ADVANCES IN GENOME BIOLOGY

GENES AND GENOMES

Editor: RAM S. VERMA

Institute of Molecular Biology and Genetics
SUNY Health Science Center
Brooklyn, New York

VOLUME 5A • 1998

 JAI PRESS INC.

Greenwich, Connecticut   *London , England*

Copyright © 1998 JAI PRESS INC.
55 Old Post Road No. 2
Greenwich, Connecticut 06830

JAI PRESS LTD.
38 Tavistock Street
Covent Garden
London WC2E 7PB
England

ISBN: 0-7623-0079-5

Printed and bound by CPI Group (UK) Ltd, Croydon, CR0 4YY

Transferred to Digital Print 2011

ADVANCES IN GENOME BIOLOGY

Volume 5A • 1998

GENES AND GENOMES

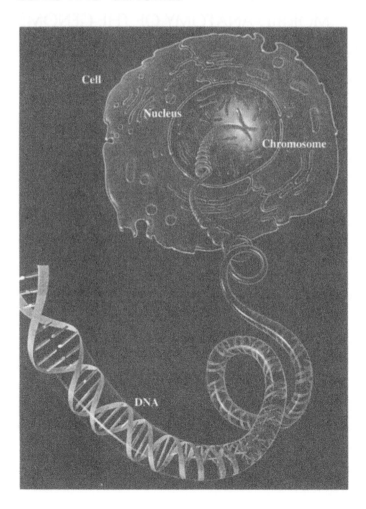

ADVANCES IN GENOME BIOLOGY

Editor: RAM S. VERMA

Institute of Molecular Biology and Genetics
SUNY Health Science Center
Brooklyn, New York

DEDICATED

TO

Americo J. Varone

CONTENTS (VOLUME 5A)

CONTENTS (VOLUME 5B)

LIST OF CONTRIBUTORS

Linda Bonen

Biology Department
University of Ottawa
Ottawa, Canada

Sandro J. de Souza

Department of Molecular
and Cellular Biology
The Biological Laboratories
Harvard University
Cambridge, Massachussetts

Elena Giulotto

Dipartimento di Genetica
e Microbiologia
Università degli Studi di Pavia
Pavia, Italy

Manyuan Long

Department of Ecology and Evolution
The University of Chicago
Chicago, Illinois

Chiara Mondello

Dipartimento di Genetica e Microbiologia
Università degli Studi di Pavia
Pavia, Italy

Christopher E. Pearson

Department of Biochemistry
and Biophysics
Texas A&M University
Houston, Texas

Joana Perdigão

Centro de Citologia Experimental
da Universidade do Porto
Porto, Portugal

Vladimir N. Potaman

Center for Genome Research
Institute of Biosciences
and Technology
Houston, Texas

Joaquim Roca   Departamento de Biologia Molecular
                                    y Celular
                                Consejo Superior de Investagaciones
                                    Cientificas
                                Barcelona, Spain

Richard R. Sinden   Department of Biochemistry
                                    and Biophysics
                                Texas A&M University
                                Houston, Texas

Adrian T. Sumner   Consultant
                                East Lothian, Scotland

Claudio E. Sunkel   Instituto de Ciências
                                Biomedical Abel Salazar
                                Universidade do Porto
                                Porto, Portugal

David W. Ussery   Department of Pharmacology,
                                    Microbiology, and Food Hygiene
                                Norwegian College of Veterinary
                                    Medicine
                                Oslo, Norway

Ram S. Verma   Institute of Molecular Biology
                                    and Genetics
                                SUNY Health Science Center
                                Brooklyn, New York

Alan P. Wolffe   Laboratory of Molecular Embryology
                                National Institute of Child Health
                                    and Human Development, NIH
                                Bethesda, Maryland

Andrei O. Zalensky   Department of Biological Chemistry
                                School of Medicine
                                University of California, Davis
                                Davis, California

PREFACE

The laws of inheritance were considered quite superficial until 1903, when the chromosome theory of heredity was established by Sutton and Boveri. The discovery of the double helix and the genetic code led to our understanding of gene structure and function. For the past quarter of a century, remarkable progress has been made in the characterization of the human genome in order to search for coherent views of genes. The unit of inheritance termed factor or gene, once upon a time thought to be a trivial and imaginary entity, is now perceived clearly as the precise unit of inheritance that has continually deluged us with amazement by its complex identity and behavior, sometimes bypassing the university of Mendel's law.

The aim of the fifth volume, entitled *Genes and Genomes*, is to cover the topics ranging from the structure of DNA itself to the structure of the complete genome, along with everything in between, encompassing 12 chapters. These chapters relate much of the information accumulated on the role of DNA in the organization of genes and genomes per se. I have commissioned several distinguished scientists, all preeminent authorities in each field to share their expertise. Obviously, since the historical report on the double helix configuration in 1953, voluminous reports on the meteoric advances in genetics have been accumulated, and to cover every account in a single volume format would be a Herculean task. Therefore, I have chosen only a few topics which in my opinion, would be of great interest to molecular geneticists. This volume is intended for advanced graduate students who would wish to keep abreast with the most recent trends in genome biology.

I owe a special debt of gratitude to the many distinguished authors for having rendered valuable contributions despite their many pressing tasks. Almost 500 pages reflect professionalism and scholarship with their own impressive styles. The publisher and the many staff members of JAI Press deserve much credit. I am very thankful to all the secretaries who have typed the manuscripts of the various contributors.

Ram S. Verma
Editor

DNA: STRUCTURE AND FUNCTION

Richard R. Sinden, Christopher E. Pearson,

Vladimir N. Potaman, and David W. Ussery

Advances in Genome Biology
Volume 5A, pages 1-141.
Copyright © 1998 by JAI Press Inc.
All rights of reproduction in any form reserved.
ISBN: 0-7623-0079-5

I. INTRODUCTION TO THE STRUCTURE, PROPERTIES, AND REACTIONS OF DNA

A. Introduction

DNA occupies a critical role in the cell, inasmuch as it is the source of all intrinsic genetic information. Chemically, DNA is a very stable molecule, a characteristic important for a macromolecule that may have to persist in an intact form for a long period of time before its information is accessed by the cell. Although DNA plays a critical role as an informational storage molecule, it is by no means as unexciting as a computer tape or disk drive. Rather, DNA can adopt a myriad of alternative conformations, including cruciforms, intramolecular triplexes, left handed Z-DNA, and quadruplex DNA, to name a few. Local variations in the shape of the canonical B-form DNA helix are most certainly important in DNA–protein interactions that modulate and control gene expression. Moreover, the ability of DNA to adopt many alternative helical structures, the ability to bend and twist, and the ability to modulate the potential energy of the molecule through variations in DNA supercoiling provide enormous potential for the involvement of the DNA itself in its own expression and replication. This chapter will focus on alternative structures of DNA and their potential involvement in biology. For more detail on some subjects, see books by Sinden[1] and Soyfer and Potaman.[2]

B. The Structure of Nucleic Acids

3. Bases

Two different heterocyclic aromatic bases with purine heterocycles, adenine and guanine, exist in DNA (Figure 1). Adenine has an amino group ($-NH_2$) at the C6 position, whereas guanine has an amino group at the C2 position and a carbonyl group at the C6 position. Two pyrimidine bases, thymine and cytosine, are commonly found in DNA. Thymine contains a methyl group at the C5 position, with carbonyl groups at the C4 and C2 positions. Cytosine contains a hydrogen atom at the C5 position, with an amino group at C4. Uracil, which is used in place of thymine in RNA, lacks the methyl group at the C5 position. Uracil is not usually found in DNA, but can result from cytosine deamination. The purines and pyrimidines are excellent candidates for informational molecules. The specific placement of hydrogen bond donor and acceptor groups provides unique structural identity. The hydrogen atoms of amino groups provide hydrogen bond donors, and the carbonyl oxygen and ring nitrogens provide hydrogen bond acceptors.

2. Deoxyribose Sugar

β-D-2-Deoxyribose is a flexible and dynamic part of the DNA molecule (Figure 2A). A shift in the positions of the C2' and C3' carbons relative to a flat plane through all carbon atoms results in various twist forms of the sugar ring. Several sugar conformations are found in DNA, the most common of which are the C2' endo and C3' endo forms (Figure 2B).

3. Nucleosides and Nucleotides

Nucleosides (adenosine, guanosine, thymidine, and cytidine) are composed of a base and a deoxyribose sugar. Nucleotide refers to the base, sugar, and phosphate group. The phosphate group is attached to the 5' carbon of the deoxyribose (Figure 3). One, two, or three phosphate groups on a sugar are designated as α, β, and γ, for the first, second, and third, respectively (Figure 3). A phosphate group can also be attached to the 3' or 5' carbon of deoxyribose.

The glycosidic bond is the bond between the sugar and the base. In the α configuration, the bond is on the 3'-OH side of the ribose sugar. This is in contrast to the β, where it is on the 5'-OH side. The base can rotate around the glycosidic bond, but generally it exists in one of two standard conformations: *syn* and *anti*. The *anti* conformation reflects the relative spatial orientation of the base and sugar

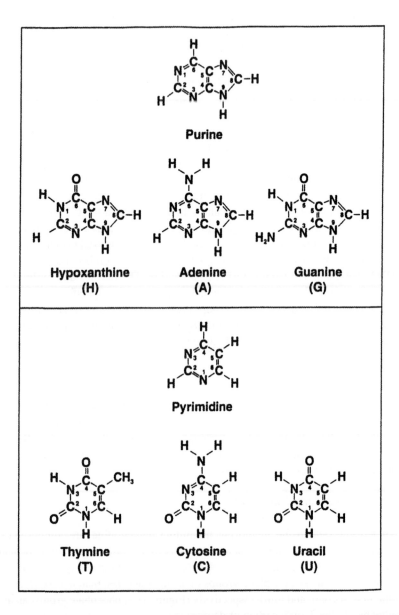

Figure 1. Purine and pyrimidine bases. (**Top**) The two-member purine aromatic ring consists of fused six- and five- member rings, each composed of carbon and nitrogen. The structures and position numbers for the basic purine ring, hypoxanthine (H), adenine (A), and guanine (G), are shown. (**Bottom**) The aromatic pyrimidine ring is composed of six carbon and nitrogen atoms. The basic ring structures, thymine (T), cytosine (C), and uracil (U), are shown.

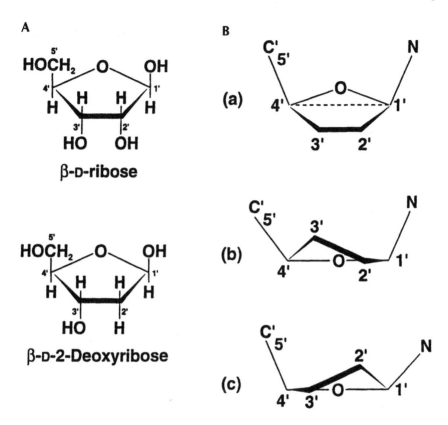

Figure 2. Sugars associated with DNA. (**A**) β-D-Ribose is found in RNA molecules. β-D-2-Deoxyribose, found in DNA, lacks a hydroxyl at the 2' position. The positions of the carbon atoms in the ribose ring are numbered with primes (e.g., 2'). (**B**) The sugar residue can adopt many different twist conformations. (**a**) Representation of an envelope conformation of a ribose sugar. (**b**) Representation of the C3' endo conformation of the ribose sugar. (**c**) Representation of the C2' endo ribose sugar conformation.

as found in most conformations of DNA, in which the ring is away from the ribose. The *syn* conformation, in which the ring is spatially over the ribose, is found in the Z-form DNA.

4.   The Phosphodiester Bond

In DNA (and in RNA) nucleotides are joined by a 3'- 5' phosphodiester bond that connects the 3' sugar carbon of one nucleotide to the 5' sugar carbon of the adjacent nucleotide through the phosphate (Figure 4). (This is termed the 3'-5' phosphodiester bond.) At the physiologically important pH 7, the ionized phos-

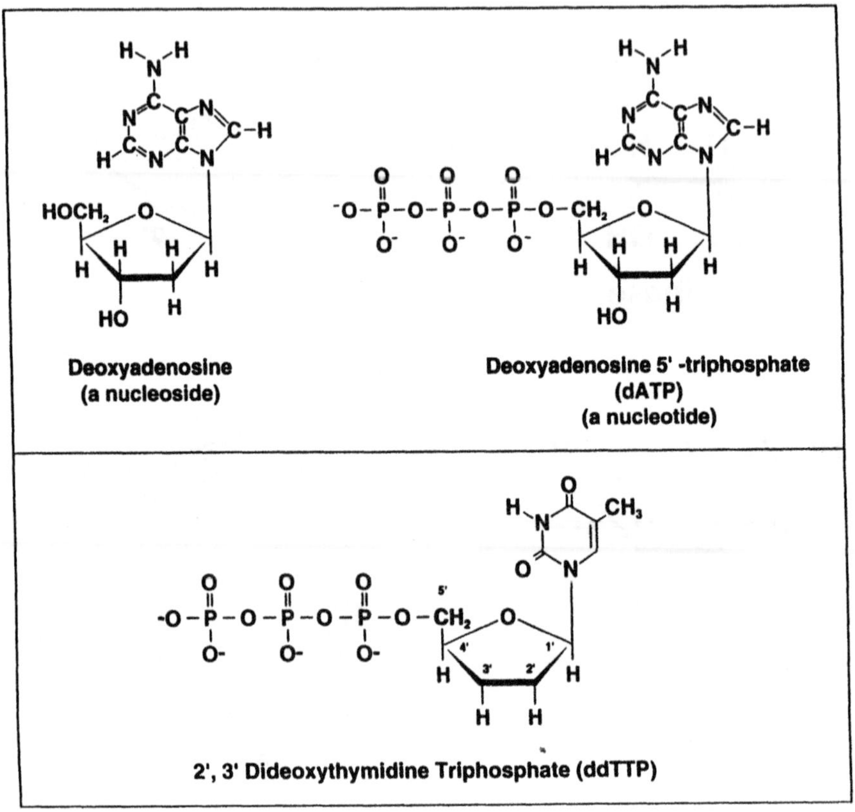

6

Figure 3. Nucleosides and nucleotides. The nucleoside deoxyadenosine consists of adenine linked through the C—N glycosidic bond to the C1' position of a 2' deoxyribose sugar. The nucleotide deoxyadenosine 5'-triphosphate (dATP) consists of adenine linked to a deoxyribose 5'-triphosphate. 2', 3'-Dideoxythymidine triphosphate (ddTTP) contains no hydroxyl group on either the 2' or 3' positions. Dideoxy-nucleotides are used for DNA sequencing reactions, since DNA polymerases require a 3'-OH for the addition of the next deoxyribonucleotide.

phate groups have one negative charge per nucleotide, which creates repulsive forces between complementary polynucleotide strands.

An important point regarding the structure of a polynucleotide is that it has two distinct ends called the 5' and 3' ends. These different ends define a polarity to the individual strands of DNA. Frequently, a hydroxyl group exists at 3' ends (3'-OH) and a single phosphate group at 5' ends (5'-PO$_4$). DNA replication and transcription occur by the addition of nucleoside 5' triphosphates to the 3' hydroxyl group of the terminal nucleotide of the polynucleotide.

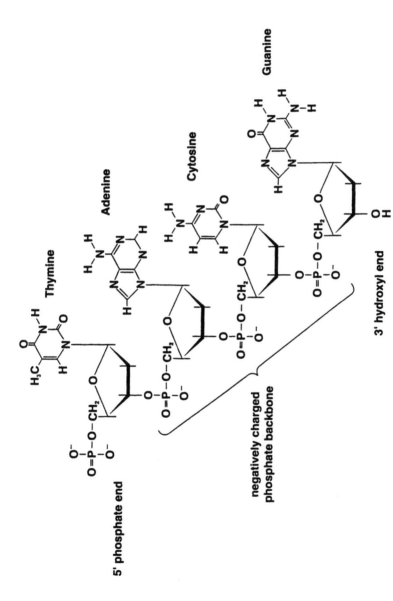

Figure 4. The phosphodiester bond. A single strand of DNA consists of nucleosides linked through a phosphate covalently bound to their 5' and 3' carbons. The charged phosphate groups are responsible for the negative charge on DNA. The two ends of the DNA chain are chemically distinct, typically having a 5'-PO$_4$ and 3'-OH.

A

C·G

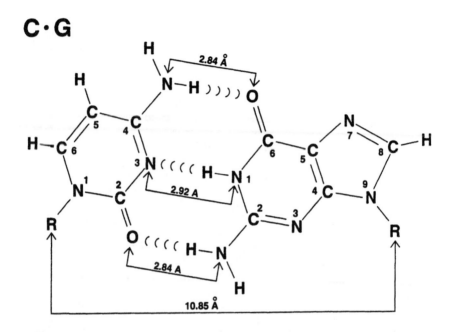

A·T

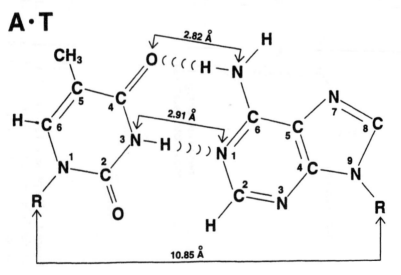

B

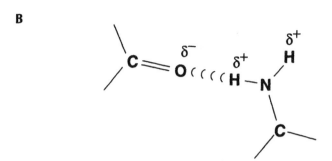

Figure 5. Hydrogen bonds in Watson–Crick base pairs. (A) The interatom distances between the C1' position of the ribose sugars are indicated. The distance between nitrogens or nitrogen and oxygen involved in hydrogen bonding are also shown. The curves in the direction of the hydrogen bond acceptor (N or O atoms) represent the hydrogen bonds. (B) Schematic of a hydrogen bond in which two electronegative atoms (N and O) partly share a proton of the NH_2 group. In this scheme the nitrogen atom is a donor and the oxygen atom is an acceptor of the hydrogen bond.

C.   The Structure of Double-Stranded DNA

The structure of the DNA described by Watson and Crick in 1953 is a right handed helix of two individual *antiparallel* DNA strands. Hydrogen bonds provide specificity that allows pairing between the complementary bases (A·T and G·C) in opposite strands. Base stacking occurs near the center of the DNA helix and provides a great deal of stability to the helix (in addition to hydrogen bonding). The sugar and phosphate groups form a "backbone" on the outside of the helix. There are about 10 base pairs (bp) per turn of the double helix.

1. Hydrogen Bonding and Base Stacking

Hydrogen bonds are short, noncovalent, directional interactions of 2.6–3.1 Å between a H atom (a donor), with a partial positive charge and a negatively charged acceptor atom, usually a carbonyl oxygen (-C=O) or nitrogen (N:) (Figure 5). In the DNA double helix, the N and O atoms involved in hydrogen bonding are separated by about 2.8–2.92 Å. An A·T base pair has two hydrogen bonds separated by 2.82 and 2.91 Å, whereas a G·C base pair has three hydrogen bonds separated by 2.84–2.92 Å (Figure 5).[3,4] Hydrogen bonds have an energy of about 3–7 kcal/mol. By contrast, covalent bonds are equivalent to 80–100 kcal/mol. Hydrogen bonds can be deformed by stretching and bending. The energy of hydrogen bonds in DNA is about 2–3 kcal/mol which is weaker than most hydrogen bonds. This is due to geometric limitations within the double helix that preclude an optimal directional alignment of the bonds.

Hydrophobic and Van der Waals interactions are involved in stacking between the aromatic planar bases. Stacking interactions are estimated to be about 4-15

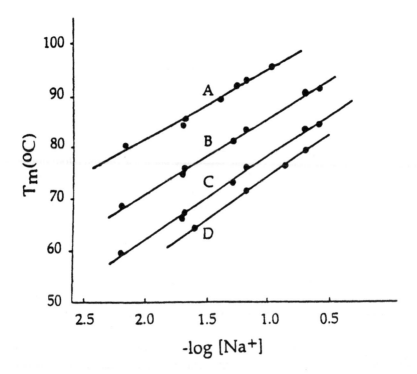

Figure 6. Dependence of DNA melting temperature (T_m) on the concentration of sodium ions in the medium. In accordance with the higher number of hydrogen bonds in the G·C pair as compared with the A·T pair, the four lines also show the dependence of T_m on the percentage of G·C pairs. (**A**) %G·C = 72; (**B**) %G·C = 50; (**C**) %G·C = 33; (**D**) %G·C = 24. Adapted from Frank-Kamenetskii.[5]

kcal/mol per dinucleotide. Base stacking provides energies of stabilization similar to those provided by hydrogen bonding.

Differences between the base stacking and hydrogen bonding energies for individual dinucleotides contribute to the heterogeneity of the B-form DNA helix. The overall energy of hydrogen bonding depends predominantly on base composition. That is, individual hydrogen bonds in A·T or G·T base pairs have relatively the same geometry and strength of hydrogen bonding. On the other hand, base stacking energies depend on the sequence of the DNA. For example, a 5' GT and 5' TG dinucleotides have very different stacking energies of 10.51 kcal/mol and 6.78 kcal/mol, respectively (see Sinden,[1] Table 1.2).

Hydrogen bonding and base stacking contribute to the stability of the DNA double helix. Since the energies for both stacking and hydrogen bonding are greater for G+C-rich DNA, it may not be surprising that the melting temperature (T_m) of the DNA is a function of G + C content of the DNA (Figure 6).[5]

2. Non–Watson-Crick Bonds

There are many ways in which two bases can form hydrogen bonds. Several of these are shown in Figure 7. Reverse Watson–Crick base pairs have one nucleotide rotated 180° with respect to the complementary nucleotide, relative to the Watson–Crick structure. In this structure, the glycosidic bonds are in a *trans* rather than *cis* orientation. Because of symmetrical hydrogen bonding potential at the C2-N3-C4 positions, thymine can rotate at the N3-C6 axis to form a reverse Watson-Crick A·T base pair. Hoogsteen base pairs utilize the C6-N7 face of the purine for hydrogen bonding with the N3-C4 face of the pyrimidine.[6] In Hoogsteen base pairing, the N7 position of purine is base-paired, altering the chemical reactivity of this position, relative to that expected for a Watson–Crick base pair. In a reverse Hoogsteen base pair, one of the bases is positioned 180° with respect to the other base compared to the normal Hoogsteen bond. A number of other base pairing schemes are possible.

3. Keto-enol Tautomerizations Can Result in Non–Watson–Crick Base Pairs

The C6 keto (C=O) position of guanine and the C4 keto of thymine can undergo a tautomerization to an enol form (C-OH). For this tautomerization to occur, the double bond must shift from the carbonyl group to the nitrogen–carbon bond in the ring. In a similar fashion, an amino nitrogen (-NH$_2$) can undergo a transition to an imino form (=NH). This can occur at the C6 position of adenine or the C4 position of cytosine. The imino or enol forms of the bases each have two isomeric forms that can exist.

The chemical equilibrium between the alternative tautomeric forms favors the keto and amino forms by about 10^4. Tautomerization leads to a reversal of the polarity of hydrogen bonding, which can result in mispairing. Enol-G can pair with T, keto-T can pair with G, imino-A can pair with C, and imino-C can pair with A. There is little evidence so far to suggest that keto-enol and amino-imino tautomerizations occur during the synthesis of DNA *in vivo*. Mispairs *in vivo* may arise from ionized and wobble base pairs.

When bases become ionized, their hydrogen bonding properties are changed. This can lead to many non–Watson–Crick base pairs. Adenine is prone to protonation at low pH, which can lead to the formation of an A$^+$·C wobble base pair. Cytosine is also very prone to protonation which can lead to a C$^+$·G Hoogsteen base pair. The ionized form of thymidine can form a T·G base pair.

4. B-form DNA

The structure of B-form DNA was determined from X-ray diffraction analysis of the sodium salt of DNA fibers at 92% relative humidity.[7,8] B-form DNA is shown schematically in Figure 8. There are about 10.0 bp per right-handed helical

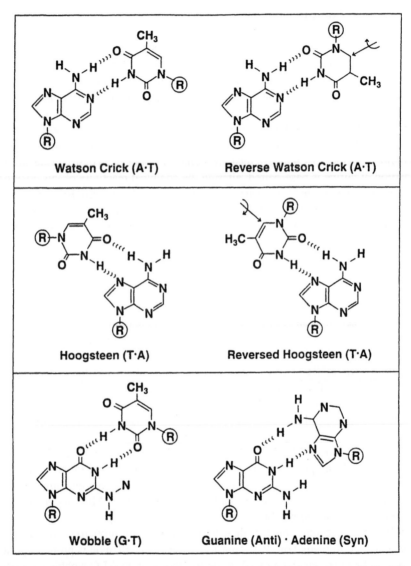

Figure 7. Base-pairing schemes. The A·T Watson–Crick base pair (top left) differs from a reversed Watson-Crick pair by the 180° rotation of the pyrimidine base (top right). T and A can also form Hoogsteen and reversed Hoogsteen base pairs (middle), with hydrogen bond formation between the pyrimidine and the N1, C6, N7 face of the purine base. A 180° rotation of the pyrimidine is required for the formation of the reversed Hoogsteen base pair. In the wobble G·T base pair (bottom left) the pyrimidine is shifted up vertically. The G(*anti*)·A(*syn*) base pair (bottom right) involves pairing between two purines, using the Watson–Crick surface of G (in the typical *anti* conformation) and the Hoogsteen surface of A (in the *syn* conformation).

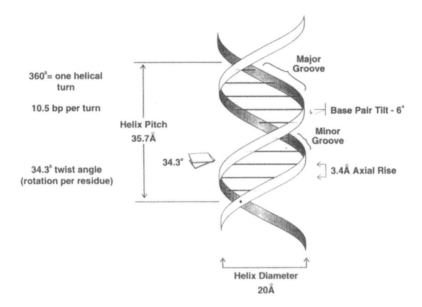

Figure 8. Structural parameters of B-form DNA. B-form DNA contains about 10.5 bp/helical turn, a 34.3° twist angle, a helix pitch of 35.7 Å, an axial rise of 3.4 Å, a tilt angle of about -6°, and a helix diameter of about 20 Å. The major and minor grooves are indicated.

turn in B-DNA in fibers. Helix parameters are defined in Table 1 and Figure 9 and listed in Table 2. The form of the ribose sugar is C2' endo (Figure 2). The term "B-form DNA" will be used to refer to the right-handed helical form commonly found for DNA in solution, where the helix repeat is 10.5 bp.

A major feature of B-form DNA is the presence of two distinct grooves, a major and a minor groove, shown in Figure 8. These two grooves provide very well defined surfaces with different shapes and geometries of hydrogen bonding potentials. The grooves are used for differential protein interaction. Certain DNA binding proteins and chemicals interact with either the major or minor groove. The Watson–Crick hydrogen bonding surfaces are not available to solvent or proteins, since they participate in hydrogen bonding with each other at the center of the double helix. The Hoogsteen hydrogen bonding surface of purines is accessible through the major groove in B-form DNA.

5. A-form DNA

A-form DNA was originally identified using X-ray diffraction in DNA fibers at 75% relative humidity.[9] The grooves in A-DNA are not as deep as those in B-DNA,

Table 1.   Definitions of Helical Characteristics

Characteristic	Definition
Helix sense	The helical rotation of the double helix. The structure described by Watson and Crick is a right-handed (clockwise) helix. Most helical forms of DNA are right-handed. Left-handed DNA, called Z-DNA, is discussed below.
Residues per turn	The number of base pairs in one helical turn of DNA, i.e., the number of bases needed to complete one 360° rotation. DNA in solution contains about 10.4–10.5 bp/turn, although this value can vary considerably as a function of base composition.
Axial rise	The distance between adjacent planar bases in the DNA double helix. In B-form DNA there are about 3.4 Å between adjacent base pairs.
Helix pitch	The length of one complete helical turn of DNA. In B-form DNA in solution, one helical turn of 10.5 bp is completed in 36 Å.
Base pair tilt	The angle of the planar bases with respect to the helical axis. The tilt angle is measured by considering the angle made by a line drawn through the two hydrogen-bonded bases relative to a line drawn perpendicular to the helix axis. For a base pair perpendicular to the helix axis, the tilt angle is 0°. In B-form DNA the bases are tilted by -6°. In A-form DNA the base pairs are significantly tilted at an angle of 20°.
Base pair roll	The angle of deflection of a base pair with respect to the helix axis along a line drawn between two adjacent base pairs relative to a line drawn perpendicular to the helix axis. Compare this to the tilt in which the line angle is measured along a line drawn through the base pair. The roll and tilt angles are offset by 90°.
Propeller twist	The angle between the planes of two paired bases. A base pair is rarely a perfect flat plane with each aromatic base in the same plane. Rather, each base has a slightly different roll angle with respect to the other base. This makes the two bases look like an airplane propeller.
Helix diameter	Diameter of the helix refers to the width in Å across the helix. B-DNA has a diameter of 20 Å.
Rotation per residue (twist angle)	The angle between two adjacent base pairs. Consider the angle between lines drawn through two adjacent base pairs. In B-form with 10 bp in one 360° helix turn of DNA, the rotation per residue is 36°. For B-form in solution with 10.5 bp/turn, $h = 34.3°$.
Shift	The displacement of two bases in the direction of the major or minor groove.
Slide	The displacement of two bases in the direction of the phosphate backbone.

and the bases are tilted to about 20°. Moreover, in A-DNA sugar pucker is C3' endo compared to C2' endo for B-DNA. Runs of homopurine/homopyrimidine DNA sequence [poly (dG)·poly (dC), for example] seem to set up an A-like helix, as determined by characteristic circular dichroism (CD) spectra.[10] Therefore, within a B-like DNA molecule, specific regions may exist in an A-DNA form. Many regions

Spatial Relationship Between Adjacent Base Pairs

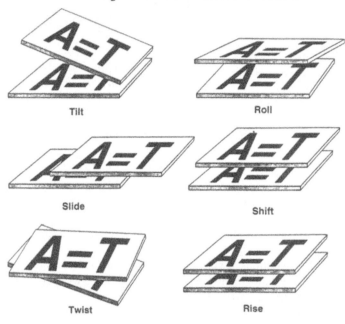

Tilt

Roll

Slide

Shift

Twist

Rise

Spatial Relationship Between Bases in a Base Pair

Propellor Twist

Figure 9. Spatial relationships between adjacent base pairs. Representations of the parameters described in Table 1.

of RNA molecules, including transfer RNA (tRNA), ribosomal RNAs (rRNA), and in parts of messenger RNAs (mRNA), exist in a double-helical A-form.

6. Sequence-Dependent Variation in the Shape of the DNA

A canonical textbook B-form helix is not likely to exist in nature. The actual shape of DNA will depend on its base composition, the local flanking sequence,

Table 2.   Helix Parameters

Parameter	A-DNA	B-DNA	Z-DNA
Helix sense	Right	Right	Left
Residue per turn	11	10.5	12
Axial rise	2.55 Å	3.4 Å	3.7 Å
Helix pitch	28 Å	36 Å	45 Å
Base pair tilt	20°	-6°	7°
Rotation per residue	33°	34.3°	-30°
Diameter of helix	23 Å	20 Å	18 Å
Glycosidic bond configuration			
dA, dT, dC	*anti*	*anti*	*anti*
dG	*anti*	*anti*	*syn*
Sugar pucker			
dA, dT, dC	C3' endo	C2' endo	C2' endo
dG	C3' endo	C2' endo	C3' endo
Intrastrand phosphate-phosphate distance			
dA, dT, dC	5.9 Å	7.0 Å	7.0 Å
dG	5.9 Å	7.0 Å	5.9 Å

and environmental conditions. Evidence for local structural variation in DNA comes from X-ray crystallography.[11,12] The twist angle between adjacent base pairs can vary considerably, from 32° to 45°.[11] Because of flanking sequence effects, not all dinucleotides will have the same twist angle. Since individual twist angles are different, the actual helical repeat and therefore the exact shape of a 10.5-bp helical turn of DNA can vary considerably.

Polymeric regions of a single base in one strand can adopt unusual helical forms. Poly(dA)·poly(dT) forms an unusual structure called heteronomous DNA,[13] with a helix repeat of 10 bp/turn. In heteronomous DNA the deoxyribose sugar in the d(A) strand is C3' endo, whereas the deoxyribose in the d(T) strand is C2' endo. The helical changes associated with phased runs of A in DNA, which are responsible for DNA bending, are another example of a variation in the DNA structure. A-form-like tracts of DNA form in runs of poly(dG)·poly(dC). Runs of (dG)·(dC) greater than 20 bp can form triple-stranded and four-stranded structures, as discussed below. Triplet repeats can form an unusual helix structure that may be more flexible than canonical B-form DNA.[14,14a]

II.   DNA CURVATURE AND BENDING

A.   Introduction

In some respects DNA is a simple molecule. A plasmid DNA of known sequence can be cut into a number of different sized pieces by a restriction enzyme. Molecules of equal length would be expected to exhibit similar flexibility. Pieces of DNA shorter than the persistence length of 150–200 bp will behave as rather stiff rods that cannot be easily bent into a circle (see refs 15 and 16 for a discussion of persistence length). Larger pieces adopt a "random coil" shape in solution. DNA molecules of defined size behave in a very predictable way when run on agarose or acrylamide gels. Shorter molecules migrate faster than larger molecules. In both agarose and acrylamide gels there is a linear relationship between the log of the distance migrated and the length of the DNA in base pairs.

Certain DNA fragments of known length do not run at their expected position on an acrylamide gel. One of the most striking examples of such anomalous migration was a 414-bp piece of kinetoplast DNA from *Crithidia fasciculata*.[17,18] This DNA migrated as if it were twice as long (e.g., it ran as if it were about 830 bp) in an acrylamide gel, but migrated at its proper position (414 bp) in an agarose gel. The anomalous migration of the kinetoplast DNA in polyacrylamide gels was attributed to the kinetoplast DNA being either stably curved or kinked (Figure 10). Although the migration through acrylamide gels is not completely understood in physical terms, the migration is believed to depend on the ability of DNA to "snake" through the gel matrix.[19] A relatively straight piece of DNA that is rather flexible can easily snake through the gel. On the other hand, DNA containing a permanent bend or kink is not as flexible and can get hung up in the gel matrix. For DNA of equal sizes, a curved fragment will take longer to snake through the small pores or matrix of the acrylamide gel than a noncurved DNA fragment. (This does not occur in agarose gels because the matrix or pore size is believed to be larger.)

The term "curved DNA" will be used to describe the phenomenon of intrinsic DNA curvature, and the term "bent DNA" will refer to axis deflection introduced at one site in DNA (frequently introduced by some external agent such as thermal motion, proteins, drugs, etc.). The investigation of behavior of DNA fragments in polyacrylamide gels has become a principal instrument in the studies of curved DNA, although this is by no means the only method used to measure DNA curvature. Indeed, as will be seen below, various techniques give different results, which in turn are interpreted in terms of very different models. However, one important observation that all techniques (including, perhaps surprisingly to some, gel electrophoresis) agree on is that nearly any DNA sequence is capable of exhibiting some curvature, under the appropriate conditions, and that in fact it is quite rare to find a piece of DNA that is not curved to some extent. Most DNA found in organisms has a gentle left-handed writhe, due to slight intrinsic curvature[20,21] (see Travers[22] for a review). Perhaps this is not surprising, when one considers that the DNA in all organisms, from bacteria to humans, must be condensed by about 1000-fold to fit inside the cell. For bacteria, this means that a typical bacterial operon, if it were not curved, but stretched out in its B-DNA conformation, would be longer than the bacterial cell! Thus DNA curvature plays an important role in controlling the condensation as well as the expression of genetic information.

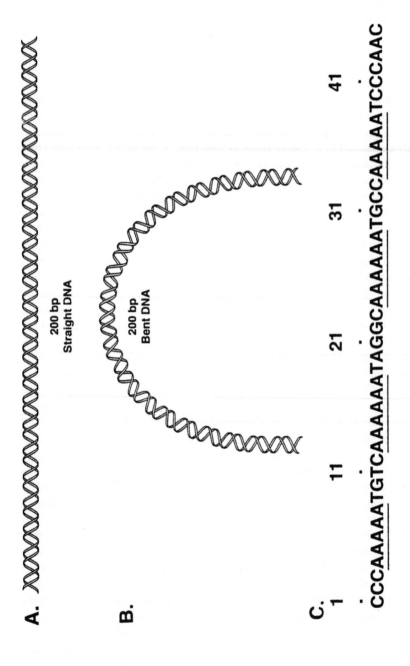

A.

200 bp
Straight DNA

B.

200 bp
Bent DNA

C.

1   11   21   31   41

CCCAAAAATGTCAAAAAATAGGCAAAAAATGCCAAAAATCCCAAC

Figure 10. Curved DNA. (**A**) A 200-bp straight DNA molecule. (**B**) A 200-bp piece of curved DNA. (**C**) Phased A tracts are preset at the site of bending in a *Crithidia fasciculata* DNA sequence.

B. DNA Sequence Organization Required for Curvature

Wu and Crothers devised a clever way to map the site of curvature in the kinetoplast DNA.[23] They cloned the bent kinetoplast DNA fragment as a dimer of a 241-bp *Hind*III DNA fragment. By cutting the dimer with a series of different restriction enzymes, DNA fragments with the curve located at various positions within the 214-bp fragment were produced. The sequence organization (i.e., the location of the curve) of each fragment will be different from that of the original fragment. Such a set of DNA molecules is called *circularly permuted*.

Circularly permuted molecules do not migrate at the same rate because the center of curvature is at different locations within the molecule. This creates different end-to-end distances. The end-to-end distance is shortest when the curve is positioned at the center of the molecule. For molecules of the same size, the end-to-end distance is important in determining gel mobility. The molecule with the shortest end-to-end distance will migrate most slowly in an acrylamide gel. This pattern was observed by Wu and Crothers.[23]

The DNA sequence shown in Figure 10B was found at the center of the bend. There are five runs of four or five A's, which, in each case, are preceded by a C and followed by a T. In addition, the runs of A are phased by 10 bp. The DNA curvature hypothesis suggested that the runs of A and the 10-bp phasing were important in curvature.[23,24] Crothers also suggested that bending may occur at a junction resulting from the interruption of B-form DNA by an A tract that adopts a non–B-DNA helix.

The 10-bp phasing is required for curvature associated with A tracts. For DNA to contain a region of stable curvature and produce the anomalous gel mobility, the small individual curves associated with A tracts must be oriented in the same direction or same plane. If the A-tracts are placed every 1.5 turns of the DNA helix (15–16 bp), the curvature resulting from the individual A tracts will have a zigzag shape, and the DNA will essentially migrate as nonbent DNA on an acrylamide gel.

The phasing hypothesis was tested by Hagerman using oligonucleotides of sequence GA_3T_3C, $G_2A_3T_3C_2$, and $G_3A_3T_3C_3$, which phases the A_3 block in one DNA strand by 8-, 10-, and 12-bp intervals.[25] Only the $G_2A_3T_3C_2$ polymer with A tracts phased at 10 bp exhibited the pattern of electrophoretic migration diagnostic for curved DNA.

Koo, Wu, and Crothers[24] also synthesized a large number of oligonucleotides containing various lengths of A tracts that were phased at different lengths. Polymers with $A_{4–9}$ were bent, with curvature being optimal for A_6. A continuous run of A's is required for curvature, since interruption of an A_5 tract with C, G, or T destroys the curvature. There is no particular sequence requirement for the base 5' or 3' to an A tract for curvature, although flanking sequences can influence curvature.

DNA sequences that do not contain runs of A's can also be curved. The curvature observed in DNA lacking phased A tracts is usually not as large as A tract curvature (in the absence of divalent cations). However, based on gel mobility experiments Brukner et al. have found that, in the presence of Mg^{2+} or Ca^{2+}, certain sequences without A tracts (e.g., $N_4G_3C_3$) can exhibit strong macroscopic curvature.[26,27] Furthermore, Brukner et al. have postulated that the direction of curvature was opposite that of the A tract curves.

C. Models for Curvature

1. The Wedge Model for DNA Curvature

Trifonov proposed a model for curvature, in which a wedge angle is associated with the AA dinucleotide and curvature is attributed to the summation of the wedge angles of the AA dinucleotides.[28,29] The sum of wedges pointing in the same direction, a condition met by the 10-bp phasing, leads to the curvature of DNA. Ulanovsky et al. used measurement of the efficiency of ligation of small DNA molecules into circles to calculate the wedge angle of an AA dinucleotide.[29] As a short piece of DNA with a defined curvature is ligated together into increasingly long polymers, at some length the total angle of curvature will result in the formation of a circle of DNA. By determining the length at which the DNA forms a circle and by knowing the number of AA dinucleotides responsible for the 360° curvature, the individual AA wedge angle can be determined, assuming no contributions from the other DNA sequences present. The optimal size for circle formation in a polymer studied by Ulanovsky et al. was 126 bp. This is much less than the dynamic persistence length of DNA of about 230 bp estimated by electron miscrocopy.[16] For a discussion of static versus dynamic persistence length and their relationship to DNA curvature, see Trifonov et al.[30] Thus efficient circularization is an indication that this sequence can readily adopt a structure in which it is curved by 360°. Within the 126 bp molecule there are 66 AA dinucleotides, and the sum of their individual wedge angles was assumed to be responsible for the 360° curvature. Trifonov estimated a total wedge angle (of both tilt and roll components) of 8.7° for each AA dinucleotide. This angle probably represents an upper limit of the wedge angle, and a later study estimated the AA dinucleotide wedge to be close to 1.1°,[20] in the absence of Mg^{2+} (Mg^{2+} is necessary for the ligation experiments) a value of roughly twice this was obtained for the same sequence, based on ligation experiments.[31]

2. The Junction Model for DNA Curvature

Wu and Crothers[23] proposed the junction model for DNA curvature (Figure 11), which suggests that there is a bend at the junction of B-form DNA and a non-B DNA helix associated with A tracts. A tracts can adopt a non–B-DNA helix called heteronomous DNA.[13] In addition, modeling studies suggested that DNA would

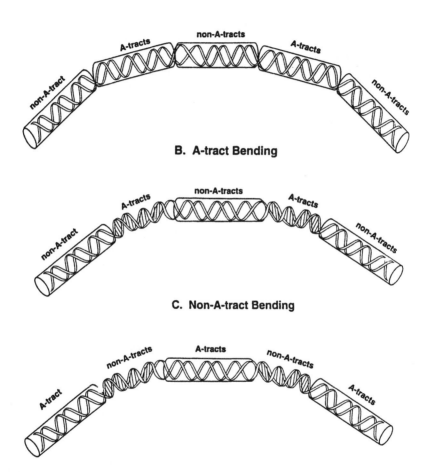

A. Junction Model

B. A-tract Bending

C. Non-A-tract Bending

Figure 11. The wedge model and junction models for DNA bending. (**A**) The wedge model suggests that curvature results from the sum of the wedge angles present between adjacent AA dinucleotides within an A tract. (**B**) The junction model suggests that bending occurs at the junction between an A tract (which has a non–B-DNA helix) and mixed-sequence DNA that forms a B-DNA helix. (**C**) The general sequence model suggests that bending results from curvature present in mixed-sequence B-form DNA interspersed with A tracts that, in this model, are considered to be straight.

bend at the junction between A-DNA and B-DNA.[32] There are two junctions, a 5' and a 3' junction, associated with each A-tract, and the site of curvature could be associated with either or both junctions. Data suggested that curvature was probably localized at the 3' end of the A tract.[23,24]

3. The General Sequence (or Non–A Tract) Model of DNA Curvature

With the exception mentioned above on non–A tract DNA curvature, it might seem pretty conclusive that the A tracts have been shown to be clearly responsible for most DNA curvature observed in the gel electrophoresis. Indeed, it is possible to predict the anomalous mobility of a particular DNA fragment on an acrylamide gel, based on the sequence, using any one of several algorithms. Most of these models assume that the curvature is due to phased A tracts, as explained above. Thus, when Goodsell et al.[33] proposed a model for curved DNA in which the A tracts are not curved at all, it seemed pretty heretical to many people in the field. Essentially the reasoning was that, since of the hundreds of DNA sequences that have been crystallized, A tracts are always rigid and never curved, the situation must be the same in solution.[33] This model proposes that the DNA curvature observed in gel electrophoresis experiments is due not to the curvature of A tracts, but to the comparative rigidity of the A tracts. The curvature arises not from tilt (remember that in the wedge model, the "wedges" consist of a combination of both tilt and roll), but almost exclusively from base pair roll, toward the major groove. The A tract has little roll (close to 0°); most base steps have, on average, a roll toward the major groove of about 3°, and certain sequences (notably GC) have a roll (again toward the major groove) of about 6° per step. Thus, it is the relative curvature of the intervening sequences flanking the A tracts that gives rise to the observed curvature, according to this model. This model predicts that the "opposite" curvature observed by Brukner for the GGGCCC curves, compared to the A tract curves, is relative to the average roll of about 3° found for most sequences. In this case the central GC roll of about 6° would be in the opposite direction (+3°) than for the A tracts (–3°), compared to most of the other steps.

D. A Tract DNA Adopts a Unique Double Helical Conformation

To better understand the DNA bending models, a bit more background about A tract DNA is needed. Part of the reasoning for the junction model was that A tract DNA could adopt a unique conformation, as discussed above. The general sequence model goes a bit further than the junction model, by saying that the A tracts are not responsible for curvature at all, but the idea that the unique B' helix of the A tract is somehow responsible in part for the observed curvature is the same. Over the years, much information has been obtained about A tract DNA.

1. Rigidity of A Tracts

Spectroscopic studies showed that short A tracts (of at least 4 bp) can adopt a structure similar to that of poly(dA)·poly(dT); this structure was independent of whether the A tracts were phased with the pitch of the helix.[34] The structure is somewhere between an A-DNA and B-DNA helix (sometimes referred to as a B' helix), and has a rigid (and straight) structure, in which the base pairs have a high propeller twist (about 20°; B-DNA has a propeller twist near 0°). This conformation is stabilized by strong base-stacking interactions.[35] The AT base pair has a high propeller twist such that the A share an hydrogen bond with the T below, resulting in "bifurcated" hydrogen bonding.[36] The helical repeat of A tract DNA is only 10.0 bp/turn, in contrast to most other DNA sequences which have a helical repeat of about 10.5 bp/turn; this indicates that A tracts adopt a type of helix distinctly different from that of normal "generic" DNA.[37,38]

2. A Tracts Have a Narrow Minor Groove

Burkhoff and Tullius[39] obtained results that were inconsistent with both the wedge and junction models. A hydroxyl radical footprinting technique was used to analyze the width of the minor groove. Since hydroxyl radical attacks ribose from the minor groove, the reactivity of DNA is dependent on the width of this groove. The hydroxyl radical footprinting pattern for B-DNA showed uniform cutting at each base. Kinetoplast bent DNA showed a reduction in cutting in the A tracts. This result was interpreted as reflecting a narrowing of the minor groove along the A tract.[39,40] The polymers studied by Hagerman, $(GA_4T_4C)_n$ and $(GT_4A_4C)_n$, showed very different results. The hydroxyl cleavage pattern in the $(GA_4T_4C)_n$ bent polymer showed periodic reduction in the cleavage along the A tracts. There was no reduction in cleavage along A tracts in the $(GT_4A_4C)_n$ polymer. These results suggest that in the 5'-A_4T_4-3' sequence the run of A's can adopt a helix different from that of B-form DNA, a helix in which a gradual narrowing of the minor groove occurs. This narrowing does not occur in the A tracts in the 5'-T_4A_4-3' sequence. Probing with diethylpyrocarbonate also indicates that the minor groove is more narrow in A tract DNA; furthermore, addition of the minor-groove binder distamycin abolished a conformation of the A tracts that was $KMnO_4$ sensitive only at lower temperatures.[41]

The results from chemical probe analysis are consistent with the crystal structure of poly(dA)·poly(dT), in which the minor groove is more narrow than average B-DNA.[35] The narrowing of the minor groove for short A tracts associated with DNA curvature has also been found in crystal structures (see ref. 42 for a review) as well as in solution, based on nuclear magnetic resonance (NMR) experiments.[43,44]

3. Cooperativity in A Tract B' Helix Formation

There is evidence for cooperativity in the formation of the A tract DNA structure. Leroy et al.[45] presented imino-proton-exchange NMR experiments to show that the structures of A_nT_n and T_nA_n sequences were quite different. Long proton exchange times for A·T base pairs were associated with longer lengths of A tracts in the A_nT_n but not the T_nA_n orientation. The shorter times in T_nA_n oligonucleotides were similar to lifetimes found in B-form DNA. Long proton exchange times were found for all sequences that exhibited anomalous migration or poly-acrylamide gels.[45] Nadeau and Crothers[43] have also confirmed that cooperative structural changes in helix structure occur in a run of A's as the length of the tract is increased. Three A's begin to set up the "A tract helix" responsible for the observed curvature. By the time a length of A_{6-7} is reached, the transition from a B-form to a different helix, an A tract structure, is complete.

4.   Temperature Dependence of A Tract B' Helix

Diekmann[46] showed a strong temperature dependence of curvature; he found that at 40°C, the curvature was greatly reduced. Curved sequences have been shown to undergo a "premelting" transition, with a midpoint near 30°C.[47] In fact, further studies have indicated that this premelting transition is due to the loss of a curved component from an equilibrium at lower temperatures.[48] This same pre-melting transition is seen for poly(dA)·poly(dT), which is known not to be curved, which means that the rigid high-propeller-twisted conformation of the A tracts is not directly responsible for curvature. The A tract regions from kinetoplast DNA were found to be sensitive to $KMnO_4$ at the 5' end of the A tracts, but only at lower temperatures—in this case there was a slight drop in reactivity at 23°C (compared to 14°C) and a dramatic reduction in reactivity at 43°C.[41]

It is certainly possible (indeed likely) that this conformation is indirectly responsible for the observed macroscopic curvature, by providing a frame of ref-erence from which to see the curvature of the rest of the DNA. If this were true, one would expect the anomalous migration of A tract curved DNA in gels to be less significant above the transition temperature than below it, and this is exactly what is observed experimentally. There is little change in migration between 5°C and 25°C, but a drastic reduction in anomalous mobility if the gel is run at 35°C.

Obviously, such a temperature-dependent transition within this range could have strong biological implications. Many sequences that have been characterized at room temperature as being "curved" may in fact not be curved at all inside a cell at 37°C; furthermore, this difference in curvature could be utilized as a type of "environmental sensor." In support of this is the finding of large regions of signif-icant curvature upstream of virulence-related genes in many different species of bacteria. It is possible that, when the bacteria are floating free in the environment, these genes are not transcribed, partly because of the curvature of the upstream region. However, at 37°C, when the bacteria have invaded a host, the upstream region is no longer as curved and might more readily allow transcription of the gene.

5.   Static versus Dynamic Curvature

So far DNA curvature has been talked about as if it were a stable, static bend in the double helix. Molecular mechanics modeling of curved DNA shows a type of flexing equilibria, such that on average a piece of DNA is curved, but it is in fact in equilibrium, or flexing, between two different conformations.[49] The models predict the flexing to be in the picosecond range. Thus the DNA appears, when averaged over time, to have a "static" curve of a given angle. This is consistent with NMR evidence, in which the "curved" conformation is due to flexing of the DNA flanking the A tracts (which are straight and have a high propeller twist).[33,50]

6. Summary

Short A tracts can form a distinct type of structure, in which the helix is rigid, the minor groove is more narrow, and the bases are stacked with a high propeller twist. The formation of this structure is cooperative, with about six consecutive A's needed for the most stable conformation to occur. This conformation is temperature dependent and appears to be stable at temperatures below 30°C. However, the presence of this structure does not always result in curvature, since the A_5 tract in the A_5N_{10} sequence can also adopt this conformation, which is not curved. Furthermore, poly(dA)·poly(dT), which is also not curved, can adopt this conformation. The high propeller twist conformation, with the bifurcated hydrogen bonds, has been shown to be unnecessary for DNA curvature.[51] Thus this structure is associated with curvature, but apparently is not the sole determinant.

E.   Which Model Best Explains DNA Curvature?

The wedge and junction models have become refined, so that now they both give essentially the same predictions.[52] Both of these models do allow for some curvature from non-A tracts (but the dominant contribution is assumed to come from phased A-tracts), and are supported by experiments using polyacrylamide gel electrophoresis (usually run in the presence of EDTA).

None of the models provide a completely accurate explanation of the physical basis for DNA curvature in solution. On the other hand, all of the models do a reasonable job of predicting the curvature (as determined by anomalous migration in polyacrylamide gels) of many DNA sequences containing runs of A's. However, there are some sequences that exhibit curvature not predicted by either the wedge or junction models. Furthermore, the wedge and junction models do not readily account for the effects of Mg^{2+} and temperature on DNA curvature.

There is compelling evidence to support the general sequence model from many different physical methods. A comprehensive review of DNA curvature from 114 different crystal structures has found the A tracts to be "relatively straight," with the curvature localized in flanking regions.[42] In addition to the X-ray crystal structures, the gel electrophoresis experiments of G_3C_3 motifs,[26,27] as well as sequence-specific flexibility, as determined by DNase I digestion patterns,[53,54] lend credence to this model. This model is also consistent with studies in which the locations of particular dinucleotide steps were compared in DNA that was wrapped around nucleosomes or into minicircles. The GC, CG, and GG steps were all found to prefer positions where the minor groove was on the outside of the circle, whereas the AA, AT, and TA dinucleotides were found to be positioned where the minor groove was facing inside[55,56] (see Wolffe and Drew for a discussion).[57] Furthermore, theoretical modeling of curved DNA has also lent support to the general sequence model.[58] The thermodynamic evidence of a premelting transition in which the rigidity of the A tracts is lost also fits well with this model and is difficult to explain in terms of either the wedge or junction model.[48] Thus it seems that much evidence is currently pointing toward the model of DNA curvature in which the A tracts are not curved at all, most other steps are curved a bit because of a roll toward the major groove, and the GC step can exhibit strong curvature. However, one criticism of the conclusions from X-ray crystallography is the use of dehydrating agents and the frequent high concentrations of divalent cations needed for crystallography,[59] an issue that has been considered by Dickerson and colleagues.[60]

The issue of the correct model for DNA bending remains to be resolved, and further experimental evidence will clearly be needed before the physical basis for bending is clearly resolved. It should be kept in mind that many assays for bending and DNA structure utilize very different protocols and procedures. For example, acrylamide gels can be run in the presence of many different buffers and in the absence of divalent cations, whereas circular ligation studies require Mg^{2+}. X-ray crystallographic buffers are very different and contain dehydrating agents. Therefore, the assay conditions can influence the DNA structure, and this complicates direct comparison of results obtained using different methods. Finally, the environmental milieu surrounding the DNA in living cells is not well defined, adding another level of uncertainty to the structure of DNA in cells.

F. Environmental Influences on DNA Curvature

The environment surrounding DNA can have a profound influence on its structure. As has already been mentioned, at 37°C most A tract curved DNA will run on gels close to normal, and is likely to be hardly "curved" at all. Other factors known to affect DNA curvature are divalent cations, such as Mg^{2+}, which will enhance the magnitude of curvature of many sequences. On the other hand, addition of NaCl to about 300 mM will greatly reduce DNA curvature. The addition of spermine (or spermidine), which is abundant in most cells, will again enhance curvature. Supercoiling of the DNA will also affect the manifestations of local curvature. Of course, the binding of a protein to DNA will have a large effect on the curvature, but this is DNA bending, the subject of the next section.

G. Proteins That Bind and Bend DNA

The DNA inside of most living organisms is compacted at least 1000-fold, and it may not be surprising that just about every DNA-binding protein will bend or wrap the DNA around itself when it binds. There are many proteins that will bind preferentially to DNA with intrinsic curvature, as well as proteins that bind to DNA and then bend it. This is something of an artificial distinction, in that most of the proteins that bind preferentially to curved DNA will, in all likelihood, bend the DNA even more. Before the proteins themselves are discussed, it is important to distinguish between DNA sequences that are intrinsically curved and sequences that are flexible.

1. Intrinsic Curvature versus Flexibility

"Flexible" sequences contain steps (usually pyrimidine-purine, such as TA or CA) that are easily deformable.[61] When a curved DNA fragment runs anomalously slow on the gel, there are two components to the observed "curvature": one is the actual "static" or intrinsic curvature, as has been discussed so far in this section. The other component is the "bendability" or flexibility of the DNA sequence. Some DNA sequences are quite readily deformable, whereas others (like the A tracts) are quite rigid. In fact, one aspect of the general sequence model for DNA curvature is that, because the A tracts are rigid and do not bend, the comparative flexibility of the adjacent sequences contributes to the observed curvature. In some crystal structures, a motif has been found to be bent in one sequence context, but not in another; this has been explained as "a demonstration of the bendability of the helix."[60]

The bendability of various DNA sequences can also be determined by locating the relative positions of dinucleotides in DNA wrapped around nucleosomes or in tight minicircles.[55,56] The result of several such studies is that GC dinucleotides are found preferentially where the minor groove is facing out, or away from the protein–DNA complex, whereas the AA dinucleotide was found preferentially in the opposite orientation—that is, with the minor groove facing in, or toward the center of the complex.[57] Thus the flexibility of the GC step will allow it to roll more toward the major groove.

2. Proteins That Bind to Curved DNA

Proteins that bind nonspecifically to DNA, yet show a preference for sequences containing curved DNA, are often chromatin-associated proteins and are responsible for maintaining the chromosome. Travers has called these proteins "DNA chaperones," since one of their main roles seems to be facilitating DNA compaction and other protein–DNA interactions.[62,63] Three different categories will be considered: the bacterial proteins, eukaryotic "curved DNA-binding" proteins, and eukaryotic nucleosomes.

Curved DNA-binding proteins in bacteria. Both of the major chromatin-associated (HU and H-NS) proteins in *Escherichia coli*, as well as the *E. coli* RNA polymerase, have been found to bind preferentially to curved DNA, with the H-NS protein showing the greatest affinity for curves.[64] The H-NS protein was isolated from *E. coli* cell extracts, based on its preference for curved DNA.[65] H-NS binding to DNA is sensitive to intercalating agents, such as distamycin,[65,66] which will remove intrinsic curves from DNA.[67]

Curved DNA sequences seem to occur within the sites of H-NS interaction in genes repressed by H-NS—at least in the cases of three H-NS dependent genes studied to date: the *proU* operon of *Salmonella*,[68] the *rrnB* gene of *E. coli*,[66] and the *hns* gene itself.[69–71] Furthermore, *in vivo* experiments seem to imply that H-NS can show strong and specific repression of a gene with a curve upstream of the promoter region.[72] It is worth noting in this last experiment that a different curve resulted in no repression by H-NS *in vivo*. The surprising thing about this was that this curve (A_5N_5) was the same curve that had been used to identify H-NS as a "curve binding" protein.[65,73] Thus it is possible that H-NS might show a strong binding preference for a particular type of curve, but exhibit different properties in binding DNA inside the cell. In fact, there are many curves (both synthetic and naturally occurring) that do not show a preference for H-NS binding. Upstream of the *E. coli proU* operon, there is a strong curve region, as found by random cloning of curved DNA fragments.[74] Although H-NS affects the regulation of the *proU* gene, it appears this occurs through a region downstream of the promoter, and deletion of the upstream curve has essentially no effect on H-NS regulation.[75] In summary, there is a correlation between the sites of H-NS interaction and DNA curvature, but not all curves will necessarily show a strong preference for binding of the H-NS protein.

Curved DNA-binding proteins in eukaryotes.   Several eukaryotic proteins have been reported to bind preferentially to curved DNA sequences. One such protein is the mouse Kin17 protein.[76] The domain of the protein responsible for binding to curves is not known, although mutational analysis has shown that it does not act through the zinc finger region.[76] A protein that binds specifically to rat mitochondrial DNA has been shown to exhibit part of this preference through binding to curved DNA.[77] Finally, the chromatin associated HMG proteins exhibit an "architecture preference"—that is, they will bind to DNA of a particular type of flexibility.[78]

Nucleosomes.   Nucleosomes can be "phased" or positioned by certain sequences containing curved DNA.[74,79] However, attempts to phase nucleosomes using synthetic curves so far have been unsuccessful.[64] It is likely that the phasing of nucleosomes will require many elements, including properly positioned flexible regions and flanking regions of "straight" DNA (see Wolffe and Drew[57] for a more detailed discussion). Other alternative DNA structures can also affect nucleosome position (see Section IV).

3. Protein-Induced DNA Bending

There are many examples of DNA bending upon protein association. In many cases, the DNA is actually wrapped around a protein core—for example, the wrapping of DNA around DNA gyrase, nucleosomes, or the bacterial HU protein. Many other proteins bind to DNA and introduce a bend that is similar in magnitude to the curvature introduced by A tracts in DNA. The catabolite activator protein (CAP) provides one example. Wu and Crothers,[23] using gel mobility and circular permutation analysis (see above), demonstrated that CAP binding introduces a significant bend into DNA. The X-ray cocrystal structure of the CAP–DNA complex has subsequently shown a bend angle of 90°.[80]

H. The Biology of DNA Curvature

Nature has found many uses for DNA curvature. One is the control of access to promoters, the switch regions that turn genes on by bending or looping the DNA. Another is the control of initiation of DNA replication, which, since it represents a major commitment for the cell, must be very carefully regulated. A third use is site-specific integration of one DNA molecule into another. A fourth use is in DNA repair, where DNA often becomes bent upon the binding of many chemicals or following UV irradiation. Finally, the compacted organization of DNA into cells requires DNA to be wrapped very tightly around DNA-packaging proteins.

1. DNA Curvature, Bending, and Gene Regulation

DNA bending and open-complex formation.   The RNA polymerase must bind and bend the DNA, and this bending (and subsequent torquing) is responsible for melting the DNA and the formation of an open complex.[81,82] The minor groove at the center of the −10 site must be placed on the inside of an intrinsic curve for optimal promoter activity.[55,83,84] Mutations that affect the curvature at the −10 site also affect the sensitivity of the promoter to DNA supercoiling.[85]

Location of DNA curves in promoter regions.   Much of the curved DNA in *E. coli* has been found to be localized in and upstream of promoter regions.[74,86] The curved regions around promoters can be grouped into three different areas: upstream, downstream, or within the promoter.

Upstream curves.   Upstream curves can have a strong influence on transcription, as has been demonstrated in many studies.[87–89] A comparison of 43 different promoters in *E. coli* found a strong correlation between the presence of upstream curves and promoter strength.[90] The promoter region of the β-lactamase gene of plasmid pUC19 has an upstream curved region, and although no effect of changing the spacing of the curve was seen on transcription *in vitro*, there was a strong dependence on rotational phasing of the curve with respect to the promoter *in vivo*.[91] This particular curve was stable, even at 60°C, in contrast to most A + T-rich curves, which often exhibit little curvature, even at 40°C, as discussed above.[46] Kim et al. have recently shown that upstream curves can enhance the effects of transcription factors.[92] There are many examples of upstream curves influencing transcription,[86] but perhaps one of the most striking examples of the importance that DNA structure can play was the experiments by Goodman and Nash, in which they replaced the binding site of a protein that was known to bend DNA, with a synthetic curve of approximately the same magnitude, and showed function in the absence of the protein.[93]

Tanaka et al. have found that the curve upstream of the *proU* promoter must be positioned properly to get high levels of reporter gene activity.[94] There was a very strong correlation between the relative gel mobility of the spacer inserts (reflecting the orientation between the upstream curve and the promoter) and β-galactosidase expression, which varied about 100-fold, in a "face of the helix" manner, depending on the spacing of the insert.[94]

Curves within the promoter region. Within the promoter region itself, curved DNA plays an important, but perhaps bit more subtle role. One of the most obvious functions is the formation of a proper RNA polymerase binding site. Ross et al. have shown that there is a "third recognition element" in some bacterial promoters, called the UP element, which is located between -60 and -40; this interacts with the C-terminal of the α-subunit of RNA polymerase.[95,96] DNA "curvature" is important, at least in the sense of relative orientation of this region of DNA with the α-subunit of RNA polymerase. The consensus sequence is A + T rich, which could reflect the importance of flexibility in bending or wrapping the DNA about the protein.[95]

In addition to this UP element, curves within the spacer element between the -35 and -10 sites can have a determinant role in promoter strength.[84,89] The geometry and flexibility of the promoter are important in determining efficient binding and initiation of RNA polymerase. For example, it has been suggested recently that the suboptimal spacing between the -35 and -10 regions of the *proU* promoter must be compensated for by an increase in flexibility with this spacer region.[97] This increased flexibility allows the -35 and -10 regions of this promoter to adopt a more favorable position for binding of the RNA polymerase σ-factor.

Downstream curves. Most endogenous curves have been found upstream or within the promoter regions.[74] In comparison, only a few curves have been characterized downstream of the transcription start site. The few reported examples of curves downstream from promoters seem to have properties distinctly different from those of the upstream curves, perhaps reflecting different roles in transcription. For example, Schroth et al. have found that there are two distinct curves flanking the promoter of an rRNA gene from *Physarum polycepahlum*; one curve is centered roughly 160 bp upstream, and the other is about 150 bp downstream of the transcription start site. Although both are curved by the same magnitude (approximately 45°), the curves exhibited different behaviors in terms of temperature dependence of the curve, the effects of EtBr, and relative mobility when the percentage of acrylamide was varied.[79] Thus these two curves would be expected to change in different ways under various environmental conditions.

III. STRUCTURE AND FUNCTION OF SUPERCOILED DNA

A. Introduction

DNA normally exists in a supercoiled form in most biological systems. Supercoiling makes structural variations in shape and helix structure particularly dynamic, with a wide variety of conformations possible through variations in twisting and writhing. Supercoiling appears to be very critical for viability in bacterial systems. *E. coli*, for example, can only tolerate mutations that change supercoiling levels by less than 25%, otherwise lethality ensues. Supercoiling is important for gene expression, DNA replication, and recombination. Although supercoiling exists in mammalian genomes, its biological significance is not yet appreciated.

Historically, various "forms" of DNA molecules have been identified, representing different topological conformations. A plasmid molecule purified from a bacterial cell will exist as a naturally occurring covalently closed circular DNA molecule that is negatively supercoiled, historically called form I DNA. DNA containing a single nick in one of the strands will lose all supercoils. The nicked molecule (also called open circular DNA) is called form II DNA. DNA that contains breaks in both phosphate backbones at the same point (or nearly the same point) along the helical axis will form a linear DNA molecule, known as form III. Form IV refers to denatured DNA (and closed circular catenated). Form V DNA is formed when two (not catenated) circular single strands anneal. Half of the DNA can form right-handed B-form DNA, and for topological reasons, half of the DNA must form left-handed turns.

B. Supercoiled Forms of DNA

1. *Lk = Tw + Wr*

Supercoiled DNA is characterized by a topological property called the linking number, *Lk*. *Lk* is defined as the number of times one strand crosses the other when oriented in a plane. *Lk* must be an integer. The relaxed DNA molecule shown in Figure 12A has a linking number *Lk* = 20. The linking number can only change when the phosphodiester backbone is broken by chemical or enzymatic cleavage.

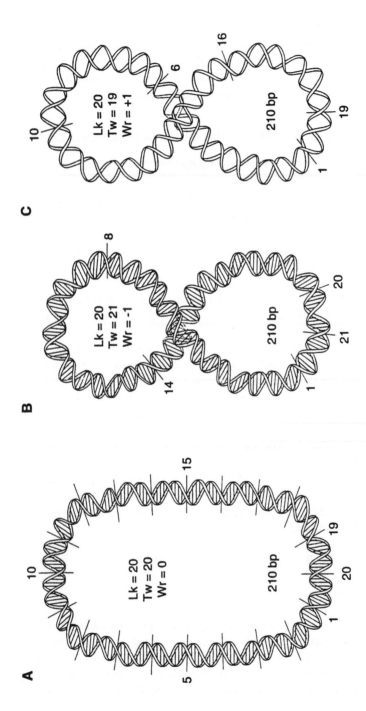

Figure 12. Positive and negative supercoils in DNA. (**A**) A covalently closed, relaxed DNA molecule containing 210 bp of DNA is shown. This molecule contains 20 helical turns, with $Lk = 20$, $Tw = 20$, $Wr = 0$. (**B**) This molecule contains one right-handed supertwist or writhe ($Wr = -1$; a negative supercoil). Because Lk cannot change, the twist must change by $+1$ ($Tw = 21$). (**C**) This molecule contains one left-handed supertwist (or positive supercoil; $Wr = +1$), and consequently $Tw = 19$.

In a covalently closed molecule, although Lk cannot change, the number of twists or turns of the double helix, as well as the number of supercoils or writhes of the double helix, can change. DNA topology is described by the simple equation

$$Lk = Tw + Wr \tag{1}$$

where Lk is the linking number, Tw is the number of helical turns in the DNA, and Wr is the writhing number of DNA. Wr describes the supertwisting or coiling of the helix in space.

The introduction into a DNA molecule with $Lk = 20$ of a negative supertwist that is a right-handed coil changes the value of Wr by -1 ($Wr = -1$) (Figure 12B). Since $Lk = 20$ and Lk cannot change, Tw must increase by +1 to a value of 21. Conversely, a decrease in the twist number by -1 to a value of 19 would require a compensating introduction of a left-handed or positive supertwist ($Wr = +1$), as shown in Figure 12C.

2. Relaxed DNA

On average, the helical repeat of DNA is about 10.5 bp per helical turn of DNA. In linear or nicked DNA, where the ends of the molecule are free to rotate, the DNA will adopt a preferred helical repeat. The preferred helical repeat of a nicked or linear DNA molecule represents the lowest energy form of the molecule. When this state of helical twist exists in a covalently closed molecule, the molecule is relaxed and contains no supercoils. In relaxed DNA, as shown in Figure 12A, the linking number equals the twist number ($Lk = Tw = 20$ and $Wr = 0$). The linking number of relaxed DNA, Lk_0, is defined as

$$Lk_0 = N/10.5 \tag{2}$$

where N is the number of base pairs in the DNA molecule and 10.5 refers to the helical repeat.

3. Negatively Supercoiled DNA

Negatively supercoiled DNA has a deficiency in the linking number compared to relaxed DNA, or $Lk < Lk_0$. Negatively supercoiled DNA is underwound with respect to helical turns, i.e., it contains fewer helical turns than the molecule would contain as a linear or relaxed molecule. This underwinding in the number of helical turns results in more base pairs per helical turn compared to B-DNA and in a decrease in the angle of twist (or the rotation/residue) between adjacent base pairs. This underwinding creates torsional tension in the winding of the DNA double helix.

The topology of negatively supercoiled DNA is illustrated in Figure 13. In Figure 13A, the 210-bp relaxed molecule has 20 helical turns, and $Lk_0 = Lk = 20$. In

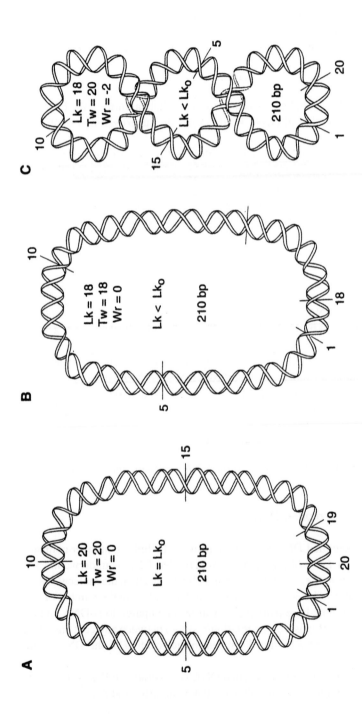

Figure 13. Introduction of negative and positive supercoils into DNA. (**A**) In this relaxed DNA molecule $Lk = 20$, $Tw = 20$, $Wr = 0$. (**B**) In this negatively supercoiled DNA molecule, the linking number deficit is 2 ($Lk = 18$), resulting in an increase in the number of base pairs per turn, a decreased angle of rotation, and $Lk < Lk_0$. (**C**) In this molecule two negative supercoils are introduced into the DNA such that $Lk = 18$, $Tw = 20$, $Wr = -2$.

Figure 13B, with $Lk = 18$, this 210-bp DNA has only 18 helical turns. This will change the average rotation per residue from 34.29° [(20 x 360°) ÷ 210] to 30.86° [18 x 360°) ÷ 210], which represents an unfavorable winding of the DNA double helix.

The molecule in Figure 13B is negatively supercoiled, since $Lk < Lk_0$, but it does not contain the familiar interwound supercoil in the molecule of Figure 13C. The linking number deficit is manifested as a twist deficit in the molecule shown in Figure 13B. It is the torsional strain in the winding of the DNA helix, a result of the decreased twist angle of DNA, which drives the supertwisting of the DNA into the supercoils shown in Figure 13C. Negatively supercoiled DNA forms a right-handed (or clockwise) supercoil. Since Lk_0 remains the same ($Lk_0 = 20$), the DNA is supercoiled by the amount $\Delta Lk = (Lk - Lk_0) = -2$. In winding in these two negative supercoils ($\Delta Wr = -2$), two additional helical turns are wound into the helix ($\Delta Tw = +2$).

4. Positively Supercoiled DNA

Although most DNA isolated from natural sources is negatively supercoiled, DNA can exist in a positively supercoiled form. DNA is said to be positively supercoiled when $Lk > Lk_0$. Positively supercoiled DNA is overwound in terms of the number of helical turns, resulting in fewer base pairs per helical turn and an increase in the winding angle between adjacent base pairs.

Overwinding a 210-bp DNA molecule by two turns to $Lk = 22$ creates the average rotation per residue from 34.29° (in relaxed DNA) to 37.71° [(22 X 360°)/210] (Figure 13). This, like the situation for negatively supercoiled DNA, represents an unfavorable state. The tension in the winding of the helix is relieved by the positive supercoiling of the DNA forming a left-handed (counter clockwise) supercoil (Figure 13D). As two positive supercoils form ($\Delta Wr = +2$), two helical turns are removed returning the number of helical turns to $Tw = 20$, which is the preferred conformation for the DNA double helix.

Positively supercoiled DNA has been isolated in what appears to be a naturally occurring form. A bacteriophage-like plasmid molecule from a *Sulfolobus* species, an archebacterium living at high temperature and low pH, has been shown to contain positive supercoils.[98] Positive supercoiled DNA would resist unwinding of the helix by heat and acid. Packaging DNA in a positively supercoiled form may be one mechanism for protecting the genetic information from denaturation.

5. Supercoils: Interwound or Toroidal Coils

In addition to existing as interwound supercoils, negative supercoils can exist as left-handed toroidal coils (Figure 14). Toroidal coils topologically satisfy the requirement for Wr, although in a toroidal coil the helix does not cross itself in a plane, as is the situation for interwound supercoil. The DNA duplex does cross

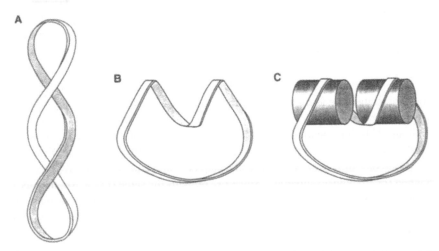

Figure 14. Right-handed interwound negative supercoils and left-handed toroidal coils. (**A**) A molecule with two interwound negative right-handed supercoils. (**B**) A molecule with two left-handed toroidal coils. (**C**) Two toroidal coils can be wrapped around a protein (represented by the cylinders).

itself in the plane of the toroidal coil. In solution, supercoils are distributed in part by a decreased angle of twist and a mixture of interwound and toroidal coils.

Knots and catenanes can have a negative or positive sign, depending on the order of the strand crossings (also called nodes). The convention for determining the sign of nodes is described in the legend to Figure 15. DNA knots and catenanes are found *in vivo* as the product of certain topoisomerases and site-specific recombination enzymes.[99–101]

6. Superhelical Density and the Specific Linking Difference of DNA

Two terms, *superhelical density* (σ) and *specific linking difference* (σ_{sp}), are frequently used to describe a level of supercoiling. Superhelical density, σ, is defined as the average number of superhelical turns per helical turn of DNA:

$$\sigma = 10.5 \ \tau/N \qquad\qquad (3)$$

where τ is defined as the titratable number of superhelical turns, N is the number of base pairs in the molecule, and 10.5 represents the average number of base pairs per turn. τ rather than Wr is used for historical reasons (see ref. 1). Since there are now many ways to measure the number of supercoils (electron microscopy, two dimensional agarose gels, as well as the original solution titration methods), τ will be defined as the number of measurable supercoils.

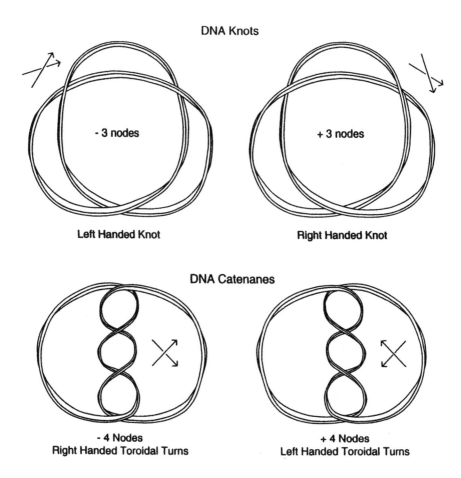

DNA Knots

- 3 nodes

+ 3 nodes

Left Handed Knot

Right Handed Knot

DNA Catenanes

- 4 Nodes
Right Handed Toroidal Turns

+ 4 Nodes
Left Handed Toroidal Turns

Figure 15. DNA knots and catenanes. (**Top**) Shown are a left-handed and a right-handed knot with negative and positive nodes, respectively. By convention, a negative node occurs when an arrow drawn along the top strand is rotated <180° in a clockwise direction to align with an arrow drawn in the same direction along the bottom double helix. When the top arrow must be rotated counterclockwise <180° to align with the bottom arrow, the node is positive. (**Bottom**) Shown are left-handed and right-handed antiparallel catenanes, each with four nodes. The positive nodes are formed by left-handed (or negatively supercoiled) toroidal turns, whereas the negative nodes are formed by right-handed (or positively supercoiled) toroidal turns.

The specific linking number difference, σ_{sp}, a term that is also used to describe the level of superhelical density, is defined as

$$\sigma_{sp} = (Lk - Lk_0)/Lk_0 \qquad (4)$$

7. The Energetics of Supercoiled DNA

The free energy of supercoiling is proportional to the square of the linking number difference according to the following relationship:

$$\Delta G = (1100 \ RT/N)(Lk - Lk_0)^2 \tag{5}$$

where R is the gas constant, T is the temperature in degrees Kelvin, and N is the number of base pairs in the DNA molecule. The free energy in supercoiled DNA can be used to drive biological reactions. Transcription and DNA replication require energy to open or unwind the DNA double helix to allow access of the bases at the center of the DNA helix. Some of this energy comes from supercoiling.

8. Effects of Temperature on DNA Supercoiling

The twist of the DNA double helix is sensitive to changes in temperature. As temperature decreases, the twist angle in DNA increases, resulting in fewer base pairs per helical turn. This results in an increase in Lk_0. In a covalently closed DNA molecule where Lk cannot change, the temperature induced change in helical winding changes Lk_0, which concomitantly affects $Lk - Lk_0$, or the superhelical density. The temperature induced change in Lk_0 is $\Delta Lk = 0.012$ helical degrees (°)/bp/°C.[102,103]

C. The Biology of Supercoiled DNA

1. Topological Domains of DNA: A Requirement for Supercoiling

Most DNA, including bacterial chromosomes and plasmid, as well as linear chromosomal DNA and small circular molecules in human cells, is negatively supercoiled. Covalently continuous circular molecules are topologically closed and thus have a defined linking number. DNA molecules must exist in a topologically closed form to be supercoiled. The *E. coli* chromosome exists as a large (2.9 × 10^6 bp) closed circle that is subdivided into 45 ±10 independent topological domains *in vivo*. A topological domain is defined as region of DNA bounded by constraints on the rotation of the DNA double helix. The nature of the molecules responsible for the topological domains *in vivo* remains to be established. Domains may be created by the attachment of DNA to the bacterial membrane through specialized binding proteins, DNA gyrase, or other topoisomerases (such as Topo IV).

Linear DNA must become organized into one or several topological domains to exist in a supercoiled state. Following infection, the ends of the linear 120-kb bacteriophage T4 chromosome are prevented from rotating, thus organizing the entire chromosome into a single topological domain. The model for the organization of human (eukaryotic) chromosomes involves independent loops, likely formed by the interaction of specific regions of DNA with defined proteins that attach to the nuclear matrix.

2. DNA Topoisomerases

Topoisomerases transiently break and reseal phosphodiester bonds in DNA, allowing one strand to pass through the other strand. This effectively provides a swivel to allow one strand to rotate around the other. It is topoisomerases that are responsible for changes in the linking number of DNA in living cells. In bacterial cells, negative DNA supercoils are introduced by DNA gyrase, and they are removed by DNA topoisomerase I. These enzymes act by breaking the phosphodiester bond and forming a covalent bond between a specific tyrosine in the protein and a phosphate group on the 5' or 3' end of the DNA. In some cases the enzymes require ATP as an energy source (see Sinden[1] for details and specific properties).

Type I and type II topoisomerases. Type I topoisomerases break one strand of the DNA, and type II topoisomerases break both strands of the DNA. The linking number of DNA, *Lk*, will correspondingly change in increments of 1 or 2 for type I and type II topoisomerases, respectively. *E. coli* topoisomerase I, the first DNA topoisomerase to be discovered, relaxes negatively supercoiled DNA. Eukaryotic type I topoisomerases can relax negative or positive supercoils. ATP is not required for type I topoisomerase activity. DNA gyrases, type II topoisomerases, are the only enzymes that can catalytically introduce negative supercoils into relaxed DNA in an ATP-dependent process. In the absence of ATP, gyrase can relax DNA.

The biological importance of DNA topoisomerases. Topoisomerases maintain a precise level of supercoiling in *E. coli*. Mutations in DNA gyrase can result in a decrease in the level of supercoiling in living cells. Similarly, mutations in topoisomerase I can result in an increase in the level of supercoiling in cells. The topoisomerase genes encoding DNA gyrase and topoisomerase I are essential. The cell will only tolerate mutation resulting in changes in linking number of about 25%. A deletion mutation of topoisomerase I was viable when a compensatory mutation in DNA gyrase occurred, reducing the ability of this enzyme to supercoil DNA. A number of excellent reviews on DNA supercoiling and the regulation and genetics of the level of supercoiling in *E. coli* are available.[104–109]

All biological reactions involving DNA likely require DNA topoisomerases. For example, following replication, two circular chromosomes will become catenated. A type II topoisomerase activity is required to transiently break the DNA double helix, allowing a duplex from the second chromosome to pass through the duplex of the first chromosome. Without this strand-passing reaction following replication, the two chromosomes cannot physically separate and the cell cannot divide.

As discussed later in this section, the movement of RNA polymerase can generate superhelical tension. Topoisomerases are needed to relax positive and negative supercoiling generated by proteins tracking through the DNA. Without these enzymes, transcription and replication would likely cease.

3. Mechanisms of Supercoiling in Cells

Supercoiling in bacterial cells is driven by the activity of DNA gyrase. DNA gyrase was the first enzyme identified that can negatively supercoil DNA. Purified DNA gyrase introduces negative supercoils to a superhelical density of about $\sigma = -0.1$ in a reaction requiring ATP hydrolysis. In living cells superhelical density is regulated by the opposing activities of DNA gyrase and topoisomerase I. The balance of supercoiling by DNA gyrase and relaxation by Topo I keeps the DNA at a finely tuned level of supercoiling in living cells. The ATP/ADP ratio is also very important in maintaining the level of supercoiling *in vivo* as gyrase activity is influenced by this ratio.[110]

Supercoiling in eukaryotic cells may be introduced by organization of DNA into nucleosomes and topoisomerase activity. In eukaryotic cells, DNA is wrapped in a left-handed fashion around a histone octamer, forming a nucleosome. The left-handed toroidal coiling of DNA around a protein, which restrains a negative supercoil, introduces a positive supercoil into the nonnucleosomal DNA in the same topological domain, with no change in Lk. Either eukaryotic type I or type II topoisomerases can relax the positive supercoil, resulting in the net introduction of a negative supercoil. Note that wrapping DNA twice around the nucleosome ($\Delta Wr = -2$) results in the introduction of a single positive supercoil. This linking number paradox, where $\Delta Lk = -1$ and $\Delta Wr = -2$, can be resolved by a compensatory change in Tw (equal $\Delta Tw = +1$). The helix repeat changes from about 10.5 to <10.5 in DNA that is wrapped around a nucleosome.

Transcriptional effects on supercoiling. In 1986, Pruss and Drlica discovered that the negative superhelical density of a plasmid was twice that normally found *in vivo* when the plasmid was purified from cells containing a mutation in topoisomerase I.[111] The plasmid sequences responsible for this unusually high supercoiling included the promoter region for the tetracycline resistance gene. Moreover, the unusually high supercoiling was dependent on transcription from the tetracycline gene. In 1983, Lockshon and Morris showed that positively supercoiled plasmid DNA could be purified from *in vivo* cells treated with novobiocin, an antibiotic that inhibits the activity of DNA gyrase.[112] Liu and Wang then argued that the movement of an RNA polymerase during transcription would generate negatively supercoiled DNA behind the RNA polymerase while generating positively supercoiled DNA in front of the RNA polymerase (Figure 16).[113] Within a topological domain of DNA when RNA polymerase moves through the

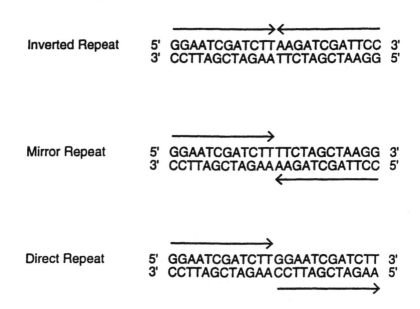

Figure 16. Transcription-dependent DNA supercoiling. Transcription can introduce DNA supercoiling as described in the text. (**A**) The single topological domain, which contains 210 bp of DNA, is divided into two subdomains by an RNA polymerase molecule. (**B**) Movement of RNA polymerase by 20 bp without rotation of the RNA polymerase results in a decrease in helical turns ($Lk < Lk_0$) behind the polymerase and an increase in helical turns ($Lk > Lk_0$) in front of the polymerase.

DNA without rotation, it divides the domain into two topologically separate subdomains. The movement of RNA polymerase (when Lk is invariant) changes the relationship of $Lk - Lk_0$, since Lk_0 behind RNA polymerase increases, whereas Lk_0 in front of RNA polymerase decreases. This creates two subdomains in which $Lk < Lk_0$ behind the polymerase and $Lk > Lk_0$ in front of the polymerase, introducing negative and positive supercoiling, respectively. To introduce negative supercoiling into DNA during transcription, topoisomerase activity must relax the positive supercoils ahead of RNA polymerase.

DNA supercoiling and gene expression. The state of DNA supercoiling can influence the regulation of gene expression, as reviewed by Drlica,[104] Esposito and Sinden,[114] and Freeman and Garrard.[115] The conformation of DNA can also influence gene expression. The very different structure of cruciforms, Z-DNA, or triple-stranded regions might provide a switch that could turn a gene on or off. A protein, for example, might bind to a B-form DNA sequence to turn on a gene, and it could not bind if the sequence existed in the alternative configuration. A model system using an inverted repeat that can form a cruciform has been shown to modulate gene expression *in vivo.*[116]

The energy from DNA supercoiling can be used to drive the opening of the promoter region, thus facilitating RNA polymerase binding. Different promoters are "tuned" for optimal transcription at different levels of supercoiling. This can be accomplished in a number of ways. For example, changes in the A + T content of the promoter could influence melting. The spacing of the -35 and -10 regions allows enormous variation in the promoter characteristics. Supercoiling differentially affects five or six kinetically distinct steps in transcription initiation, as reviewed by McClure.[117]

The regulation of the *E. coli* DNA gyrase and topoisomerase genes are a classic example of DNA supercoiling-dependent gene regulation. Transcription from the DNA gyrase gene is turned on by a low level of supercoiling and is turned off when the level of supercoiling is high. Moreover, the topoisomerase I gene is turned on when supercoiling levels are high. Thus while the opposing enzymatic activities of DNA gyrase and topoisomerase I regulate the level of supercoiling, the levels of these enzymes are also regulated by DNA supercoiling.

A supercoiled DNA template is required for precise initiation of DNA replication at the *E. coli* DNA *oriC*. The *oriC* DNA is wrapped into a toroidal coil around a complex of DnaA proteins. In an ATP-dependent process, three 13-bp AT-rich direct repeats are stably unwound, exposing this region to recognition and binding by a DnaB–DnaC complex. Other proteins then assemble on DNA to initiate replication from *oriC*.

IV. CRUCIFORM STRUCTURES

A. Introduction

Within the DNA there are defined, ordered sequences (dosDNA). Such elements include inverted repeats, mirror repeats, and direct repeats (Figure 17), which can form cruciforms, intramolecular triplex DNA, and slipped mispaired structures, respectively.

An inverted repeat (or palindrome) is a DNA sequence that "reads" the same in either direction (from the 5' to 3' in either strand). Inverted repeat sequences are widely distributed in the chromosomal DNA of many eukaryotes, including plants, yeast, *Neurospora*, *Physarum*, *Drosophila*, mouse, *Xenopus*, and human (for a review, Pearson et al.).[118] In all cases their distribution is nonrandom and is clustered at or near regions of genetic regulation. Cruciform structures may transiently form under physiological conditions to serve as recognition signals for regulatory proteins of transcription, recombination, or replication (see above). Some possible roles that inverted repeats may play are described below.

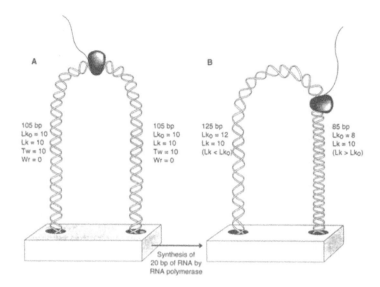

Figure 17. Defined ordered sequence DNA (dosDNA). Inverted repeats, mirror repeats, and direct repeats represent defined ordered sequence DNAs (dosDNAs). The arrows above and below the base sequences show the organization of symmetrical complementary sequences in DNA. In inverted repeats (or palindromes) the two strands of the DNA are complementary to each other, as well as being self complementary. A mirror repeat has identical base pairs in one strand surrounding a center of symmetry. Arrows show complementary base pairs in the mirror repeat sequence. Certain mirror repeats can form intramolecular triplex structures (as described in Section VI). Direct repeats contain a particular sequence that is repeated or duplicated. Direct repeats can be adjacent (as in the example shown), or they can flank an intervening sequence. Direct repeats can form slip-mispaired structures (as described in Section VII).

B. Formation and Stability of Cruciforms

1. Cruciform Formation Requires DNA Supercoiling

Energy from DNA supercoiling is involved in melting the center of the inverted repeat, allowing the intrastrand nucleation required for cruciform formation. For cruciform formation about 10 bp must unwind at the center of symmetry (Figure 18A).[119–121] This provides a region in which intrastrand base pairing can occur. Following nucleation, the inverted repeat extrudes as a cruciform. Cruciforms form in supercoiled but not relaxed DNA.[122–124] A very specific minimal level of superhelical energy is required for cruciform formation, σ_c.[125] DNA supercoiling is also required for cruciform stability. Stability comes from the relaxation of negative supercoils as cruciforms are formed. Cruciform formation results in the relaxation of one negative supercoil per 10.5 bp of DNA that forms the cruciform. The relaxation of supercoils reduces the free energy (ΔG) of supercoiling.

A. S-type Kinetics   **B. C-type Kinetics**

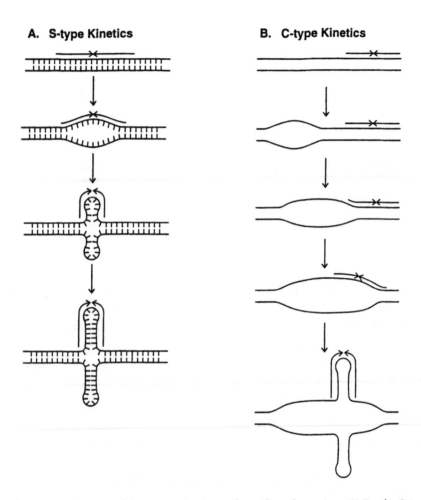

Figure 18. S-type and C-type mechanisms of cruciform formation. (**A**) For the S-type cruciform mechanism, 10 bp at the center of symmetry must unwind. Nucleation occurs with intrastrand hydrogen bond formation flanking the center of symmetry. Following nucleation, branch migration occurs in the extrusion process. (**B**) For C-type kinetics an A + T-rich region of DNA flanking the inverted repeat breathes, forming a denaturation bubble. The bubble becomes enlarged, encompassing the inverted repeat. Within this unwound region nucleation can occur, resulting in formation of a cruciform structure.

2.   Effect of Base Composition on the Formation of Cruciforms

The rate of unwinding the center of an inverted repeat depends on its base composition as well as temperature, ionic strength, and superhelical density of the DNA. The thermal stability of the central 10 bp is important in defining the rate of cruciform formation. In general, there is a correlation between the T_m of the entire inverted repeat and the rate of cruciform formation: DNAs with lower T_m values form cruciforms more easily than sequences with higher T_m. As expected, there is a cooperative relationship between superhelical density and temperature for cruciform formation.

3. C-Type Cruciform Formation

The characteristics of cruciform formation discussed above pertain to solutions of physiological ionic strength. This type of transition has been called the S-type (where S refers to salt-dependent). The S-type transition is dependent on supercoiling, temperature, ionic conditions, as well as divalent cation.[126–128] A second mechanism, called the C-type, occurs in solutions lacking salt[129,130] (Figure 18B). Under C-type conditions, the rate of cruciform formation is independent of base composition.

C-type behavior is due to A + T rich DNA sequences, called C-type inducing sequences, located within several hundred base pairs of the inverted repeat.[130,131] Under conditions of low ionic strength, the A + T-rich C-type inducing sequence unwinds, forming a large unwound region that probably includes the inverted repeat.[132] Nucleation leading to hairpin formation can occur within this unwound region (Figure 18B). A C-type transmitting sequence must be present between the C-type inducing sequence and the inverted repeat for C-type cruciform formation.[131] The presence of a G_4C_4 block between the A + T-rich C-type inducing sequence and the inverted repeat will prevent C-type cruciform formation. Presumably, the high thermal stability of G + C-rich DNA resists melting and prevents unwinding through the inverted repeat.

4. The Removal of Cruciforms from Supercoiled DNA

There are three ways in which cruciforms can be converted back to the linear form: by introduction of positive DNA supercoils, heating above the T_m of the inverted repeat, and removal by transcription or DNA replication.

Introduction of positive supercoils. The introduction of positive supercoils into DNA will drive cruciforms back into the linear form, since they require negative supercoiling for stabilization. Positive supercoils can be introduced by the binding of ethidium bromide, which intercalates into DNA. Upon the removal of ethidium bromide, the DNA becomes reequilibrated with negative supercoils, and if ethidium bromide is removed at 0–4°C, this low temperature prevents the formation of cruciforms (except for $(AT)_n$ inverted repeats).

Heating above the T_m. Cruciforms are stable in supercoiled DNA below the T_m of the hairpin arm. Upon incubation above the T_m, the hairpin arm is melted, and upon cooling, the inverted repeat returns to the linear form, presumably because two nucleation events are provided by the plasmid DNA flanking the inverted repeat. Reannealing from the ends of the inverted repeat results in rapid formation of linear DNA. It is probably the kinetics of this process, compared to the slower rate of intramolecular base pairing, that drives the formation of linear DNA, despite the fact that cruciforms are thermodynamically stable in supercoiled DNA.

The removal of cruciforms by transcription and DNA replication. Transcription by RNA polymerase through a cruciform will reconvert an inverted repeat into the linear form. As RNA polymerase traverses the cruciform, it will necessarily unwind the base pairing in the hairpin arms, and the duplex DNA behind the RNA polymerase provides a nucleation site for hybridization of the complementary strands, as demonstrated experimentally by Morales et al.[133] Transcription and replication are probably responsible for removing cruciforms from DNA *in vivo*.[134]

C. Cruciform Structure

Cruciforms have two structural characteristics, the four-way DNA junction and the stem loops.

1. Four-Way Junctions

The three-dimensional structure of four-way DNA junctions (Holliday structure) has been a field of intense investigation (Figure 19). Most of the structural analysis of four-way DNA junctions has been modeled on small stable junctions composed of four synthetic oligonucleotides (reviewed by Lilley and Clegg[135] and Wemmer et al.[136]). Using these stable junctions, it was found that (1) four-way junctions are normally fully base paired[136]; (2) all four branches are in the right-handed B-conformation[137,138]; (3) there is twofold symmetry,[139] which is in agreement with the stacked X-conformation[138] (see below), and the most likely isomer is the antiparallel conformer.[140,141] The structure depends critically on the DNA sequence, mainly at the junction, which determines the distribution of the stereoisomers,[142,143] and on the type and amount of counterion used in the solutions, which determes the geometry of the helices.[144] In the absence of salt, the junction is in an extended conformation, probably planar with fourfold symmetry, with unstacked bases at the junction.[145] Micromoles of Mg^{2+} enable the four-way junction to adopt a more compact, X-shaped structure with twofold symmetry, with pairwise coaxial stacking of helices and apparently no unpaired bases.[145] In the presence of Na^+ (≥ 50 mM), the structure is similar to that in the presence of Mg^{2+}, in that it is also compact (X-shaped)[135,146]; however, the structure has only imperfect twofold symmetry,[147] and the junction bases are still unstacked.[144,145]

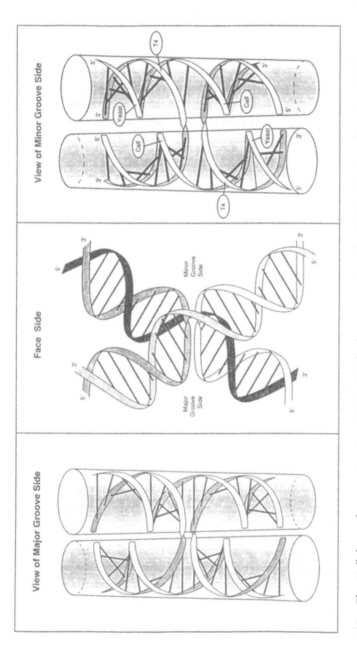

Figure 19. The Holliday or four-way junction. (**Center**) Helical representation of a Holliday junction. The four individual strands have different shadings. (**Left**) This representation shows the major groove side of the two DNA helices. In a Holliday junction, no Watson–Crick hydrogen bonds are broken, and all base pairs remain stacked. (**Right**) This representation shows the minor groove side of a Holliday junction. Also shown are the sites of cleavage of the yeast, calf, and T4 Holliday structure resolvases. (Figure modified with permission from Lilley and Clegg.)[135] Reproduced with permission from the *Annual Review of Biophysics and Biomolecular Structure, Volume 22.* ©1993 by Annual Reviews, Inc.

It has been reasoned that the more physiological structure is the more compact (X-shaped), rather than the extended conformation.[135,146,148]

The structural model of four-way DNA junctions has recently had several interesting developments, namely that the given structure of a particular set of sequences does not have a stable steady-state structure.[149] It is interesting to note that cruciforms can induce DNA bending,[150] and that four-way junctions are relatively flexible.[151] Regardless of the geometric conformation, each four-way DNA junction possesses both a minor groove face and a major groove face. On one face the four base pairs at the point of strand exchange all present minor groove edges, and on the other side, the corresponding major groove edges are presented.[145]

2. Stem Loops

Single-stranded loops at the tip of cruciform arms or DNA hairpins are sensitive to single-strand nucleases. For direct inverted repeats, with no intervening sequence between the repeats, in the cruciform conformation the loops contain two or three unpaired bases.[152,153] For nondirect inverted repeats, the loop size is dependent upon the length of the intervening sequence. Hairpin loops can involve base stacking,[154] non–Watson–Crick base pairing,[155,156] and, possibly, extrahelical bases.[157] Hairpin loop conformation and dynamics are exquisitely sensitive to small changes in the bases in the loop and adjacent stem sequences.[154,158] Similarly, the presence of hairpin loops can affect the stem structure[159] and may affect the flexure, writhe, and torsional strain experienced by the molecule. Intrahelical torsion may severely affect the conformation of a four-way junction.[151] Hence, for cruciforms with short arms, the presence of a hairpin loop, as opposed to a free end, may affect the overall conformation of the four-way junction. The presence of loops could impart a certain stiffness to the DNA structure, which may be important for protein recognition of the target, such as a cruciform/hairpin as opposed to a Holliday junction, a DNA cross-over or a looped (wrapped) DNA molecule.[160]

3. Four-Way DNA Junctions, Holliday Junctions, and DNA Cross-Overs

Holliday structures, believed to occur as recombination intermediates,[161] contain a four-way DNA junction. In the case of the Holliday structure, it is a junction between two separate DNA molecules (whereas a cruciform is an intramolecular structure). DNA cross-overs are regions where two separate DNAs cross, as would occur at points where two double helices intersect by the process of supercoiling, looping, or folding, and at nucleosome linkers where DNA strands enter and exit the nucleosome.[162] DNA cross-overs also resemble four-way DNA junctions, except that there is no covalent interaction between the two helices. Both Holliday junctions and DNA cross-over structures, although similar, are biologically distinct from cruciforms.

4. Branch Migration of Four-Way DNA Junctions

Both the extrusion of cruciforms from inverted repeats as well as the formation of Holliday junctions during genetic recombination require branch migration. Branch migration involves the stepwise breakage and reformation of hydrogen bonds between base pairs as one strand is exchanged for another in the two arms of a four-way junction. The point of exchange occurs at the junction. This is a key step in most models of homologous recombination and is implicated in site-specific recombination. The number of hydrogen bonds formed equals the number broken, and the process can occur spontaneously. It occurs between regions of homology; in the case of cruciforms, branch migration is limited to the boundaries of the inverted repeat being extruded. Panyutin and Hsieh have shown that single base mismatches, such as those that occur in imperfect inverted repeats, can impede the process of branch migration.[163] Recent evidence indicates that the process of branch migration of DNA in four-way junctions, long thought to occur spontaneously at a rapid rate, occurs at a very slow rate.[164]

The process of spontaneous branch migration has been modeled as a one-dimensional random walk in which there is an equal probability of moving forward or backward at each step. The rate of spontaneous branch migration is exceedingly sensitive to metal ions. The rate of branch migration increases dramatically at concentrations of $MgCl_2$ below 500 μM. In the absence of any metal ions, branch migration is exceedingly fast, with a step time of 50 μsec.[165] This increase in branch migration rate coincides with the loss of base stacking in the four-way junction over similar concentrations of the divalent ion. As described above, a constrained (immobile) four-way DNA junction adopts different conformations, depending on the metal ion present. In the absence of metal ions, the junction adopts a square-planar conformation. In the presence of multivalent ions, such as Mg^{2+}, the junction assumes a stacked X structure, in which base stacking is retained through the cross-over point. It seems that the presence of multivalent metal ions at concentrations that induce base stacking at the junction also results in relatively slow rates of spontaneous branch migration.[165] This indicates that the structure of the Holliday junction is a critical determinant of the rate of spontaneous branch migration. The rate of branch migration in the presence of Na^+ alone is similar to that observed in the presence of low concentrations (100 μM) of Mg^{2+}, which probably reflects a loss of base stacking in both instances.

It has been proposed[165,166] that the Holliday junction undergoes an isomerization at each migratory step, from a folded, stacked X structure to a square-planar conformation. This structural isomerization may involve the displacement of Mg^{2+} in the vicinity of the cross-over, leading to the disruption of base stacking at the junction. Such conformational changes are likely to require energy, and therefore the process is slow.

Branch migration is known to affect binding, protein-induced DNA structural distortion, and endonucleolytic activity of phage, bacterial, and yeast Holliday junction resolvases.[167–173] There are cellular proteins that can drive branch migration[174] (reviewed by Adams and West).[175]

D.   Assays for Cruciform Structures in DNA

Many assays for cruciforms have been developed. Many of these are very similar to assays applied to the analysis of Z-DNA, intramolecular triplex DNA structures, as well as other non–B-DNA structures. Table 3 summarizes the basics of these assays.

1.   Application of the Psoralen Crosslinking Assay
for Quantitating Cruciforms

Psoralen is an excellent probe for *in vivo* studies because it is freely permeable into eukaryotic cells and reasonably permeable into bacterial cells.[182–184] Because of the low solubility of many psoralen derivatives and the relatively low binding constant to DNA, low levels of psoralen bind to DNA by intercalation in its "equilibrium dark-binding" mode, even when a solution is saturated with psoralen. On absorption of 360 nm light, covalent cyclobutane rings are formed between one or both ends of the psoralen molecule and pyrimidine bases. The extent of DNA cross-linking can be precisely controlled by varying either the light dose, the psoralen concentration, or both. Using the psoralen crosslinking assay, Zheng et al. demonstrated the superhelical density-dependent existence of a series of 106-bp cruciforms with A + T-rich centers in *E. coli*.[134]

2.   Nuclease Assays for Cruciforms in Cells

If a nuclease existed in cells that specifically recognized and cut cruciforms, then identification of specific cutting at inverted repeats should provide an indication of the existence of cruciforms. Panayotatos and Fontaine cloned the T7 endonuclease VII gene on a plasmid under transcriptional control of a regulated promoter.[186] When the gene was turned on and the nuclease was expressed in cells, cuts were made at the inverted repeats. In addition, cuts were made throughout the entire chromosome, resulting in complete destruction of the intracellular DNA. This experiment is certainly consistent with the interpretation that cruciforms existed in cells. It is difficult to rule out, however, that the nuclease was also recognizing other structural features in chromosomal DNA.

Table 3. Assays for Cruciform Structures

Assay	Description	Representative Reference
Single-strand nuclease cutting of cruciform loops	S1 nuclease is a single-stranded endonuclease that does not cut relaxed plasmid DNA containing the linear form of an inverted repeat. Digestion of supercoiled plasmid DNA, however, results in the introduction of a double strand break at the center of the inverted repeat. The location of the cut can be mapped by digestion with a restriction endonuclease, which cuts the plasmid DNA once at a known site. From analysis of the sizes of the two fragments on agarose gels, the S1 nuclease-sensitive site can be determined. The analysis of restriction-S1 fragments on DNA sequencing gels can identify the loop of a cruciform with base pair resolution.	Lilley[176] Panayotatos and Wells[123]
Chemical modification at the loop of hairpin stems	The reactivity of the unpaired bases at the tips of hairpin arms to certain chemicals, including chloroacetaldehyde, osmium tetroxide, and diethylpyrocarbonate, provides an assay for cruciforms. Chemical cleavage (after modification) can be used to identify the site of modification with base pair resolution. Alternatively, sites of modification can be identified by primer extension procedures, since DNA polymerization will stop at the modified base. The products of either protocol are analyzed on a DNA sequencing gel.	Palecek[177] Furlong et al.[178]
Analysis using cruciform resolvases	Resolvases, which cut cruciforms at the base of the arms, can be used to identify cruciforms in DNA.	Lilley and Kemper[179]
Psoralen interstrand cross-linking assay for cruciforms	Psoralen forms interstrand cross-links between the two strands of a DNA duplex on exposure to 360-nm light. The inverted repeat can be covalently locked into either the cruciform or linear form. Because the cross-linked hairpin arm is half the mass of the cross-linked linear form, these DNA species from cloned inverted repeats migrate very differently on a polyacrylamide gel, permitting analysis of the extent of cruciform formation. This assay is applicable *in vivo*. Cross-linking a cruciform also locks a bend, inherent in the four-way junction into DNA, that provides an additional assay for cruciforms.	Sinden et al.[180] Zheng and Sinden[120]
DNA topological analysis	Cruciform formation results in the relaxation of one negative supercoil for about every 10.5 bp extruded into a cruciform. The change in electrophoretic mobility on an agarose gel provides a very sensitive assay for cruciform formation.	See Sinden[1]
Alteration of restriction cleavage	When cruciforms contain a restriction enzyme site at the center of the inverted repeat, digestion will occur in the linear but not the cruciform form.	Mizuuchi et al.[124]
Cruciform-binding antibodies	Cruciform-specific antibody binding, which can be detected by a variety of methods, provides a sensitive assay for cruciforms.	Frappier et al.[181]

3. Detection of Cruciforms in Cells Using Chemical Probes

Certain chemicals that modify the unpaired bases at the loop of a hairpin arm can enter cells and preferentially modify these regions of DNA. Following modification, the DNA can be purified and analyzed to identify the site of modification. A disadvantage of this approach is that the cells are killed following treatment with many chemical probes. Osmium tetroxide (OsO_4), for example, severely damages living cells. This approach has been used by McClellan et al. to modify the loops of cruciform arms in a series of $(TA)_n$ inverted repeats.[187] Sequences greater than 30 bp were chemically modified, but 24-bp sequences were not reactive. Moreover, the level of chemical modification was higher with increasing lengths of the inverted repeats. Since the level of negative supercoiling required for cruciform formation increases with shorter $(TA)_n$ inverted repeats, the longest inverted repeats would be expected to exist as cruciforms at the highest levels in cells.

4. Analysis of DNA Supercoiling as an Indication of Cruciform Formation in Cells

The formation of a cruciform in a plasmid results in the relaxation of one negative supercoil for about every 10.5 bp of DNA extruded into a cruciform. *E. coli* maintains a precise level of negative supercoiling in cells. Therefore, the relaxation of negative supercoils in cells by the formation of a cruciform should theoretically induce a decrease in the linking number in plasmids containing extruded cruciforms and lead to the restoration of the original level of supercoiling by the action of DNA gyrase. Following plasmid purification from cells, providing that no change in the linking number occurred during purification, a population of topoisomers should be present that have more negative supercoils than plasmids in which the cruciform did not form. The increase in the number of supercoils upon cruciform formation *in vivo* (ΔLk_c) should be approximately equal to $\Delta Lk_c = N/10.5$, where N is the number of base pairs in the inverted repeat that have extruded into the cruciform. Typically, the formation of cruciforms *in vivo* does not occur in all topoisomers of a plasmid population. In the case where cruciforms formed in 50% of the topoisomers, rather than a single Gaussian distribution of topoisomers there would be two overlapping Gaussian distributions. The higher negatively supercoiled population, resulting from cruciform formation, would be shifted by ΔLk_c. Haniford and Pulleyblank used this system to present one of the first indications of formation of cruciforms in *E. coli*.[188]

5. Application of Cruciform Antibodies to the Analysis of Cruciforms in Eukaryotic Cells

Monoclonal antibodies have been raised and isolated that recognize structural features of cruciforms, but not the DNA sequence of specific inverted repeats. *In vitro* antibody binding to DNA can be detected by the DNA filter binding or the retardation of cruciform-containing DNA during gel electrophoresis.[181] When antibodies are applied to isolated nuclei, they can be detected by binding a second fluorescent antibody to the anticruciform antibody. The localization of the fluorescence by microscopy reveals the sites and relative abundance of cruciforms.

In eukaryotic cells, cruciforms are detected most strongly at the G_1/S boundary of the cell cycle, just before the beginning of the period of DNA synthesis (S phase). Ward et al. estimated that there might be as many as 3×10^5 cruciforms per eukaryotic nucleus and suggested that inverted repeats may form cruciforms as a prerequisite for the initiation of DNA replication.[189,190] As discussed below, inverted repeats are quite common at origins of DNA replication. Perhaps the formation of a cruciform triggers initiation of DNA replication at a specific origin. Alternatively, cruciforms may simply accumulate at the G_1/S boundary, during the period when no replication is occurring. As soon as synthesis begins, cruciforms may be converted by replication through the inverted repeat back into the linear form.

Caution should be exercised in interpreting the cruciform antibody (and Z-DNA antibody, see Section V) studies of alternative conformations in eukaryotic cells. These analyses require the isolation and purification of nuclei to allow the antibody to enter the nucleus. The cells must be gently lysed, removing the nuclei from their natural environment. It has been shown by Cook et al. that even the gentlest isolation procedures result in significant changes in the chromatin organization of DNA.[191] In eukaryotic cells, negative supercoiling in the bulk of chromosomal DNA is restrained by the organization of DNA into nucleosomes. Relaxed DNA should not support the formation of cruciforms (or Z-DNA or triplex structures). However, if the chromatin structure is disrupted and nucleosomes are lost, unrestrained supercoils may be introduced into DNA as an artifact and may not reflect the natural *in vivo* situation of the DNA. It is interesting to note that in addition to these studies on cruciforms, there are several reports of other non–B-DNA structures that appear to be dynamically regulated throughout the eukaryotic cell cycle, including triplex DNA,[192] Z-DNA,[193,194] and single-stranded DNA.[195–198]

E. The Biology of Inverted Repeats

1. Protein Binding Sites

Many short inverted repeats (4–20 bp) represent the binding site for specific proteins. Restriction endonuclease cleavage/modification sites range from 4 to 10 bp in length. However, recognition and cleavage of these sites occur in the linear form, since most restriction sites are too short to form a cruciform (which requires a 3–4-bp loop at the end of the hairpin stem). Many DNA operator sequences have quasipalindromic symmetry, in which the inverted repeat is not perfectly symmetrical. The binding of dimeric repressor proteins usually involves recognition of a 6–10-bp DNA sequence by each monomer repressor.

2. Inverted Repeats and Other Sequence Elements Associated with Eukaryotic Replication Origins: Regulation of Replication Initiation

A pressing question in cellular and molecular biology is how the cell limits DNA synthesis to one round per cell cycle. The mechanism that inhibits reinitiation is not known, although chromatin conformation, chromatid pairing, and nuclear membrane permeability may be involved in the process. Replication of genomic DNA is restricted to the S phase of the cell cycle, and control mechanisms ensure that the entire genome is replicated only once per cell cycle.

Replication origins and replicators, the specific sequences that control the initiation of DNA replication, are poorly defined in mammalian cells.[199,200] Unlike the small simple genomes of prokaryotes and viruses, the multichromosome genome of a mammalian cell is more complex. With this increased complexity one might expect increasingly complex or more numerous modes or levels of initiation regulation. However, despite the lack of a particular consensus sequence, there are certain types of sequences that are common to many replication origins of parasitic, prokaryotic, eukaryotic, and mammalian organisms, some of which are described below.

A- l T-rich sequences.   The helical axis of DNA can be curved unidirectionally if there exists a series of A tracts that are phased in their spacing (reviewed by Hagerman[201]; also see Section II above). Curved DNA is known to occur at prokaryotic, yeast, viral, and mitochondrial replication origins. Curved DNA has been associated with the DHFR and c-*myc*-associated origin regions, as well as monkey and human origin enriched sequences *(ors)*.

DNA unwinding elements.   Common to most origins are sequence elements that facilitate the unwinding of the DNA. DNA unwinding elements (DUEs) are specific but not unique sequences. They are usually A + T-rich, and their function is dependent upon the base stacking interactions.[202,203] DUEs were first recognized as regions of single-strand nuclease hypersensitivity contained within the *E. coli*[204] and yeast[202,205] replication origins. Mutations in the DUE of the yeast autonomously replicating sequence (ARS) that inhibit DNA unwinding *in vitro* also inhibit ARS activity *in vivo*.[202]

DUEs are believed to function by unwinding the two strands to permit the entrance of the replication machinery to initiate replication, some of these regions extend to lengths up to 100 nucleotides. Recently the DUE of SV40 was found to be within the G + C-rich early region adjacent to the central inverted repeat.[206] This corresponds directly to the site of initial primer synthesis.[207,208] Cruciforms may play a role in DNA melting, for the easily melted regions of pBR322 contain inverted repeats.[209,210]

The Epstein-Barr virus replication origin has its DUE located at the origin of replication, oriP.[211] Recent evidence indicates that both the dyad symmetry element and the family of inverted repeats are sensitive to single-strand nucleases.[211] In duplex DNA the structure of the dyad symmetry element is a large single-stranded bubble containing a stem loop formed by the 65-bp dyad, whereas the family of repeats are in the cruciform conformation. Williams and Kowalski concluded that the intrinsic ability of the oriP elements to form alternative structures may be important in the initiation process, specifically facilitating the access of the replication machinery to the parental DNA strands.[211] The unwound single-strand bubble containing a base paired hairpin is reminiscent of the single-strand initiation (*ssi*) signals of plasmids and phages. It is tempting to speculate that the unwound oriP is recognized by replication factors in a fashion similar to that of the *ssi* signals.

Inverted repeats.   Inverted repeats are common sequence elements in many prokaryotic and eukaryotic replication origins (reviewed by Pearson et al.)[118] and are important for the initiation of DNA replication in phages, plasmids, prokaryotes, and viruses of both prokaryotes and eukaryotes.

Hairpins in single-stranded bacteriophage genomes.   Inverted repeats in single-strand DNA can form hairpins, and in single-stranded chromosomes these are important for replication. A region containing three inverted repeats of 44, 21, and 20 nt is required for the initiation of replication in bacteriophage G4 (see Table 4.2 in Sinden).[1] The hairpin structures of the 44-bp and 21-bp inverted repeats are required for this region to function as the origin. Following infection, the DnaG primase protein binds to the hairpin formed within the 20-bp inverted repeat as a prerequisite for initiation of synthesis of an RNA primer.

Eukaryotic viruses.   Inverted repeats are associated with many eukaryotic viral origins of replication. For example, the SV40 origin of replication contains a perfect 27-bp inverted repeat and an imperfect 15-bp "early palindrome." The linear duplex forms of the inverted repeats are required for replication. Herpes simplex virus contains ori_{L1}, a 144-bp A + T-rich inverted repeat with a perfectly symmetrical central 20-bp region,[212] and ori_{L2}, a 136-bp inverted repeat that is similar to ori_{L1}.[213]

Plasmid pT181: Is a cruciform cut to begin replication?   Plasmid pT181 contains a small inverted repeat that constitutes the origin of replication. This inverted repeat can form a cruciform *in vivo,* and it may be involved in the initiation of replication.[214] RepC, an initiation protein, binds to the origin of single-stranded or double-stranded DNA and introduces a nick at the center of the inverted repeat.

Replication begins at the nick. The binding of the RepC protein is enhanced by the formation of an unwound structure, possibly the cruciform, at the origin.

Inverted repeats in eukaryotic genomes. To investigate the sequence/structure requirements of higher eukaryotic replication origins, numerous laboratories isolated libraries of early replicating sequences that are enriched in replication origins which are activated at the onset of the S phase. The nascent fragments, ranging from several hundred base pairs to over 1 kb, were cloned, generating a library of early replicating sequences. These sequences, by the nature of their isolation, should contain replication origins at or near their center and thus are termed "origin-enriched sequences. These were isolated from avian, mouse, monkey, and human cells (reviewed by Pearson et al.).[118] Some of these cloned sequences were capable of autonomous replication upon transfection into mammalian cells. Sequence analysis did not reveal a single major consensus sequence among the clones or between the libraries. However, in each of the above-mentioned libraries the sequences were enriched with both short inverted repeats as well as A + T-rich tracts. In addition, the origin of bidirectional replication of the dihydrotolate reductase (*DHFR*) gene and the c-*myc*-associated replication origin are both known to contain inverted repeats (reviewed by Pearson et al.).[118]

3. Mammalian Cells: The Effect of Anti-Cruciform–DNA Antibodies

Monoclonal antibodies have been produced with unique specificity to cruciform DNA structures.[181,215] These antibodies recognize conformational determinants specific to DNA cruciforms and do not bind linear double-stranded DNA, linear single-stranded DNA containing a stem loop structure or tRNA. The binding site of these antibodies has been mapped to the four-way (elbow-like) junction at the base of the cruciform.[215,216]

Introduction of the anticruciform DNA antibodies into a permeabilized cell system, which is capable of carrying out DNA replication, resulted in a 2- to 11-fold enhancement of DNA synthesis.[217] An enhanced replication of known early replicating sequences such as *ors*8, *DHFR,* and c-*myc* was also detected.[217] Taken with the above mentioned precaution of *in vivo* analysis (see above), this effect was apparently caused by the antibody supposedly stabilizing cruciforms encountered near the origins of replication and allowing multiple initiations to occur at these sites.

Using the anti-cruciform DNA antibodies in the same system, Zannis-Hadjopoulos and colleagues[189,190] were able to quantify the number of cruciform structures in living cells by fluorescent flow cytometry. Two major populations of cruciforms were observed throughout the cell cycle, each with an estimated 0.6×10^5 and 3×10^5 cruciforms per cell, respectively. Cruciforms were observed in a bimodal distribution throughout the S phase, their numbers reaching a maximum at the G_1/S boundary. The second wave occurred at 4 hours into S phase, but at a lower level than that observed at the G_1/S boundary. The timing of these waves was coincident with both the maximum rate of DNA synthesis and the relative enhancement of DNA synthesis by the antibody.[217] A limited number of cruciforms were detected in G_2/M nuclei. These data suggest that the formation of cruciforms is cell-cycle regulated. It also suggests that both cruciforms and active replication origins are grouped closely together in discrete regions within the nucleus in early S phase. The above observations support the hypothesis that certain inverted repeats may represent potential initiation sites for DNA replication.

4. Inverted Repeats and Gene Amplification

Very long inverted repeats in mammalian cells have been associated with gene amplification of oncogenes (e.g., N-*myc*, c-*myc*, *erb*-B, *sis*, *ras*) and of genes involved in drug resistance, such as the adenylate deaminase (AMPD), adenine phosphoribosyltransferase (APRT), carbamoyl-phosphate synthetase/aspartate carbamoyl transferase/dihydroorotase (CAD), and dihydrofolate reductase (DHFR) genes (reviewed by Fried et al.[218] and Windle and Wahl[219]). Treatment of mammalian cells with replication inhibitors and certain drugs can induce gene amplification, giving rise to inverted and tandem repetitions of genes and their surrounding regions. Such genetic amplification allows for multiple copies (>1000 in the case of DHFR) and overexpression of the gene product. The palindromic arms of the amplified mammalian inverted duplications are extremely large, in excess of 100 kb. In each case studied, the arms are separated by a stretch of noninverted DNA of approximately 150–1000 bp.[218] The formation of the large inverted repeats has been explained by way of an extra-chromosomal double rolling circle model.[169,218] This model requires that an origin of replication be in the vicinity of the DNA region to be amplified. Interestingly, sequence analysis of the sites of the recombination joints of inverted duplications in amplified DNA reveals that the DNA contained A + T-rich sequences as well as short inverted repeats with the potential to form hairpin (cruciform) structures, both of which are common to origins.[169,220,221]

The long inverted repeats of amplified DNA do not form cruciforms *in vivo*.[221] However, the long inverted repeats do form cruciforms *in vitro* following DNA isolation.[221] Processes of genomic DNA isolation can induce topological perturbations in the DNA[222]. The amplification and synthesis of long (plasmid-length) inverted repeats can be mimicked in an *in vitro* T antigen/SV40 origin-dependent replication/amplification assay by using extracts from carcinogen-treated HeLa cells.[223-225]

5. Inverted Repeats, Cruciforms, Z-DNA, and Nucleosomes

Nucleosomes interfere with the binding of initiation factors to promoters[226] and origins of replication.[227,228] A yeast ARS placed within the nucleosome has severely reduced replication activity compared to its normal location in the linker region.[228] Histones bind poorly to inverted repeats[229,230] or cruciform DNA structures,[231–233] and it is likely that cruciform structures exist in the spacer region between nucleosomes.[231] Cruciforms may play a role in nucleosome phasing, such that they expose nucleosome-free DNA sequences, making them accessible to DNA binding proteins specific for transcription, recombination, and/or replication.[231–234]

The formation of Z-DNA may also play a similar role in nucleosome phasing.[235–239] The SV40 viral replication origin region, in addition to the central inverted repeat and adjacent AT tract, contains 21-bp and 72-bp repeats. Both the 21-bp and the 72-bp repeats enhance the replication efficiency, in an orientation-independent manner. The position of the repeats is important.[240,241] The 21-bp and the 72-bp repeats are believed to enhance replication by maintaining a nucleosome-free region at the SV40 origin.[242–244] Interestingly both the 21-bp and 72-bp repeat regions have been reported to adopt the Z-DNA conformation.[245–249]

6. *Inverted Repeats, Stem Loops, Cruciforms, and Transcription*

Inverted repeats are present in a number of regulatory regions of genes, ranging from simple plasmid borne antibiotic resistance genes[250] and viral genes[251] to well-studied eukaryotic genes.[252–255] DNA sequences that can potentially form secondary structures are frequently localized to promoter regions, suggesting their role in gene regulation[250,256,257] (reviewed by Horwitz and Loeb).[258]

Involvement of an inverted repeat in N4 virion RNA polymerase promoter recognition.   The bacteriophage N4 RNA polymerase, which is packaged into the virion, is required for early transcription. Early transcription requires a supercoiled template, on which the *E. coli* SSB protein acts as a transcriptional activator.[259] The promoters of these early genes contain inverted repeats necessary for transcription. Remarkably, the inverted repeat symmetry and not a defined base sequence is required for transcription.[257] A model has been proposed in which DNA is supercoiled by DNA gyrase, driving the formation of a short cruciform structure stabilized by a particularly stable hairpin loop in one strand, containing the sequence CGAAG. The SSB protein removes any potential alternative secondary structure in the strand containing the sequence CTTCG. The N4 RNA polymerase then binds to the strand containing the hairpin stem.[257] This system represents probably the most convincing evidence for a system in which a cruciform structure is involved in the regulation of gene expression.

Possible involvement of cruciforms in transcription regulation.   The twin-domain model of transcriptional supercoiling[113] predicts the formation of positive supercoils in front of the transcription complex, and the generation of negative supercoils behind it.[260,261] Supercoil waves generated by transcription may cause the reversal of negative supercoil-induced altered structures to the B-form. Based on the proposition that supercoiling can be affected by protein tracking mechanisms such as transcription, several researchers investigated the effect of cruciforms on such processes. There have been several reports of cruciforms regulating DNA transcription: examples of cruciform extrusion driven by transcription[262]; cruciform absorption by transcription[133]; and transcription inhibition by the presence of cruciforms.[263-265] There is a report of a cruciform-inhibiting transcription and the relief of inhibition by the addition of HMG1.[173] HMG1 binds to DNA cruciforms[266] (and see below). Cruciforms on negatively supercoiled templates stalled the progress of RNA polymerases. The addition of HMG1 to the reaction relieved the block by binding to the cruciform and altering the DNA conformation to permit the progression of the RNA polymerase. Apparently HMG1 was able to bind and force the reabsorption of the extruded cruciform to the linear form inverted repeat, thus allowing the passage of the transcription complex. HMG1 can also remove the transcriptional block caused by left-handed Z-DNA on a supercoiled template.[267]

ADP-ribosyltransferase, implicated in DNA repair, binds specifically to DNA four-way junctions.[268] It has been proposed that this structure-specific mode of binding by the enzyme may be linked to its autoregulation of expression.[269] The promoter structure of the ADP-ribosyltransferase gene contains several noteworthy inverted repeats, which could form cruciform structures. The functional significance of these inverted repeats was revealed by deletion analysis; removal of one inverted repeat resulted in diminished promoter function, whereas removal of the other increased promoter activity. Overexpression by a heterologous promoter of ADP-ribosyltransferase led to repression of the endogenous promoter, suggesting that the ADP-ribosyltransferase gene product acts as a negative modulator of it own promoter.

7. Other Biological Phenomena Associated with Cruciforms

Cruciform extrusion and regulation of superhelicity and protein–DNA interaction. The extent of supercoiling is known to affect transcription, recombination, and replication.[261,270] Since cruciform extrusion causes an effective relaxation in DNA distal to it (the number of helical turns relaxed is approximately equal to the number of helical turns contained in the stem-loop of the cruciform)[271] the process of cruciform formation may be a mechanism regulating local superhelicity. In this manner, cruciform extrusion or melting may regulate the recognition and binding of proteins to specific sequences proximal to the cruciform.[264] A particular instance of this was reported by Pearson et al.[272] It appears that the presence of a cruciform stabilizes the sequence-specific binding of a protein(s) at a site proximal to it. Presumably the cruciform altered the structure of the target sequence such that it was more efficiently recognized and/or bound by the protein. Cruciforms are known to affect structural alterations in the flanking sequences.[150,273-275] This mode of transmission along the DNA molecule has previously been observed (refs. 130 and 275, and references therein) and is referred to as "telestability."[276] This transmission has been demonstrated to occur over long distances and can affect protein–DNA interactions such as the interaction of RNA polymerase with promoters,[277] and nuclease specificity.[55] Horwitz found that the resulting decrease in superhelicity by cruciform extrusion at one plasmid based promoter is below the optimum for expression from another promoter on the same plasmid, indicating that cruciform formation can act at a distance.[264]

Cruciform-specific proteins. It has long been suggested that cruciform structures may form transiently under physiological conditions and serve as the recognition site for initiator or other protein factors[278] (reviewed by Pearson et al.).[118] Presented below is a discussion of proteins that recognize and bind DNA in a structure-specific fashion and are likely to be involved in transcription or replication.

Prokaryotic four-way junction resolvases. There are several proteins with nuclease activity that interact with four-way junctions. Resolvases are ubiquitous enzymes that cut the four-way junction.[124,279–284] These enzymes cut 4–5 bp up the stem on either the 3' or 5' end of a cruciform arm (see Figure 19). Cuts are always made on opposite sides of the four-way junction to resolve the recombinant into two symmetrical molecules. Several of these have been analyzed in great detail by the use of hydroxyl radical cleavage of the DNA complexed with the proteins: the bacteriophage encoded resolvases T4 endonuclease VII[285] and T7 endonuclease I,[286] and the *E. coli* RuvC resolvase.[172] The T7 endonuclease I and the T4 endonuclease VII contact the DNA backbone without detectably altering its structure. On the other hand, by hydroxyl radical analysis, the RuvC resolvase[172] reveals only structural alterations, apparent as increases in radical cleavage intensity, but no protection footprint. Since a sequence dependence of the cleavage reaction for the RuvC resolvase was demonstrated, this protein might not interact directly with the DNA backbone.[172] In the hydroxyl radical cleavage patterns of the bacteriophage endonuclease–cruciform complexes, only two diametrically opposed (T4) or all four junction strands (T7) are protected. T4, yeast, and calf thymus resolving enzymes interact with the minor groove face of junctions.[287]

Mammalian cruciform–DNA-binding proteins: HMG1, HMG2, and the HMG box. A number of mammalian cellular proteins have been reported that recognize the Holliday-like four-way junctions of DNA (reviewed by Duckett et al.)[288]; among them is the ubiquitous HMG1.[289] HMG1 is an abundant nuclear protein with >10^6 molecules of HMG1 per cell.[290–292] HMG1 has been demonstrated to bind to gene regulatory regions, such as promoters, and replication origins with greater affinity than to "junk DNA."[293] The high mobility group protein, HMG1, binds to four-way junctions.[289] It has been reported that HMG1 can protect the single-stranded tips of cruciform stems from S1 nuclease digestion,[173] but not from T4 endonuclease VII, which interacts at the junction.[287] Not only is HMG1 able to bind to specific DNA structures, it is also capable of distorting the conformation upon binding. It has been reported to form beaded structures[293,294] as well as unwind supercoiled DNAs.[295–298] Recently several groups reported the ability of HMG1 to bend DNA such that there is facilitated T4 ligase-mediated circularization.[299,300] Such circularization assays have been carried out for several proteins, including the bacterial HU protein.[301] Several authors have suggested that although HMG1 may not be similar to the bacterial HU protein at the sequence level, it certainly may be the mammalian functional equivalent.[299,302] It has been proposed that the cellular function of HMG1 and HMG2 is to bind to folded regions of DNA, and thus aid in chromatin compaction.[293,294,303] HMG1 and HMG2 proteins have been implicated in both transcription[173,267,304–307] and replication processes.[290,308–310]

A protein–DNA binding motif known as the HMG box,[78,311,312] which is present in many different proteins, binds different forms of DNA, such as single-stranded DNA,[313] sharp angles in DNA,[314] and four-way DNA junctions.[289,315,316] Clearly, a wide variety of altered DNA structures can be recognized and bound by HMG box proteins. In addition to binding to four-way DNA junctions, HMG1 binds to B-Z DNA junctions.[266] Both posttranscriptionally modified HMG1 and 2 were isolated as Z-DNA-binding proteins. By using a brominated nucleic acid as a probe,[317] it was later discovered that these proteins were preferentially binding to brominated Z-DNA and not to nonbrominated DNA.[318,319] Recently Lippard and co-workers isolated HMG1[151] as well as other HMG box proteins that specifically recognize cisplatin DNA adducts.[320]

The HMG box is a domain of positively charged protein sequences characterized by one of the internal repeats contained in HMG1. The HMG1 protein is composed of three regions, consisting of two internal homologous repeats (HMG boxes) and an acidic tail.[321] The internal repeats consist of about 80 amino acids that are rich in basic and aromatic residues. The acidic tail is composed almost entirely of about 30 aspartates and glutamates. Although there is no apparent absolute conservation of the sequence for the HMG boxes, it is likely that they all have a similar structure. The HMG box is necessary and sufficient to bind DNA.[321] Both the whole HMG1 and its individual HMG boxes bind in a structure-specific fashion that is sequence independent. All proteins that contain HMG boxes bind to DNA. Some HMG box proteins recognize specific sequences that are generally A + T-rich, whereas others are indifferent to sequence (see discussion below).

The family of HMG box proteins includes proteins that may play key roles in transcription,[322,323] replication,[171,324] DNA repair,[151,318,319] V-D-J immunoglobin gene recombination,[325] and sex determination (SRY).[314,326] That the HMG box, a single motif, can be used for so many varied roles indicates a new class of nucleic acid binding proteins. Of the many proteins that contain HMG boxes, there are several classes: those that contain one HMG box, such as SRY[315]; those that contain two, such as HMG1; and the human upstream binding factor (hUBF)[322] and the *Xenopus* UBF,[327] which contain four and five HMG boxes, respectively. The pressing questions are: (1) What is it that gives some HMG box proteins the ability to recognize in a site-specific manner? and (2) Is this recognition mediated by the rest of the polypeptide? A general trend is emerging—proteins that contain a single HMG box seem to bind with strong sequence specificity, and those with multiple HMG boxes bind in a sequence-tolerant fashion.[328] The HMG1 protein itself contains two HMG boxes; this class of proteins binds solely in a structure-specific fashion. The human and *Xenopus* UBF proteins bind with little sequence specificity, and the *Xenopus* UBF, unlike other HMG box proteins, can bind to tRNA as well as DNA four-way junctions.[327] The one-box class binds with significant sequence specificity as well as structure specificity. The sex-determining factor, SRY, contains one HMG box and binds specifically to DNAs containing its target sequence, AACAAAG. However, when SRY binds to its target it produces a sharp bend in the DNA; thus the interaction does contain some structure specificity.[314]

The abundance of HMG1 and 2 indicates that they must be playing a role that is required continuously (such as chromatin structure), or that there are several roles within the cell, and these roles are mediated by the variety of modifications possible. More HMG box proteins will likely be discovered, and further research into the effect of modifications will be forthcoming.

Other four-way junction binding proteins. One more cruciform-specific binding protein (CBP) has been analyzed by the hydroxyl radical technique. CBP from human cells recognizes four-way DNA junctions in a structure-specific fashion[272] and is void of nuclease activity. It is apparent from the footprinting pattern of CBP[160] that the mode of interaction differs significantly from that of the bacteriophage encoded resolvases (see above) T4 endonuclease VII,[285] and T7 endonuclease I,[286] and the *E. coli* RuvC resolvase[172]: (1) CBP interacts with the junction, giving clear areas of protection and simultaneously introducing several changes in the fine structure of the cruciform. (2) Whereas the hydroxyl radical cleavage patterns of the bacteriophage endonuclease–cruciform complexes show protection on only two diametrically opposed (T4) or all four junction strands (T7), the hydroxyl radical cleavage patterns of the CBP–DNA complex reveal protection on both junction strands of stem loops as well as both strands of the branch arms.[285,286] This indicates that CBP apparently interacts at both major and minor groove faces of the cruciform junction. In addition, the fact that CBP protects both junction strands as well as both strands of branch arms indicates that CBP apparently interacts at both major and minor groove faces of the cruciform junctions; in contrast, T4, yeast, and calf thymus resolving enzymes interact with the minor groove face of junctions.[287] Apparently CBP provides a novel type of cruciform DNA–protein interaction, in that there are firm contacts with the sugar phosphate backbone (protection) as well as structural alterations of the cruciform substrate, both reflecting a putative cruciform-stabilizing function of CBP in the cell. The ability to structurally alter the DNA by binding of CBP provides a putative role for it in preparation of DNA for the processes of replication, transcription, or recombination. The asymmetric binding further predicts that there may well be a specific orientation of CBP required for an interaction with other proteins at a functional DNA element.

Several proteins have been implicated in binding specifically to DNA cross-overs, which occur at points of DNA looping and folding, and at nucleosome linkers.[162] The histone H1, like HMG1, can also bind specifically to DNA cross-overs.[328,329] Interestingly, the proteins are known to associate with each other,[330,331] and it is known that HMG1 modulates the histone H1-induced condensation of DNA[332] and facilitates nucleosome assembly.[333]

Several reports indicate that eukaryotic DNA topoisomerases may specifically recognize DNA cruciforms.[334,335] The vaccinia virus topoisomerase I has been shown to resolve Holliday junctions.[336] This enzyme has a specificity for the site CCCTT; four-way junctions with these sites in two juxtaposed arms are recognized and cut by the enzyme. The cutting is essentially identical to that of type I topoisomerases in that it takes place by concerted transesterifications at the two CCCTT sites rather than by hydrolysis.

Topoisomerase II actually cleaves DNA hairpins,[337] such that the cut site is one nucleotide 3' of the hairpin base. Recognition of the hairpin was sequence independent and required a double-stranded/single-stranded junction. Base pairing up to the base of the hairpin on the 5' side had no effect on the scission of the hairpin, whereas base pairing on the 3' end inhibited cutting. The requirement for a single stranded region was delimited to two to four nucleotides on the 3' side of the hairpin. Topoisomerase II has recently been demonstrated to specifically recognize and bind to Z-DNA[338]; the relationship between these different binding capabilities is not known. It may be that topoisomerase recognizes such structures as a means for regulating supercoil tension during processes of transcription, replication repair, or recombination.

Hairpin and stem loop binding proteins. In the human enkephalin gene enhancer there is an imperfect palindrome of 23 bp[339] that contains both cAMP-responsive elements (CRE-1 and CRE-2). *In vitro* studies reveal that short oligonucleotides of this enhancer are able to undergo reversible conformational transitions from duplex to the hairpin structures.[339] Although CRE-binding protein binds weakly to the duplex palindrome, it binds specifically and with high affinity to the G·T mispair in the hairpin formed by one of the two strands.[340] Interestingly, the hairpin contains an alternative CRE binding site, which contains two G·T mispairings, in contrast to the A·T and G·C pairs found in the duplex site. Furthermore, a T in the mispaired positions is required for the high affinity recognition, as a fully base paired hairpin is not recognized. Thus the binding of CRE-binding protein is both sequence and structure specific.

There are proteins that bind specifically to single-stranded DNA that lies between two members of an inverted repeat.[341,342] Such a situation occurs at the early palindrome of the SV40 virus, as there are two different human cellular proteins that bind to this palindrome.[341] Each binds with sequence specificity to one of the single palindromic strands. One was discovered to bind in a cell-cycle-regulated fashion. The other protein, binding to the opposite strand with some sequence specificity, is the replication protein A, a single-strand-binding protein.[341,343] It has been proposed that the particular conformation of the loop structure, which is unique to each arm of a particular cruciform, can be specifically recognized by proteins.

Since each loop conformation is distinct, specific recognition of a particular loop would be determined by the loop sequence and perhaps the stem sequence (i.e., the environmental context of the loops). Alternatively, it may be the specific mismatch base pairings that occur in the stem loop(s) of a particular imperfect inverted repeat that are the recognition structures of proteins.[340]

The ubiquitous nature of short inverted repeats in the genomes of eukaryotic organisms suggests that these sequences were evolutionarily maintained. The multitude of proteins that can interact with both sequence and structure specificity with cruciform DNAs suggest that cruciforms play biological roles in the cell.

V. LEFT-HANDED Z-DNA

A. Introduction

In 1970 Mitsui et al. suggested that the polymer poly d(I-C)·poly d(I-C) could adopt a left-handed conformation under specific conditions, based on the X-ray diffraction pattern and circular dichroism (CD) spectra of poly d(I-C)·poly d(I-C).[344] However, much later Sutherland and Griffin[345] found that the polymer analyzed by Mitsui et al. adopted an unusual non–B-DNA right-handed configuration called D-DNA. In 1972, Pohl and Jovin presented the CD spectrum of a left-handed alternating copolymer poly d(G-C)·poly d(G-C).[346] This CD spectrum was very different from that for classical B-form DNA. The classical, characteristic B-form CD DNA spectrum shows a positive ellipticity (θ) peak at about 270–280 nm, a negative ellipticity at about 250 nm, and a cross-over point (where the ellipticity equals zero) at about 260 nm. The Z-DNA spectrum shows an inversion of the peaks relative to the spectrum of B-DNA. In 1979 Rich and co-workers solved the X-ray crystal structure of d(CpGpCpGpCpG).[347] Quite unexpectedly, this 6-bp oligonucleotide (the first crystal structure of a DNA molecule to be solved) existed as a left-handed helix.

Following the realization that certain alternating purine pyrimidine sequences could exist as a left-handed helix, an enormous scientific effort ensued to characterize the DNA sequences and environmental conditions required for Z-DNA formation. The discovery of Z-DNA is important because the ability of DNA to adopt non–Watson–Crick structures could have profound implications for the processes of replication, recombination, or transcription. Z-DNA can exist readily in bacterial cells, and although sequences that can form Z-DNA are widely found in human DNA, nature has yet to divulge the biological role of Z-DNA in eukaryotic cells.

B.   The Structure of Z-DNA

1.   The Left-Handed Helix

There are only a few similarities between B-DNA and Z-DNA, including the double-stranded structure, antiparallel orientation of the two strands, and Watson–Crick hydrogen bonding.[348] Apart from these similarities, there are a number of differences between B-DNA and Z-DNA. First, there is a zigzag phosphate backbone in Z-DNA compared to a smooth backbone in B-DNA. Second, B-DNA has distinct major and minor grooves. In Z-DNA, the major groove has all but disappeared into a nearly flat surface. The one visible groove is deep and narrow. This groove is structurally analogous to the minor groove in B-DNA. Third, the Z-DNA helix, at an 18-Å diameter, is narrower than B-DNA (20 Å in diameter). Fourth, the helix repeat in Z-DNA is 12 bp per turn, compared to 10.5 per turn for right-handed B-DNA. Finally, in B-form DNA the helix pitch is, on average, 36 Å, with an average rise of 3.4 Å. The helix pitch of Z-DNA is 44.6 Å, with a rise of 3.72 Å. (The helix parameters for B-DNA and Z-DNA are listed in Table 2.)

Two additional major structural differences between B-DNA and Z-DNA are the sugar pucker and the configuration of the glycosidic bond in deoxyguanosine. In B-form DNA all sugar residues exist in the C2' endo configuration. In Z-DNA the sugar pucker for dC remains C2' endo, but the pucker changes to C3' endo in dG residues. The structures of C2' endo, *anti* dG and C3' endo, *syn* dG are shown in Figure 20. The zigzag phosphate backbone is the result of the alternating C2' endo conformation in pyrimidines and C3' endo conformation in purines. One major consequence of the change in sugar pucker is the effect this has on the distance between the phosphate groups attached to the C5' and C3' ribose positions (Figure 20). Changing the sugar pucker from C2' endo to C3' endo reduces this distance dramatically. This shortening has the result of pulling the base away from the center of a B-DNA helix and closer to the phosphate backbone, which is on the outside of the DNA helix. The other significant change occurring in deoxyguanosine in Z-DNA is a 180° rotation around the glycosidic bond. The configuration goes from the *anti* configuration found in B-form (and A-form) DNA to the *syn* configuration in Z-DNA.

A major structural rearrangement required for the B-DNA to Z-DNA transition is a flipping or inversion of the bases relative to the helix axis. Within a region of Z-DNA the bases flip over 180°. This flipping is accompanied by the rotation of the glycosidic bond of the purines from *anti* to *syn* and the sugar pucker change.

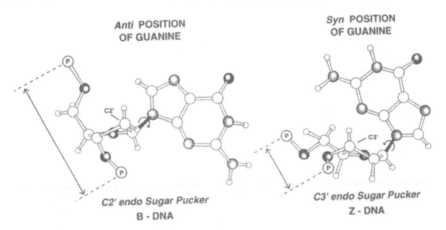

Figure 20. Conformations of guanosine in B-DNA and Z-DNA. B-DNA. The N9—C1' glycosidic bond (shown in black) is in the *anti* position, and the sugar is in the C2' endo pucker conformation. Z-DNA. The glycosidic bond is in the *syn* position (rotated 180° with respect to the *anti* position), and the deoxyribose is in the C3' endo pucker conformation. (Figure modified from Rich et al.[348] with permission.) Reproduced with permission from the *Annual Review of Biochemistry*, Volume 53. ©1984 by Annual Reviews, Inc.

For the pyrimidine nucleotides in Z-DNA, the sugar accompanies the base in its 180° rotation.

2. Base Stacking and Positioning in the Z-DNA Helix

In B-form each DNA base pair is oriented in the helix similarly with respect to its adjacent base pairs, such that all dinucleotide base pairs are structurally similar. Another prominent feature of B-DNA is that the base pairs are stacked regularly on top of one another. As shown in Figure 21, there is reasonable overlap between the GpC dinucleotides. There is somewhat less physical overlap between the bases in the CpG stack. In B-form DNA the bases are hydrogen bonded in a position almost perfectly centered within the double helix (the dot in Figure 21 denotes the center of the DNA helix). As illustrated in Figure 22, the bases in B-form DNA stack into a cylinder within the center of the double helix, with the sugars and phosphates on the outside of the helix.

At the base pair level, another major difference between B-DNA and Z-DNA is that in Z-DNA the base pairs are displaced, or sheared, with respect to their stacking positions. The orderly stacking in the center of the helix found in B-DNA is nonexistent in Z-DNA. Analysis of the CpG dinucleotide in Figure 21 shows that the two cytosines are stacked, but the guanines are positioned under and over the ribose sugars. Thus, in the CpG dinucleotide the bases are sheared by being effectively pulled away from each other with respect to the center of the helix. In the GpC dinucleotide in Z-DNA, the bases are relatively well stacked.

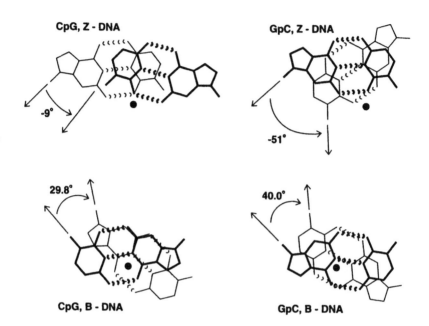

CpG, Z - DNA

GpC, Z - DNA

CpG, B - DNA

GpC, B - DNA

Figure 21. Base stacking in Z-DNA and B-DNA. The CpG and GpC base pairs are shown in Z-DNA and B-DNA. The bases in bold are stacked above the bases denoted by the thin lines. Bases are stacked well in B-DNA, especially in the GpC dinucleotide. Poor stacking overlap occurs in the CpG dinucleotide in Z-DNA. The twist angles shown are positive for B-DNA and negative for Z-DNA. (Figure modified from Rich et al.[348] with permission.) Reproduced with permission from the *Annual Review of Biochemistry,* Volume 53. ©1984 by Annual Reviews, Inc.

The position of bases within the helix also distinguishes DNA from B-DNA. In B-DNA the bases form a cylinder within the double helix, with the hydrogen bonds at the center of the helix. However, as shown in Figures 21 and 24, the bases in Z-DNA are positioned toward the outside of the Z-DNA helix such that neither the bases nor the hydrogen bonds overlap the helix. In B-DNA, the bases are protected from the solvent by their location at the center of the helix. In Z-DNA, certain ring positions are much more chemically reactive than in B-DNA. For example, the N7 and C8 positions of guanine are exposed to the solvent in Z-DNA (Figure 22).

3. A Repeating Dinucleotide in Z-DNA

Because all base pairs are positioned similarly with respect to other base pairs, B-form DNA has a base pair repeating unit of 1. In Z-DNA the repeating base pair unit is 2. The CpG dinucleotide is very different from GpC. There are 12 bp/turn

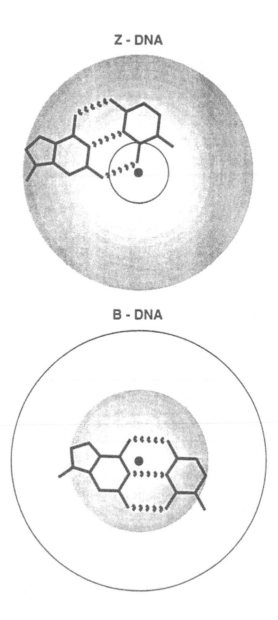

Figure 22. Location of base pairs within the B-DNA and Z-DNA helix. End-on views of the location of a G·C base pair in the Z-DNA and B-DNA helices. The shaded regions indicate the positions the base pairs occupy within the helical cylinder. The bases are centrally stacked in B-DNA, whereas they are organized toward the outside of the helix in Z-DNA.

of Z-DNA, which should produce an average helix twist of -30° (the value is negative, to denote the left-handed rotation in Z-DNA compared to a positive rotation found in right-handed helices). However, the twist angle for CpG is only -9°, whereas the twist angle for GpC is -51° (Figure 21). Consequently, in contrast to B-DNA, each base pair in Z-DNA is not oriented in the helix in a similar fashion with respect to the adjacent base pairs. However, every two base pairs are oriented within the helix in a fashion identical to the next two adjacent base pairs. The twist angle for a two base pairs relative to the next two base pairs is -60°, producing a helix repeating unit of 2.

4. The B-Z Junction

When a region of Z-DNA exists within a larger B-form DNA molecule, there must be a junction between the right- and left-handed helices. The best estimation of the physical structure of a B-Z junction is that it consists of a region of probably three or four unpaired or weakly paired bases. There is likely not a sharp transition from the left- to right-handed helix, but a region of a few base pairs that is partially unwound. The selective or differential reactivity of many chemicals at the B-Z junctions has provided sensitive assays for Z-DNA.

C. Formation and Stability of Z-DNA

1. The Sequence Requirement for Z-DNA

Z-DNA can form in regions of alternating purine-pyrimidine sequence, with $(GC)_n$ sequences forming Z-DNA most easily. $(GT)_n$ sequences also form Z-DNA, but they require a greater stabilization energy for formation than $(GC)_n$. $(AT)_n$ generally does not form Z-DNA, since $(AT)_n$ easily forms cruciforms. $(AT)_n$ can form Z-DNA under two special conditions. First, up to 10 alternating A·T base pairs must be embedded within a $(GC)_n$ or $(GT)_n$ region to form Z-DNA.[349] Second, $(AT)_n$ can form left-handed DNA at high negative supercoiling or in high salt, and in the presence of $NiCl_2$.[350]

2. Chemical Modifications Stabilize Z-DNA

Certain chemical modifications of DNA will drive the B-DNA $\rightleftarrows$ Z-DNA equilibrium in favor of Z-DNA. Bromination at the C5 position of cytosine or the C8 position of guanine allows the stabilization of Z-DNA at lower salt concentrations than required to stabilize a nonbrominated polymer. The addition of a bulky group at the C8 position of guanine favors the *syn* conformation, which places this group on the outside of the Z-DNA helix (see Figure 22). In the *anti* conformation, the bulky group would be inside the cylindrical axis of the DNA near the phosphate backbone. Methylation of the N7 position of guanine also favors the formation of Z-DNA. Bromination and, especially, methylation (or ethylation) of DNA from environmental mutagens will affect the stability of the B or Z conformation of the DNA double helix.

In bacterial and eukaryotic DNA many bases are methylated. Methylation is used in bacteria for restriction modification systems and the methyl-directed mismatch repair system. In eukaryotic cells, the C5 position of cytosine in the CpG dinucleotide is frequently methylated, as a general mechanism preventing transcriptional activity from large regions of eukaryotic chromosomes.

3. The Effect of Cations on the B-to-Z Equilibrium

The formation of Z-DNA was first observed in very high concentrations of salt (3 M NaCl). The distance between the negatively charged phosphates is much closer in Z-DNA than in B-DNA. Therefore, Z-DNA will be destabilized by the charge-charge repulsion of the negative charges on the phosphates. High concentrations of monovalent cations (Na^+, K^+, Rb^+, Cs^+, Li^+) can be required to shield the negative charges and stabilize the Z-conformation. Divalent cations (for example, Mg^{2+}, Mn^{2+}, and Co^{2+}) are much more effective at shielding negative charges than are monovalent cations and will stabilize Z-DNA at much lower concentrations. Co^{3+} will stabilize Z-DNA at millimolar concentrations. In addition, spermine, spermidine, and modified polyamines will stabilize Z-DNA.

4. DNA Supercoiling Stabilizes Z-DNA

In the early 1980s, experiments from the laboratories of Wells, Rich, and Wang demonstrated the role of supercoiling in Z-DNA formation.[351,352] Klysik et al.[353] showed that when a plasmid containing the sequence $(dCdG)_n \cdot (dCdG)_n$ was analyzed by agarose gel electrophoresis in the presence of 4 M salt, relaxation indicative of the formation of Z-DNA was observed. Subsequently, Stirdivant et al.[354] and Peck et al.[352] showed that as the level of negative supercoiling increased, the concentration of salt required to drive the Z-DNA transition decreased. In fact, the level of negative superhelical energy necessary to drive the B-to-Z transition at physiological ionic strengths was well within the level of supercoiling found in DNA purified from cells. $(dCdA)_n \cdot (dGdT)_n$ also forms Z-DNA in supercoiled plasmids.[355,356] However, $(GT)_n$ Z-DNA-forming sequences require a higher level of negative supercoiling to form Z-DNA than a $(GC)_n$ tract of equal size. The level of negative supercoiling required to form Z-DNA is a function of length of the alternating purine–pyrimidine tract. As a general rule, the longer the Z-DNA-forming sequence, the less negative supercoiling required to drive the B-to-Z transition. Table 4 shows the level of negative supercoiling required to form Z-DNA for a number of Z-DNA forming sequences.

DNA supercoiling is one physiological condition that can drive the formation of Z-DNA. Z-DNA is quite stable in supercoiled DNA, since the formation of Z-DNA effectively unwinds DNA, resulting in the relaxation of supercoils. As dis-

Table 4. Supercoiled Density Dependence for Z-DNA Formation

Z-DNA Sequence	σ_c	References
$(CG)_{16}$	-0.031	Peck et al.[352]
$(CG)_8GG(CG)_7$	-0.034	Ellison et al.[357]
GCGCGCGAGCGCGCGCGCGCTCGCGCGC	-0.042	Ellison et al.[357]
$CG(TG)_{20}AATT(CA)_{20}CG$	-0.032	Blaho et al.[358]
$(CG)_{13}AATT(CG)_{13}$	-0.025	Blaho et al.[358]
$(CG)_6TA(CG)_6$	-0.042	Sinden and Kochel[359]
$(TG)_6TA(TG)_6$	-0.057	Kochel and Sinden[360]
$(TG)_{26}$	-0.047	Haniford and Pulleyblank[355]

cussed above, conditions that unwind DNA are thermodynamically favored. This situation is analogous to that described in the preceding section for cruciforms and the formation of intramolecular triple-stranded DNA discussed in the following section. When a 12-bp region of B-DNA undergoes a transition from a right-handed helix to one complete left-handed helical turn, the number of twists in the DNA will decrease by 2.14 (1.14 from the removal of B-DNA (12 ÷ 10.5); plus 1 from the formation of Z-DNA). In terms of 10 bp units, the formation of Z-DNA removes 1.78 turns for every 10 bp that form Z-DNA ([10 ÷ 10.5] + [10 ÷ 12]).

5. Protein-Induced Formation of Z-DNA

The widespread occurrence of alternating purine–pyrimidine sequences in eukaryotic DNA, especially alternating (GT) sequences, has led to speculation that Z-DNA- and Z-DNA-binding proteins must exist. The identification of proteins that bind Z-DNA might implicate a role for the left handed conformation of DNA in biology. Proteins that bind to Z-DNA have been identified from many organisms, including *E. coli*, *Drosophila*, and higher eukaryotes. It is not yet known with certainty, however, if any Z-DNA-binding protein has a biological role that actually involves binding to left-handed DNA.

The first example of a protein that could recognize and specifically bind to Z-DNA was a Z-DNA antibody. Anti-Z-DNA antibodies recognize left-handed Z-DNA but not right-handed DNA. Lafer et al.[361] showed that some antibodies could induce the formation of Z-DNA and stabilize Z-DNA by binding the left-handed helix in relaxed (linearized) plasmid DNA. This was a significant finding, since it provided an example of a protein that could affect an equilibrium between two alternative helical forms of DNA.

D. Assays for Z-DNA

The major structural differences between Z-DNA and B-DNA allow many physical and chemical approaches to be used as assays for these conformations. Many of these approaches are also applicable for cruciform and triplex DNA structures. Table 5 describes several assays for Z-DNA.

Table 5.   Assays for Z-DNA

Assay	Description	Representative Reference
Z-DNA antibodies	Many polyclonal and monoclonal antibodies have been raised against Z-DNA. Binding can be detected by a variety of methods, including the binding of the antibody–DNA complex to nitrocellulose.	Lafer et al.[362] Moller et al.[363]
DNA topological assays	The formation of Z-DNA within a plasmid results in the relaxation of about 1.78 negative supercoils for every 10 bp of B-DNA that form Z-DNA. This relaxation can be detected on agarose (especially two-dimensional) gels.	Singleton et al.[351] Haniford and Pulleyblank[355] Peck and Wang[364] Nordheim et al.[365]
Chemical probes of Z-DNA	Many chemicals (such as diethylpyrocarbonate) react preferentially with the N7 and C8 positions of guanine when in the Z-DNA helix. Chemicals that specifically react with unpaired bases (bromoacetaldehyde (BAA), chloroacetaldehyde (CAA), osmium tetroxide (OsO_4), hydroxylamine, aminofluorene derivatives) react at the B-Z junctions. Sites of chemical modification can be mapped by S1 nuclease cleavage or primer extension analysis. Psoralen and other intercalating drugs bind poorly to Z-DNA. Psoralen photobinds in a hypersensitive fashion to certain B-Z junctions. These characteristic differences provide an assay for Z-DNA. Psoralen photobinding can be measured with base pair resolution by an exonuclease III mapping procedure or by primer extension analysis.	Johnston and Rich[366] McLean et al.[367] Rio and Leng[368] Kochel and Sinden[369] Hoepfner and Sinden[370]
Restriction/modification assays	Restriction endonucleases and methylases generally recognize B-form DNA, not Z-DNA or the B-Z junction. Differential restriction or modification is an indication of the presence of B- or Z-DNA.	Azorin et al.[371] Singleton et al.[372] Zacharias et al.[373] Vardimon and Rich[374]

E.   Z-DNA *In Vivo*

Z-DNA has been demonstrated in living cells by a number of different chemical, enzymatic, and physical approaches. One of the first *in vivo* assays for Z-DNA involved the analysis of psoralen cross-linking and photobinding to the Z-DNA forming sequence $(CG)_6TA(CG)_6$. The pattern of photobinding to this sequence in living cells was consistent with an *in vitro* superhelical density of $\sigma = -0.035$.[359,375] Jaworski et al.[376] provided evidence for Z-DNA by using an *Eco*RI methylase (MEcoRI) assay, in which the *Eco*RI site (GAATTC) was not methylated when it existed within a Z-DNA region. Haniford and Pulleyblank[376a] used a linking number assay to provide evidence for the existence of Z-DNA *in vivo*. A bimodal distribution of supercoiled plasmid containing a $(GC)_n$ Z-forming insert was not observed in cells under normal growth conditions, but was observed in cells treated with chloramphenicol, which can lead to an increase in DNA supercoiling and thus drive Z-DNA formation. Several chemical probes have also been used to detect Z-DNA *in vivo*. Rahmouni and Wells[377] applied osmium tetroxide (OsO_4), a chemical probe for unpaired pyrimidines, to detect Z-DNA *in situ*. Thymines within the *Eco*RI sites (GAATTC) between two $(CG)_n$ blocks or flanking a $(CG)_n$ block were hypersensitive to OsO_4. In these experiments, reactivity was not observed with $(CG)_5$ Z-DNA-forming sequences, but reactivity was detected with $(CG)_6$ and longer tracts, indicative of the existence of Z-DNA.

Anti-Z-DNA antibodies have also been used extensively as probes for Z-DNA in eukaryotic cells. Nordheim et al.[378] were the first to use Z-DNA antibodies to detect Z-DNA in eukaryotic cells in the polytene chromosomes of the *Drosophila* salivary gland. Because the harsh conditions initially required for the antibody binding may introduce negative supercoiling and thus Z-DNA formation,[379] several approaches have been developed to maintain as natural a chromosomal state as possible for antibody binding studies. Jackson and Cook[380] have shown that gentle purification of nuclei from cells can result in changes in the organization of DNA in chromosomes. Therefore, a protocol involving encasing eukaryotic cells in agarose beads was developed by Jackson and Cook. Cells are gently lysed in the beads, and the cytoplasm and membrane fragments can be washed out of the beads. The nucleus remains and is permeable to antibodies and certain restriction enzymes and nucleases. Using this approach, Whittig et al.[193,194] have presented evidence consistent with the existence of Z-DNA in eukaryotic cells. The level of antibody binding suggested the existence of Z-DNA in every 100 kb of DNA. To demonstrate that Z-DNA was dependent on unrestrained supercoiling, nuclei in the agarose beads were treated with DNase I to introduce nicks into DNA. This reduced antibody binding. Moreover, when a topoisomerase I inhibitor was added, a higher level of anti-Z-DNA antibody binding was observed. These results are consistent with the interpretation that the formation of Z-DNA in eukaryotic cells is a DNA supercoiling-dependent process.

F. Possible Biological Functions of Z-DNA

Z-DNA may play regulatory roles in gene expression, DNA replication, or genetic recombination. Although a definitive role for Z-DNA remains to be established, this discussion will enumerate some of the possibilities. Many of the roles of Z-DNA are similar to those described below for triplex DNA.

The formation of Z-DNA could alter the level of supercoiling and possibly control the level of expression. In bacterial cells, gene expression can be regulated by the level of unrestrained supercoiling in DNA. Although many other factors are important, the expression of certain genes is regulated by supercoiling. The topoisomerase I and DNA gyrase genes of *E. coli* are two such examples of genes regulated by DNA supercoiling. Unrestrained supercoiling can exist at active gene regions, and the energy from supercoiling may facilitate RNA polymerase binding. The level of unrestrained supercoiling will decrease within a topological domain as a region undergoes a transition from B-DNA to Z-DNA. If a region of Z-DNA returns to the B-form, the level of negative supercoiling will increase. Since a certain level of negative supercoiling may be required for optimal gene expression, the ability to alter supercoiling levels by B-to-Z or Z-to-B transitions provides one mechanism for gene regulation. DNA supercoiling also acts to compact chromosomes, which will increase the local concentration of the DNA. A change in the concentration of DNA can influence the binding of regulatory proteins. The very different helix structures of B-DNA and Z-DNA clearly provide different signals to DNA-binding proteins. The formation of B-DNA or Z-DNA within a particular DNA-binding site may present a defined substrate for specific proteins designed to recognize a particular DNA helix. Differential protein binding to the B- or-Z form of a helix may be required for gene expression acting as a switch in a positive or a negative fashion.

Several different Z-DNA-forming sequences have been cloned into the regulatory regions of several genes to examine the influence of a Z-DNA-forming sequence on transcription. To date there is no consensus for the role of Z-DNA in gene regulation. Z-DNA can have a positive, negative, or no effect on gene expression.[381–384] This result may not be unexpected, given the complexity of eukaryotic gene regulation, including the multiple transcription factors that are likely to be associated with each promoter. Each promoter certainly binds many different transcription factors, and their binding may be differentially influenced by B- or Z-DNA. The existence of Z-DNA may also affect nucleosome positioning on promoters, which could strongly influence gene expression.

Z-DNA may be involved in genetic recombination, which may involve the transient formation of a left-handed helix (for a review see Blaho and Wells).[385] When a single strand begins base pairing with a homologous duplex, a single turn of left-handed DNA may form for every helical turn of right-handed DNA formed. Within the base paired region, sequences with alternating purine–pyrimidine symmetry would adopt a left-handed helix most easily. DNA adjacent to a Z-DNA-forming region may preferentially be recombinagenic. In support of this idea, many DNA sequences with the potential to form Z-DNA have been associated with sites of genetic recombination.[386–391] Additional evidence for Z-DNA in recombination comes from the observation that many proteins involved in recombination bind to Z-DNA. RecA (from *E. coli*) and RecI (from *Ustilago*) facilitate duplex alignment and strand exchange. These proteins bind preferentially to Z-DNA.[385,390,392–394]

Z-DNA binding proteins have been purified from many different organisms, including *E. coli*, *Drosophila*, yeast, nematodes, chicken, frog, bull testis, and human cells,[246,375,395–400] although, for the most part, the role of the proteins in binding Z-DNA is not understood. *Drosophila* topoisomerase II binds preferentially to Z-DNA.[338] Moreover, binding of GTP to the enzyme weakens its affinity for B-DNA while increasing the binding to Z-DNA. In negatively supercoiled DNA regions, such as genes primed for or active in transcription, Z-DNA formation may act to facilitate topoisomerase II binding.

VI. TRIPLE-STRANDED NUCLEIC ACIDS

A. Introduction

The triple-stranded helices (triplexes) of nucleic acids were first described in 1957[401] and have been studied for more than 30 years.[1,2,402–406] Intermolecular triplexes attract much attention because of their potential therapeutic and biotechnological applications.[2,407–410] Intramolecular triplex/single strand structures (H- and H*-DNA) are presumed to play important roles in DNA function (see refs. 1, 2, 404, and 406 for reviews).

B. The Structure of Triplex DNA

1. Sequence Requirements

A triple helix consists of a duplex, in which the bases are paired via Watson–Crick hydrogen bonds, and a third strand, the bases of which form hydrogen bonds with one base of each base pair of the duplex. After the formation of hydrogen bonds in Watson–Crick base pairs, purine bases have potential hydrogen-bonding donors and acceptors that can form two hydrogen bonds with incoming third bases. By contrast, each pyrimidine base already involved in the duplex can form only one additional hydrogen bond with incoming third bases. The hydrogen bonds of such a type are traditionally called Hoogsteen bonds, after their discovery in adenine-thymine cocrystals by Hoogsteen in 1959. To maximize the number of stabilizing hydrogen bonds, the third strand bases bind to the purine bases of the duplex. If these purines were randomly distributed in both duplex strands, this would result in an energetically unfavorable conformation of the sugar-phosphate backbone of the third strand due to base switching from the one duplex strand to the other, and in the lack of stabilizing stacking interactions in the third strand. Therefore, a duplex appropriate for stable triplex formation contains purine bases only in one strand. Thus the precondition of triplex formation is the presence of a homopyrimidine (Py) sequence in one strand of the duplex and a complementary homopurine (Pu) sequence in the opposite strand (Py·Pu tract).

Two hydrogen bonds with the duplex purine strand can be formed by both pyrimidine and purine bases of the third strand.[2,411–413] Energetically stable triple helices are formed when the third strand is composed of either only pyrimidine bases (Py·Pu·*Py* type triplex) or mainly purine bases with a low proportion of pyrimidine bases (Py·Pu·*Pu* type triplex) (the third strands are shown in italics). Triplexes can be also divided into intramolecular and intermolecular types (Figure 23).

Typical intermolecular triplexes can be formed when the polymeric or oligomeric third strand of appropriate sequence binds (through Hoogsteen bonding) to the double-stranded Py·Pu tract.[401,414–416] A DNA sequence containing a Pu·Py tract of mirror symmetry can form an intramolecular H- or H*-DNA structure in which one half of a Pu·Py tract unwinds and then a Py (or Pu) strand bends around

A

Py•Pu•*Pu*

B

Py•Pu•*Py*

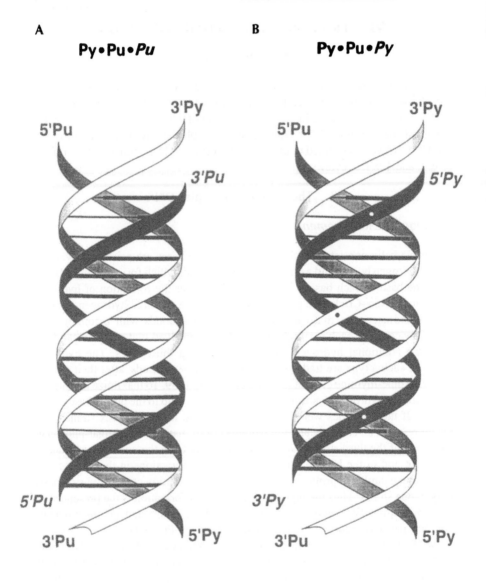

C

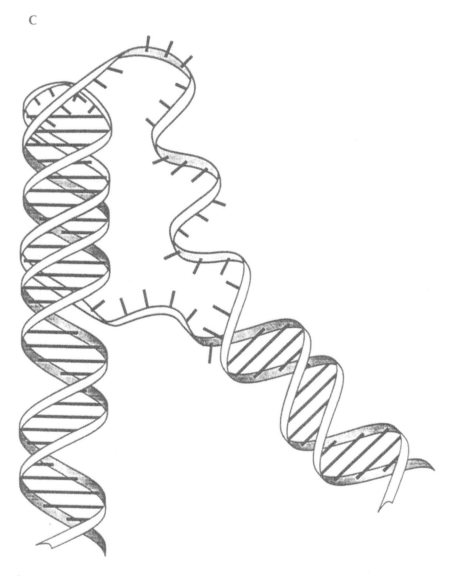

Figure 23. Intermolecular and intramolecular triple helices. (**A**) An intermolecular Py·Pu·*Pu* triple helix is shown, with the polypurine third strand organized antiparallel with respect to the purine strand of the Watson–Crick duplex. (**B**) An intermolecular Py·Pu·*Py* triple helix is shown, with the polypyrimidine third strand organized parallel with respect to the purine strand of the Watson–Crick duplex. (**C**) An intramolecular triplex DNA is shown. As with intermolecular triplex DNA, the third strand lays in the major groove, whereas its complementary strand exists as a single strand. (Figure modified from Wells et al.[594] with permission.)

A

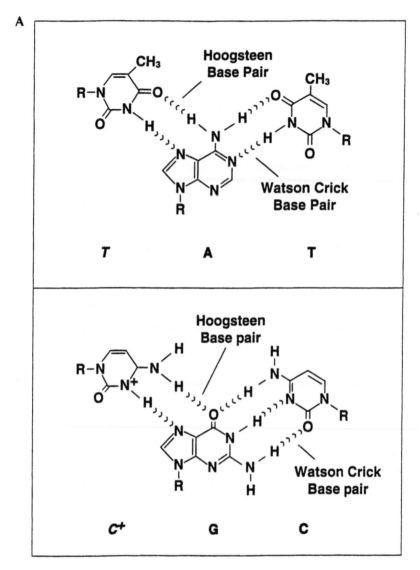

Figure 24. Base triads in triplex DNA. (**A**) The top panel shows the *T·A·T* triad, bottom panel the C·G·C⁺ triad. In the T·A·T triad, A forms both Watson–Crick and Hoogsteen base pairing to thymines. In the C⁺·G·C triad, G forms a Watson-Crick base pair and a Hoogsteen base pair with a protonated cytosine. In *Py·Pu·Py* triads the third base forms hydrogen bonds in a Hoogsteen configuration. (**B**) In the C·G·G triad the central G of the Watson–Crick base pair forms a Hoogsteen base pair with the third-strand G. In the T·A·A triad, the central A forms hydrogen bonds in a Hoogsteen base pair with an additional A. In Py·Pu·*Pu* triads, the third base forms hydrogen bonds in a reverse Hoogsteen configuration.

a point of symmetry and makes Hoogsteen hydrogen bonds with the Pu strand in another half of the Pu·Py tract. A former Pu (or Py) counterpart of the folded-back strand remains unpaired.[417,418]

2. Isomorphism of Base Triads

Figure 24 shows stable base triads that can be formed with natural bases. Two configurations of third-strand bases relative to the Watson–Crick base pair are possible: Hoogsteen base pairing,[6] which is realized in the Py·Pu·Py triplex, and reverse-Hoogsteen base pairing for the Py·Pu·Pu triplex, in which third-strand bases are rotated $180°$ relative to their positions in the previous scheme. Note that not all of the possible 16 Py·Pu·Py and 16 Pu·Pu·Pu triads[419] are stable, most of them would have one or zero Hoogsteen hydrogen bonds. Base triads in a Py·Pu·Py triplex (C·G·C^+ and T·A·T) are isomorphous, that is, they can be superimposed one over another, so that the positions and orientations of corresponding glycosidic bonds, as well as positions of C1' atoms, practically coincide. The conformation of the sugar-phosphate backbone is regularly repeated along the third strand. The Py·Pu·Pu triplexes include C·G·G and T·A·A triads, which are not truly isomorphous; so the conformation of the backbone is not perfectly regular along the third strand. The Py·Pu·Pu triplex can also accommodate a number of T·A·T triads at the expense of some distortion in optimum base pairing and/or backbone conformation.[405,420]

3. Major Groove Location and Antiparallel Orientation of the Third Strand

Base pairs of double-helical nucleic acid molecules retain hydrogen-bonding capabilities in both the minor and major grooves of the double helix (see, for review, Helene and Lancelot).[421] The third strand is placed in the major groove of the DNA, where purine bases may form two normal Hoogsteen hydrogen bonds (see, for a review, Cheng and Pettitt).[413] The major groove of DNA is deep enough and, upon third-strand binding, the resulting triplex has a diameter that is only a few angstroms larger than the 20-Å duplex diameter.[422]

An antiparallel orientation of similar strands and *anti* glycosidic bonds was observed in both the Py·Pu·Py and Py·Pu·Pu triplexes.[2,405,423] (Some researchers call Py·Pu·Py triplexes parallel because the purine and the third strand are parallel, and Py·Pu·Pu triplexes antiparallel because the duplex purine strand and the third strand are antiparallel.)

4. Strand Geometry Is Possibly Intermediate between A- and B-Forms

Depending on the sequences involved and the type of third strand, DNA triplexes may have differences in overall structures, with the geometric parameters of constituent strands in triple helix close to the B-form or intermediate between the B- and A-forms.[405,413,422,424–426]

C. The Formation and Stability of Triplex DNA

1. Kinetics of Triplex Formation

The formation of H-(H*)-DNA requires a denaturation bubble in the center of the mirror-repeated Py·Pu tract that allows the duplexes on either side to rotate and fold back, providing the formation of the first triad.[427] Subsequently, one-half of the Py·Pu tract hydrogen bonds to a single strand as it is released by the progressive unwinding of the other half of the Py·Pu tract. The rate of intramolecular triplex formation depends on the sequence of the Py·Pu tract, so that H-(H*)-DNA can form between 2 min and several hours, and can be increased by increasing the temperature.[428,429] The reverse transition, from intramolecular triplex to duplex, takes less than several minutes.[428] The association rate constants for intermolecular Py·Pu·Py triplexes were on the order of 10^3 $M^{-1}s^{-1}$, which is 10^3 times lower than the corresponding constant for duplex formation. The triplex dissociation rate constants were in the 10^{-5} to 10^{-4} s^{-1} range, resulting in a triplex lifetime of 1–10 hours at 37°C.[430,431] Dissociation constants for the complex of 21–22-mer pyrimidine oligonucleotides with their DNA target may be as low as 1–10 nM.[430-432]

2. Electrostatic Stabilization

The formation of triple-stranded nucleic acids requires a reduction of repulsion between negatively charged phosphate groups of the three strands. The repulsion between these phosphates might be screened by submolar concentrations of monovalent cations (e.g., Li^+, Na^+, K^+).[401,412,433] Much lower (millimolar) concentrations of divalent metal cations (e.g., Mg^{2+} or Mn^{2+}), which bind more tightly to phosphates, stabilize triplex *in vitro*.[401] Submillimolar concentrations of polyamines, the distributed charges of which allow them to interact simultaneously with different nucleic acid strands, are effective in stabilizing triple-helical nucleic acids *in vitro*.[434-438] Lysine-rich peptides as models of basic protein fragments have a triplex-stabilizing effect comparable to that of polyamines.[439] If triplexes contain clusters of charged bases in third strands (e.g., protonated cytosines), the repulsion between these bases may partly destabilize triple helix.[440,441] See Table 6 for a list of triplex-stabilizing factors.

3. Hoogsteen Hydrogen Bonds and Stacking Interactions

The sequence specificity of triplex formation is the major evidence of the importance of Hoogsteen hydrogen bonds. All stable triads have two Hoogsteen hydrogen bonds. The importance of two Hoogsteen hydrogen bonds explains the requirement of low pH for Py·Pu·Py triplexes where C·G·C+ triads form under acidic pH conditions because of protonation of the N3 of the third-strand cytosines (see, for a review, Frank-Kamenetskii).[444] In many cases the Py·Pu·Pu triplexes are stabilized by divalent cations.[418,445,456] Stronger Hoogsteen hydrogen bonds are possibly formed when divalent metal cations coordinate to purine bases in the third strand.[2]

Table 6. Triplex Stabilizing Factors

Type	Characterized in vitro	Availability in vivo	Reference
Electrostatic stabilization	Monovalent cations (Na[+], K[+]), up to 1 M	0.15 M	Krakauer and Sturtevant[433]
	Divalent cations Mg^{2+}, 1–10 mM Zn^{2+}, 0.1–10 mM	Up to 1 mM Bound in metalloenzymes	Felsenfeld and Miles[411] Darnell et al.[442]
	Polyamines: putrescine (2+) >1 mM Spermidine (3+) > 0.1 mM	May be as high as 1 mM, mostly in a macro-molecule-bound form. Free polyamines in a micromolar range	Hampel et al.[435] Singleton and Dervan[437] Thomas and Thomas[438] Sarhan and Seiler[442a]
	Spermine (4+) > 0.01 mM		Davis et al.[443]
	Basic polypeptides, 0.01–1 mM	Probably as protein fragments	Potaman and Sinden[439]
Hoogsteen hydrogen bonds	Similar or slightly weaker than Watson–Crick Low pH for C·G·C[+] and C·G·A[+] triads Transition metal cations for Py·Pu·Pu triplex	No No	Cheng and Pettitt[413] Frank-Kamenetskii[444] Bernues et al.[445]
Stacking	Stabilizing contribution similar to duplex DNA	Yes	Roberts and Crothers[446] Cheng and Pettitt[413]
Hydration	Spines of hydration in all three grooves	Yes	Radhakrishanan and Patel[405] Weerasinghe et al.[447]

Supercoiling for H (H*) DNA	Negative supercoiling dependence on Py·Pu length and pH	Static, transcriptionally and chromatin rearrangement-induced supercoiling	Lyamichev et al.[448] Kohwi and Kohwi-Shigematsu[418] Sinden[1]
Length of third strand	≥ 6 nt for H-DNA, ≥ 10 nt for intermolecular	One H-DNA motif in 50,000 bp of human DNA One triplex-forming sequence in 300–500 bp	Lyamichev et al.[449] Schroth and Ho[450] Horne and Dervan[451] Behe[451a]
Ligands	Intercalating agents		Pilch and Breslauer[452] Duval-Valentin et al.[453] Wilson et al.[454] Thuong and Helene[403]
Covalently bound agents	Intercalators, chemically and photochemically cross-linking agents		Thuong and Helene[403] Soyfer and Potaman[2]
Modified third strand	Non-ionic backbones, modified nucleobases		Thuong and Helene[403] Soyfer and Potaman[2] Nielsen et al.[455]

Experimental studies of mismatches and bulges in triplex structures show the general importance of stacking interactions.[431,446,457,458] The free energy penalty for introducing a single mismatch ranges from 2.5 to 6 kcal/mol, which is close to the corresponding values for DNA and RNA.

4. Hydration Effects

The NMR and molecular dynamics simulation data show that long-lived water molecules with lifetimes of >1 nsec are immobilized in all three grooves of the Py·Pu·Py and Py·Pu·Pu triple helices.[405,447] Water molecules may stabilize the triplex structure by solvating the complex, by screening repulsive electrostatic interactions between phosphate groups across the narrow groove between the third strand and duplex purine strand of the Py·Pu·Py triplex, or by bridging polar groups belonging to different strands.[405]

5. Requirement of Supercoiling for Intramolecular Triplex

Relaxation of some torsional stress in closed circular DNA may result from some non–B-form DNA structure (reviewed in refs. 1, 177, 459). Therefore, increased levels of DNA supercoiling can provide the formation and stabilization of intramolecular Py·Pu·Py triplex (H-DNA)[427,448,460] or Py·Pu·Pu triplex (H*-DNA).[418] Increased DNA superhelicity and low buffer pH may substitute for each other, and the superhelical tension necessary for H-DNA formation depends linearly on increasing pH.[448]

6. Length Dependence

Increasing the length of the Py·Pu tracts facilitates the formation of triplexes—the longer the lengths of Pu·Py tracts, the less the negative superhelix densities are required to induce the intramolecular triplex.[427,449,460] A theoretical lower limit of 15 bp for a Py·Pu tract was estimated for H-DNA formation.[449] At Py·Pu tracts that are several dozen nucleotides long, multiple conformers may form.[429,461] The length dependence for intermolecular Py·Pu·Py triplex showed an increase in triplex stability with increasing oligomer length.[462,463] In the case of Py·Pu·Pu triplex, the length of the Pu third strand seems to have an optimum because of an imperfect fit to the major groove of the duplex.[412,464,465]

D. Assays for Triplex DNA

A number of physical and chemical techniques can be used as assays for H- and H*-DNA *in vitro* and *in vivo* (Table 7).

Table 7.   Assays for H- and H*-DNA

Assay	Description	References
Two-dimensional agarose gel	The formation of H-DNA results in the relaxation of one negative supercoil for every 10.5 bp of B-DNA that forms H-DNA.	Lyamichev et al.[448]
Chemical probes	Chemicals that specifically react with unpaired bases react at the tip of triple helix and unpaired half of purine or pyrimidine strand.	Kohwi and Kohwi-Shigematsu[418] Johnston[466] Hanvey et al.[467]
	Dimethyl sulfate reacts with the N7 of guanines, and protection of guanines from modification indicates the bases involved in the triple helix. Similarly, pyrimidines protected from photochemical modification map to a triplex-forming sequence.	Lyamichev et al.[468]
	The pattern of 4,5',8-trimethylpsoralen photobinding to AT and TA dinucleotides in duplex provides a quantitative measure of triplex formation.	Ussery and Sinden[469]
Enzymatic assays	Single-strand-specific nucleases recognize and cut unpaired regions in H- and H*-DNA. Inability of restriction enzyme to cut DNA indicates the presence of triple-stranded region.	Lyamichev et al.[448] Hanvey et al.[467]

E.   Triplex DNA *In Vivo*

1.   *Search for Triplexes in Cells*

Several attempts have been made to determine whether the Py·Pu sequences do really form triple-stranded structures in the cell. Information from these studies may be separated into results from direct determinations and indirect indications of the triplexes *in vivo*.

Immunological assays.   To directly probe the triple-stranded structures in chromosomes, monoclonal antibodies were produced by immunizing mice with poly[d(Tm^{5}C)]·poly[d(GA)]·*poly[d(m^{5}CT)]* triplex, which is stable at neutral pH.[470,471] These antibodies did not bind to calf thymus DNA or other non-Py·Pu DNAs, such as poly[d(TG)]·poly[d(CA)], and did not recognize Py·Pu DNAs containing m^{6}A (e.g., poly[d(TC)]·poly[d(Gm^{6}A)]) which cannot form a triplex since the methyl group at position 6 of adenine prevents Hoogsteen base-pairing. One type of monoclonal antibody was demonstrated by numerous criteria to be specific for the T·A·*T*-rich Py·Pu·*Py* triplex DNA, whereas another type was specific for poly[d(TC)]·poly[d(GA)]·*poly[d(CT)]* triplex. Chromosomes fixed in methanol/

acetic acid were stained by antibodies in the presence of *E. coli* DNA, but not in the presence of polymer triplex. To avoid an ambiguity in interpretation that could arise as the acid fixation itself can change the structure under study,[379] nuclei were also stained after fixation in cold acetone.[470] Unfixed, isolated mouse chromosomes also reacted positively with the antibody, particularly when they were gently decondensed by an exposure to low ionic conditions at neutral pH, indicating that fixation is not mandatory for antibody staining.[192] Additional evidence that triplexes really do exist *in vivo* is provided by immunoblotting of triplexes in crude cell extracts.[472] Thus there is a growing body of immunological evidence that the triplexes are present in eukaryotic chromosomes.

Chemical probing.   Direct chemical probing techniques were also used to detect triplexes in plasmid DNA in *E. coli*. The osmium/bipyridine modification pattern characteristic for H-DNA was observed in plasmid DNA after its host *E. coli* cells were preincubated in the pH 4.5 or 5.0 media.[473] Upon cell incubation in the presence of Mg^{2+} and chloramphenicol at neutral pH, chloroacetaldehyde reactivity of DNA triplex-forming sequence $(dG)_{30}$ was similar to that for the $C \cdot G \cdot G$ triplex *in vitro*.[474] Glaser et al.,[475] using diethylpyrocarbonate, failed to detect any modification pattern consistent with the presence of H-DNA, whereas in experiments *in vitro* they were able to detect it. High levels of DNA supercoiling and proper environmental conditions are the major limiting factors in the formation and detection of H- or H*-DNA structures.

Photochemical probing.   Trimethylpsoralen photobinding to the Py·Pu tracts in plasmid DNA showed that the formation of H-DNA in *E. coli* cells was dependent on DNA superhelicity and extracellular pH.[469] When cells were grown in K media, which acidifies over time down to pH 5, the trimethylpsoralen photobinding pattern was consistent with the presence of H-DNA. The use of topoisomerase I-mutant cells with a higher level of supercoiling *in vivo* was the triplex-promoting condition in these experiments.

Indirect assays.   Several indirect studies addressed the possibility of triplex existence *in vivo*. The GATC site at the center of or adjacent to a Py·Pu mirror repeat was undermethylated in plasmid grown in *E. coli* strain JM101.[476] This result could be explained by the participation of this Py·Pu tract in H-DNA *in vivo*, since *Dam* methylase methylates the GATC site when it is in the double-stranded B-DNA form, but not in an alternative non-B conformation. However, other unknown factors might be involved in *Dam* undermethylation as partial methylation of H-DNA *in vitro* was detected, and the GATC undermethylation was observed when the plasmid was grown in JM cells, but not when grown in other types of *E. coli* cells.

Deletion analysis of the Py·Pu tracts inserted in the tetracycline resistance gene[477] showed a significant instability of those (longer) inserts that can form intramolecular triplexes (H- or H*-DNA) *in vitro*. These data may reflect the existence and mutational role of triplexes *in vivo*, provided there is a satisfactory model for the underlying process.

Sarkar and Brahmachari[478] showed that the Py·Pu tract cloned into the transcribed region of a bacterial gene significantly decreased the gene expression. Some unusual structure in the Py·Pu tract might be responsible for the premature transcriptional termination. Rao reported that the cloned $d(GA)_n \cdot d(TC)_n$ sequences that can potentially adopt triplex structures could slow down the DNA replication fork movement.[479] In both of these cases the H-DNA was suggested to be responsible for preventing polymerases from progressing along the DNA template in the cellular system.

In summary, none of the triplex searches *in vivo* has been conclusive. This is not surprising, since if triplex DNA participates in gene functions it cannot exist all the time, at least under physiological conditions, because a truly stable structure cannot regulate a changing physiological state.[480] It is hoped that further work will provide more convincing evidence of the existence of triplexes *in vivo*.

2. Factors that Could Be Responsible for Triplex Formation In Vivo

Supercoiling. Topoisomerases can create definite levels of (unrestrained) DNA supercoiling (see refs 1, 106, and 263 for reviews) in topological domains created by proteins and/or RNA molecules. Nucleosomal dissociation during transcription may convert a restrained supercoiling into an unrestrained supercoiling.[1] During the progression of RNA polymerase, a wave of negative supercoiling arises in DNA behind the RNA polymerase and positive supercoiling in front of the RNA polymerase (see Liu and Wang[113] and Sinden[1] for more detail). A number of studies experimentally confirmed the difference in the level of supercoiling upstream and downstream of the transcribed site.[134,260,262,377,481,482] An assortment of other mechanisms may create localized or transient torsional stress in eukaryotic DNA: binding of transcription factors or other proteins, activity of helix-tracking proteins, looping of DNA by protein binding at two distant locations, histone acetylation, and gyrase activity of topoisomerases.[483]

Cations. Upon triplex formation, the very high negative charge density that originates from the phosphate groups in three strands may be reduced by counterions (Table 6). Concentrations of metal cations in cells are far from stabilizing,[442] and metal cations may only be partially responsible for triplex formation and maintenance *in vivo*. Concentrations of polyamines in the nucleus may be as high as 5 mM,[442a] where they are largely bound to nucleic acids and phospholipids.[443] Thus polyamines might be a class of compounds which stabilize triplex DNA *in vivo*. Basic polypeptides stabilize triplex DNA *in vitro*.[439] A similar stabilizing effect in cells might come from the basic protein fragments.

Proteins. The H-DNA structure formed through the interaction of two distant Py·Pu tracts of the same DNA duplex was proposed to be stabilized by single strand-binding proteins that could fix an unpaired single strand.[485] About 20 proteins binding to Py·Pu sequences have been reported. Several examples include transcription factors such as Sp1[485] and PuF,[486] which bind to double-stranded DNA, and proteins that preferentially bind to homopyrimidine[487–490] or homopurine single strands.[491,492] In the absence of conclusive evidence on the role of Py·Pu tract-binding proteins, it is suggested that they may shift the equilibrium between the double-stranded conformation and H-DNA that has an unpaired strand.[487,491,492] For example, such proteins may bind to and "capture" a single-stranded region resulting from Py·Pu tract breathing or denaturation under torsional stress.[487,491]

Another option in H-DNA stabilization consists of a preferential protein interaction with the triple helix.[493] The triple helix geometry or high negative charge density might be the features recognized by triplex-binding proteins. In support of this option, experiments with basic oligopeptides as reasonable models of surface-localized protein domains showed their triplex-stabilizing effect.[439] However, there have been no direct data on triplex stabilization by proteins, and in some cases the experimental results were consistent with protein binding to Py·Pu tracts that had B-conformation but not H-DNA structure.[475,494,495]

To summarize, there is an assortment of various factors that may induce and/or stabilize the triple-helical structures in the cell. When considered separately, these factors contribute to triplex stability *in vitro*. However, many of them are available in cells simultaneously, and their combined effects are not evident. For example, the stabilizing effects of cations of various valences interfere with each other, since these cations compete for the same binding sites.[430,437] Clearly, more data are necessary to understand how triplexes could be formed and maintained in living cells.

F.   Possible Biological Roles of H-(H*)-DNA

Py·Pu tracts capable of forming triple-stranded DNA structures have been found in the genomes of various organisms. In eukaryotes, the Py·Pu tracts a few dozen base pairs long constitute up to 1% of the entire genome.[451a,496,497] Analysis of the human genome showed one H-DNA-forming sequence in every 50,000 bp.[450] H-DNA has been suggested to play a role in key biological processes such as transcription, replication, recombination, and DNA condensation (see, for a review, Mirkin and Frank-Kamenetskii,[404] Sinden,[1] Frank-Kamenetskii and Mirkin,[406] and Soyfer and Potaman[2]).

1.   Possible Regulation of Transcription

The Py·Pu tracts often occur in 5' flanking regions of eukaryotic genes, and a number of these tracts have been shown to be sensitive to the single-strand-specific nucleases (see Soyfer and Potaman[2] for a list of sequences). When cloned in plasmids, many of these Py·Pu tracts can adopt an H-DNA structure *in vitro*.[387,448,475,495,498–501] H-DNA might influence the regulation of gene expression in a number of different ways: by affecting the level of DNA supercoiling in the topological domain in which it forms; by changing the local as well as the global structure of DNA; and by influencing nucleosome organization and nucleosome phasing from the triplex region.

Indirect influence of H-DNA on transcription. Generally, the efficiency of transcription increases with increasing superhelical density of the templates.[107] However, some bacterial and eukaryotic transcription systems require an optimum superhelical density.[502–506] Thus H-DNA formation, partially relieving excessive superhelical tension, could serve to maintain the optimum template topology (as suggested in ref. 507).

Direct involvement of H-DNA in transcription. The formation of H-DNA can make the structure of the Py·Pu tract inappropriate for the transcription factor and subsequent RNA polymerase binding, thereby inhibiting gene expression (Figure 25A). For example, long triplex-forming $(G)_n \cdot (C)_n$ tracts placed 5' to the promoter inhibited transcription.[508] The gel comigration data show that the $(G)_n \cdot (C)_n$ tracts can bind an activator protein present in human cells[508] and chicken BPG1 protein.[509] The local unwinding of DNA in the promoter sequence of the human Na,K-ATPase $\alpha 2$ gene created by H-DNA formation extends into an adjacent TATA box, and thereby may down-regulate transcription by disrupting interactions between DNA and TATA-box-binding protein.[510]

Structure–function analyses for many eukaryotic genes have demonstrated a stimulatory influence of the Py·Pu sequences on promoter function. Transcription efficiency was partially lost when the Py·Pu tracts were deleted from promoter regions of human epidermal growth factor receptor,[511] c-*myc*,[512,513] *ets*-2,[514] and decorin genes[515]; rat neuronal cell adhesion molecule gene[494]; mouse c-Ki-*ras*[516] and transforming growth factor $\beta 3$[517] genes; and *Drosophila hsp26*[475] and actin[518] genes. Cloning of the H-forming Py·Pu tract upstream of the β-lactamase promoter stimulated transcription compared with the plasmid lacking the Py·Pu tract.[507] Triplex-distorting variations in the sequence reduced the transcriptional efficiency. Similarly, mutations in the repeating c-*myc* sequence motif (ACCCTC-CCC)$_4$, which would result in more mismatches within the suggested triplex, led to reduced transcriptional activity of the gene.[519] Two models of positive H-DNA influence on transcription may be consistent with these data.

A B-DNA-binding protein may interact with the promoter sequence as a repressor, preventing RNA polymerase binding (Figure 25B). When H-DNA forms, the repressor protein can no longer bind to the promoter sequence, allowing RNA polymerase access to the promoter sequence. The formation of H-DNA at the repressor binding site might be similar to the action of inducer protein, which

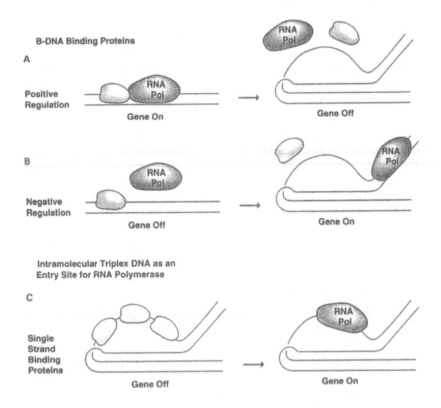

Figure 25. Models for intramolecular triplex involvement in gene regulation. (**A**) A positive acting B-DNA-binding protein may promote binding of RNA polymerase and turn on transcription (left). Intramolecular triplex formation might prevent binding of the positive activator, thus preventing binding of RNA polymerase (right). (**B**) In the negative regulatory scheme, a B-DNA-binding repressor that prevents RNA polymerase binding to the promoter (left) may no longer bind to its recognition sequence when in the intramolecular triplex form (right). RNA polymerase may be able to bind to the unwound region and begin transcription. (**C**) Single-strand-binding proteins bound to the single-stranded loop of the intramolecular triplex structure might prevent RNA polymerase binding and thus transcription (left). The unwound single-strand region or the duplex-triplex junction may provide an entry site for RNA polymerase to the promoter (right) (as in model **B**).

serves to displace the repressor and facilitate RNA polymerase binding at the adjacent site. No experimental data are available to illustrate this attractive hypothetical option in the utilization of the H-DNA by living cells.

The unwound structure of the H-DNA may be appropriate for the binding of RNA polymerase (Figure 25C), which otherwise should itself (with the aid of auxiliary factors) locally unpair the DNA duplex to initiate transcription.[484] Thus H-form could serve as the RNA polymerase entry point. Single-strand-binding proteins might transiently occupy a single-strand-repressing RNA polymerase binding when transcription is not required. The H-DNA structure appropriate for RNA polymerase entry could be formed under the influence of negative supercoiling (see above). Model studies in which RNA polymerases bound to locally unwound DNA templates and initiated[520] and elongated transcription[521] support the RNA polymerase entry-point hypothesis.

In other cases a lack of correlation between the H-forming potential of the DNA sequence and the transcriptional efficiency of the promoter has been reported. A point mutation in the Py·Pu tract of the promoter of *Drosophila hsp26* gene destabilized the H-DNA *in vitro,* but had no effect on the level of gene expression.[475] Replacement of one triplex-forming tract, $(GA)_n \cdot (TC)_n$, with another one, $(G)_n \cdot (C)_n$, reduced the transcription efficiency to the level characteristic of the promoter sequence lacking the essential $(GA)_n \cdot (TC)_n$ region. In this case, the H-DNA itself played no role in maintaining high transcriptional efficiency. A structure–function analysis of the c-Ki-*ras* promoter has shown that various point mutations in the Py·Pu tract do not affect its capability to form stable H-DNA *in vitro* and those destabilizing the triplex lead to a comparable drop in transcriptional activity relative to the original promoter.[495] It was suggested that the Py·Pu sequences contained binding sites for specific proteins, and the mutations introduced might not affect H-forming potential, but reduce the protein binding affinity for the sequence.[475,495] A similar conclusion has been drawn for the Py·Pu tract in the NCAM promoter.[494] Among possible proteins that might bind to the Py·Pu sequences upstream of promoters are GAGA[522] and Sp1 transcription factors,[485] NFκB,[523] and NSEP proteins.[512]

In addition to regulation at the initiation and termination stages, transcription in eukaryotes can be regulated at the elongation stage.[524] The possible triplex role at the elongation stage may be illustrated by the experiments in which codon degeneracy was used to engineer a 38-bp H-forming sequence into the β-galactosidase gene of the pBluescriptIISK+ plasmid.[478] An 80% lower expression of this β-galactosidase gene compared with another plasmid, where other codons that did not constitute the Py·Pu tract coded for the same amino acid sequence, was explained by the formation of H-DNA, which blocked the RNA polymerase progression along the DNA template.

In summary, the contradictory results on the importance of H-DNA in the regulation of transcription have been reported,[475,495,519,525] and the mechanisms by which H-DNA-forming sequences influence gene expression are still far from being well understood.[406] New information on mechanisms of H-DNA formation, and its stabilization and influence on conformations of adjacent regulatory sequences can advance our understanding of the potential involvement of H-DNA in biological processes.

A

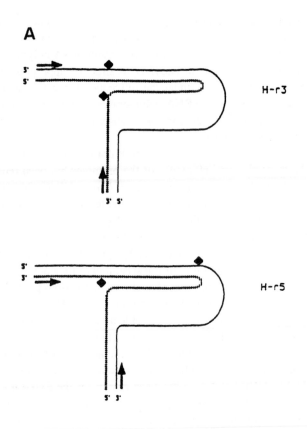

B

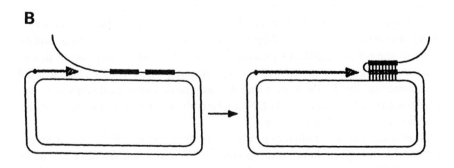

C

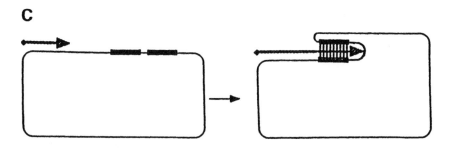

Figure 26. DNA polymerase inhibition by triplex DNA. (**A**) Inhibition by H*-DNA. The progression of a DNA polymerase along one strand of the double-stranded template may be hindered by the preexisting triple-stranded DNA structure. Depending on the H* DNA isomer, DNA polymerase advancing in the 3' to 5' direction faces the triplex either at the end or in the middle of the Py·Pu tract. (Reproduced from Mirkin and Frank-Kamenetskii.[404]) (**B**) Inhibition by H-DNA formation during strand displacement. When the new DNA chain is synthesized in a strand displacement reaction through the middle of the homopolymer template, the displaced strand containing a Py·Pu sequence may fold on itself downstream of the replication fork and form a block to replication. During the replication of open circular DNA under conditions favorable for both replication and H*-DNA formation, the T7 DNA polymerase stalled exactly in the middle of the Py·Pu tract when the purine-rich strand was displaced. (Reproduced from Samadashwily et al.[529]) (**C**) When a new DNA chain is synthesized through the middle of a single-strand homopolymer template (as occurs during synthesis of the lagging strand), the downstream half of the homopolymer sequence can fold back and form a triplex.[527] This triplex serves as a trap for the DNA polymerase, which cannot continue DNA synthesis. (Reproduced from Mirkin and Frank-Kamenetskii.[404]).

2. Possible Regulation of Replication

A phenomenon similar to transcription elongation control was observed for DNA replication. The $(GA)_n \cdot (TC)_n$ tracts pause several DNA replication enzymes *in vitro* under conditions appropriate for triplex formation.[526–528] Termination of DNA replication by the triplex-forming Py·Pu tracts has also been found in other studies[262,479,529] (see, for a review, Mirkin and Frank-Kamenetskii).[404] There are several different triplex structures that might be involved in the replication blockage.

The H*-DNA consisting of C·G·G and T·A·T triads could exist in different structural isomers dependent on the specific sequences designed.[262] The termination sites were located differently for specific H*-forming regions, but were mapped precisely by the chemical probing to the triplex-forming sequence. Figure 26A shows the sites where DNA polymerase movement may be hindered by the triple-stranded DNA structure that has been formed prior to DNA synthesis. Dependent on the H*-DNA isomer, DNA polymerase moving in the 3' to 5' direction along the template strand might stall either at the end or in the middle of the Py·Pu tract.

An active DNA polymerase itself may create a DNA structure that blocks polymerization. When the new DNA chain has been synthesized through the middle of the Py·Pu template, another portion of the Py·Pu sequence may fold back and form a triplex (Figure 26B).[527] During the replication of single-stranded DNA, a DNA polymerase is trapped in the center of the homopyrimidine or homopurine tract.[527–529] Upon double-stranded DNA replication, the DNA polymerase continuously synthesizes a strand complementary to the leading strand, and the other, lagging strand is unpaired for sufficient periods of time. The Py·Pu tract-containing lagging strand may fold on itself downstream of the replication fork. The resulting triplex presents one more type of replication block (Figure 26C). In accordance with this model, in experiments on open circular DNA, which cannot form a triplex before replication, T7 DNA polymerase stalled exactly in the middle of the Py·Pu tract when the purine-rich strand was displaced.[529] Further evidence of replication termination by the induced triplex came from a mutational analysis. H*-DNA-destabilizing mutations relieved the DNA polymerase from the polymerization block, whereas the compensatory mutations restoring the H*-DNA-forming potential restored the replication block.

The models for DNA replication blockage by triplex structures were elaborated in experiments *in vitro*. Besides DNA polymerase, the actual replication fork contains a number of accessory proteins that may destroy the preformed triplex, prevent DNA single strands from folding onto their target duplex, etc. The single strand-binding protein easily restores replication by unwinding the intramolecular triplex, but is less effective in disrupting an intermolecular triplex.[530] In *in vitro* experiments, the *E. coli* and the SV40 large T antigen DNA helicases were able to unwind model intermolecular triplexes[531] and intramolecular triplexes.[532] *In vivo* data on a role of Py·Pu tracts in replication are scarce. The $(GA)_{27}·(TC)_{27}$ tract-containing region of DNA located 2 kb from the integration site of the polyoma virus is a strong terminator of DNA replication in rat cells.[533] The involvement of this Py·Pu tract in replication termination was confirmed by the DNA polymerase pausing when a corresponding fragment was cloned into SV40 DNA.[526] Brinton et al.[534] found that an unusual cluster of simple repeats, including a Z-DNA-forming region, $(GC)_5(AC)_{21}$, and a long Py·Pu tract with a potential for H-DNA formation, has a significant effect on replication of a plasmid shuttle vector. One copy of this cluster, when cloned on either side of SV40 origin of replication, reduced the amount of DNA replicated in COS cells up to twofold. Two copies on both sides of the origin reduced replicated DNA down to 5% of that in a vector without the cluster.

3. Possible Triplex-Mediated Chromosome Folding

The fact that the Py·Pu tracts are distributed over the whole length of genomic DNA[496,497,535,536] has led to the suggestion that triplexes may promote chromosome compact packaging.[537,538] A series of experiments was designed to show DNA condensation due to triplex formation between different DNA molecules or

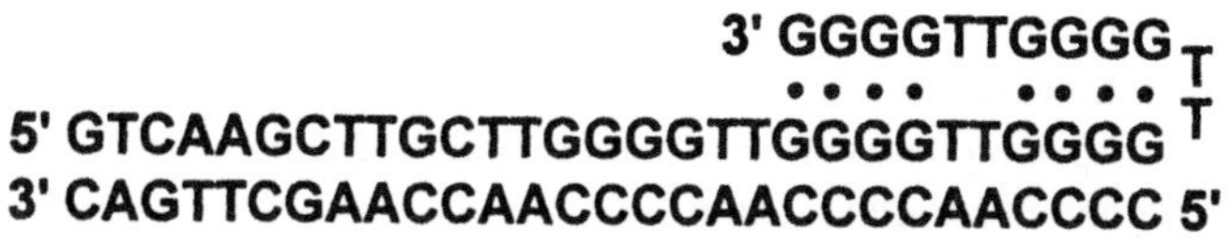

```
3' GGGGTTGGGG ┐
   • • • •   • • • • ┤ T
5' GTCAAGCTTGCTTGGGGTTGGGGTTGGGG ┘ T
3' CAGTTCGAACCAACCCCAACCCCAACCCC 5'
```

Figure 27. Triplex DNA structures can form at telomeres. Telomeres have a single-strand end of G-rich sequences that can form a variety of triplex and quadruplex structures (see Section G). The single-stranded overhang $(T_2G_4)_2$ and DNA duplex of the same sequence can form a triplex in the presence of Mg^{2+} cations at physiological pH. Since the triple-stranded structure is not an appropriate substrate for many proteins, this triplex structure of telomeres may provide a plausible explanation for the *in vivo* resistance of chromosome ends against degradation and recombination. (Reproduced from Veselkov et al.[541])

distant Py·Pu tracts of the same molecule.[472,538] When a molecule of poly[d(Tm⁵C)] formed three-stranded structures with the Py·Pu tracts of several $(GA)_{45}·(TC)_{45}$ insert-containing plasmids, this was seen in an electron microscope as the plasmid "rosettes."[472] In other experiments, linear plasmids containing single Py·Pu tracts at the ends produced linear dimers via triplex formation, and linear plasmids with Py·Pu tracts at both ends gave rise to quasicircular DNA molecules.[538] In the above-described experiments, a triplex structure is formed between the Py·Pu duplex and one strand of another unwound PyPu tract. The action of topoisomerase or other nicking-closing activity is necessary to create the structure consisting of interwound strands, which belong to different long DNA molecules.

4. Structural Role at Chromosome Ends

Telomeres are structures that stabilize the ends of eukaryotic chromosomes. They usually present a very long repeating motif consisting of six to eight nucleotides with the general sequence $(T/A)_mG_n$.[539] Single-stranded oligonucleotides modeling telomeric structures, consisting of the DNA duplex and the single-stranded overhang, form inter- and intramolecular quadruplexes in the presence of monovalent sodium and potassium cations.[540]

The study of a synthetic model of *Tetrahymena* chromosome telomeric terminus, consisting of the DNA duplex and the single-stranded overhang $(T_2G_4)_2$, showed that in the presence of Mg^{2+} and physiological pH the overhang folds back to form a triplex (Figure 27).[541] These authors suggested that the triplex structure of telomeres provides a plausible explanation for the *in vivo* resistance of chromosome ends against degradation and recombination.

5. Possible Role in Recombination

Specific types of triple-stranded structures have been suggested to mediate homologous DNA strand recombination in the presence of RecA protein.[542–545]

Such triple-stranded structures do not require the Py·Pu tracts, patterns of hydrogen bonding in them drastically differ from those in triplexes formed in the Py·Pu
tracts, the axial spacing between base pairs is stretched to 5.1 Å, and homologous
strands are aligned in parallel fashion.

The conventional intramolecular triplexes may also play roles in recombination.
The DNA rearrangement in the immunoglobulin class from IgM to IgA, IgG, or
IgE occurs in the highly repetitive regions of DNA (switch regions). For instance,
the potentially H-DNA-forming sequence $(AGGAG)_{28}$ is located in the switch
region of murine IgA.[546] The unwound structure may provide a single strand to
pair with a homologous region of a second chromosome-initiating recombination,
which should result in the splicing of antibody genes from a number of individual
gene segments. Non–B-DNA structures are possibly involved in an unequal sister
chromatid exchange, in which part of the gene is duplicated on one chromosome
and deleted from the other chromosome.[387] The recombination region contains
simple repeats of $(TC)_n$ capable of forming H-DNA followed by stretches of the
Z-forming $(TG)_n$ sequence. Formation of dimer molecules in recombinant plasmids carrying the H-forming Py·Pu tracts occurs six times more often than in a
control plasmid.[547] In human cells homologous recombination between the plasmids containing H-DNA-forming sequences occurs three times more often compared to controls.[548]

Recombination can be transcriptionally induced between two direct repeats separated by the sequence containing the H*-forming Py·Pu tract (Figure 28).[551] The
$(dG)_n \cdot (dC)_n$-containing plasmid constructs allowed efficient recombination
between homologous sequences of *lac* and *tac* promoter sequences separated by
either 200- or 1000-bp regions. The recombination was RecA independent, the
recombination rate being dependent on the length and orientation of $(dG)_n \cdot (dC)_n$
with respect to the gene. Under the active transcription conditions in *E. coli*, the
plasmids formed C·G·G-type triplexes, as shown by chemical modification. The
H*-DNA in this study was suggested to bring two remote sequences in close proximity to make recombination favorable.

The above described hypothetical models of H-DNA in genetic recombination
may include a displacement mechanism (Figure 29A), an interaction of the H- and
H*-forms (Figure 29B), and an H-form-induced duplex bend that brings homologous sequences into close proximity. Several results showing increased rates of
plasmid dimerization, transcription-driven recombination between direct repeats,
and the presence of H-forming tracts close to recombination points are indirect
evidence in favor of the triplex playing a role in recombination. More experimental and theoretical considerations are needed to fully establish this role.

G.   Control of Gene Expression

The formation of triple-stranded DNA complexes presents one of the most
important examples of how DNA function can be influenced by changing its struc-

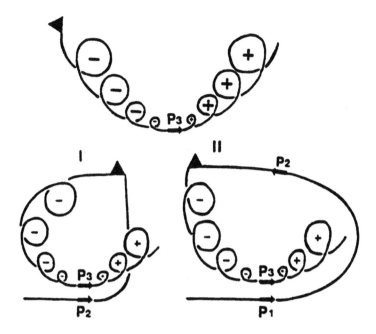

Figure 28. Transcription-stimulated genetic recombination mediated by triplex formation. Intramolecular triplex structures introduce bends into the DNA, which can influence the positioning of various DNA sequences. This model shows the triplex-induced approach of different homologous sequences in promoter regions (designated P1, P2, and P3). Once the promoter containing P3 is activated, the transcriptional process results in two supercoiled domains: positive supercoiling accumulates ahead of the transcribing RNA polymerase, whereas negative supercoiling accumulates behind RNA polymerase. Negative supercoiling-induced H*-DNA (designated by a filled triangle) significantly bends the DNA helix, bringing P2 and P3, which are separated by 200 bp, into close proximity to stimulate recombination. Similarly, P1 and P3, which are 1000 bp apart, may be brought together. (Reproduced from Kohwi and Panchenko.[549])

ture. Intermolecular triplex formation at specific regulatory sites using specifically designed oligonucleotides may be used to inhibit deleterious expression of genes related to cancer, AIDS, and so forth. Triplex-based inhibition of gene expression is feasible because of a natural abundance of Py·Pu tracts in genes and their 5' flanking regions (see Table 6.1 in ref. 2).

Morgan and Wells[550] showed that mRNA synthesis on a duplex polymer template is inhibited when the binding of a homopyrimidine third strand results in a triple-helical complex. During the past few years, triplex formation was shown to inhibit transcription and subsequent protein synthesis in a number of complicated

A. **D-Loop Formation with an Unpaired Strand**

Unwinding of theTriplex Strand to Form a Second Duplex

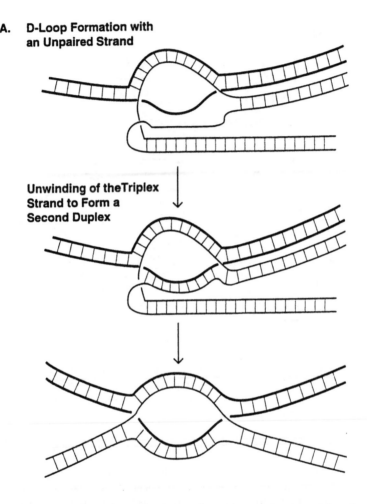

Figure 29. Hypothetical involvement of an intramolecular triplex in genetic recombination. (**A**) Possible mechanism of H-DNA-mediated genetic recombination. A free single strand of H-DNA pairs a complementary strand of a second homologous duplex (top). Following D-loop formation, the third strand of the triplex unpairs. Being complementary to the displaced D-loop strand, it initiates the formation of the Watson–Crick duplex (middle). Rotation of the bottom duplex from right to the left shows that a classical recombination intermediate has been formed (bottom). (Reproduced from Sinden, 1994.[1]) (**B**) Recombination may involve two H-DNA structures. This model suggests the formation of different (H-r5 and H-y3; or H-r3 and H-y5) isomers in different DNA molecules containing the same Py·Pu sequence. Single strands of these H and H* forms are complementary and form a Watson–Crick duplex. The unwinding of the strands forming triple helices and their pairing in the Watson–Crick duplexes result in a classical recombination intermediate. (Reproduced from Sinden.[1])

B.

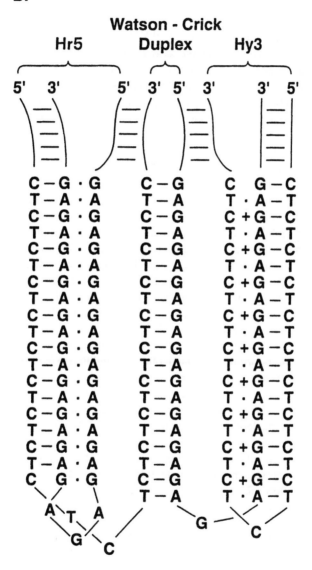

experimental cell-free systems. This inhibition is based on sufficiently high affinities of triplex-forming oligonucleotides for double-stranded DNA ($K_{diss} \approx 1$–10 nM) and the lifetimes of triple-stranded complexes on the order of several hours.[431,432,437] These values approach those typical for many sequence-specific DNA-binding proteins, so the proteins involved in transcription cannot easily displace the Py·Pu tract-bound oligonucleotide. Specificity of triplex formation is high: even one mismatch in a 15 nucleotide long triplex-forming oligonucleotide results in at least a 10-fold decrease in affinity compared to a perfect triplex.[414]

Several mechanisms are relevant to the transcription inhibition by the triplex-forming oligonucleotides (Figure 30).[403,410,551] The transcription machinery generally involves RNA polymerase and associated factors, activating proteins that bind upstream in the promoter region, and activators that bind to enhancer sequences at long distances from the RNA polymerase binding site and act via folding of the double helix. Triplex-forming oligonucleotides can (1) prevent long distance interactions between enhancer-bound proteins and RNA polymerase and associated factors by influencing the local bending ability of DNA; (2) eliminate binding of activators, acting either through long distance interactions or through factors bound adjacent to the RNA polymerase (not shown); (3) block tracking of transcription factors initially bound to a distant enhancer sequence and sliding toward the RNA polymerase machinery bound around the promoter site; (4) repress binding and interactions of basal protein factors associated with the RNA polymerase; (5) inhibit the RNA polymerase binding to the promoter site; (6) block initiation of transcription upon binding downstream of promoter but in contact with the transcription machinery; (7) inhibit elongation of transcription when bound to the Py·Pu site within the transcribed gene. Other mechanisms, such as the recruitment of inhibitory factors (e.g., proteins recognizing triple-stranded structure) or the alteration of chromatin assembly, may also be suggested. More details may be found in recent reviews.[408,410]

Table 8 lists some examples of inhibition of transcription initiation or chain elongation by triplex-forming oligonucleotides. In addition to *in vitro* inhibition of gene expression, experiments have shown that the inhibiting activity is retained when the preformed DNA-triplex-forming oligonucleotide complex is introduced into cultured cells.[552,553] Moreover, incubation of the cells in the presence of triplex-forming oligonucleotides resulted in the uptake of the oligonucleotides and a subsequent inhibition of gene expression and protein synthesis.[554–558]

In *in vitro* experiments, triplex-forming oligonucleotides, when used in excess to their target duplexes (mol oligonucleotide/mol template > 100), significantly inhibited (up to 90%) RNA polymerases.[551,558,565] In *in vivo* experiments, 50% inhibition of mRNA synthesis or cell proliferation may be obtained at oligonucleotide concentrations in extracellular milieu of up to 100 μM.[555–558] Inhibition of transcription depends on lifetimes of oligonucleotides, which in enzyme-rich cellular media are on the order of several dozen minutes[452,566]; however, a decreased level of mRNA was noticed, even 7 days after removal of natural phosphodiester oligonucleotides from cell culture.[557]

In many cases the Py·Pu tracts are available in the genes encoding the key proteins in the pathogenesis of various diseases so that triplex methodology is applicable. Table 9 lists some diseases that may be treated with triplex-forming oligonucleotides and their analogs.[407] The least complicated and most likely therapeutic applications of triplex-forming oligonucleotides will be as antiviral

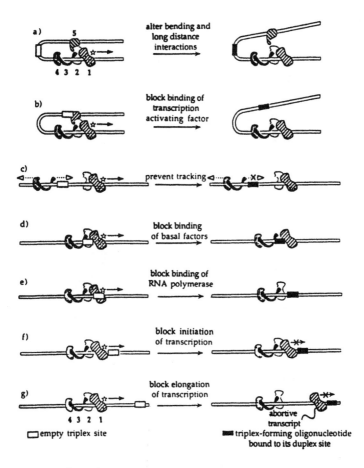

Figure 30. Hypothetical mechanisms by which triplex-forming oligonucleotides can inhibit transcription. The transcription process involves RNA polymerase and associated protein factors, activating factors that have binding sites in the promoter region, and activator proteins that bind to enhancer sequences at long distances upstream of the RNA polymerase binding site and require the double helix to fold. The star and arrow indicate the transcription start and the direction of transcription, respectively. An open box represents an unoccupied PyPu tract, whereas a filled box represents an intermolecular triplex formed on duplex DNA. Proteins: RNA polymerase (1), associated basal transcription factors (2), transcription activating factors (3,4,5). Triplex formation may result in (**a**) disruption of long-distance interactions between enhancer-bound proteins and the transcription complex; (**b**) blockage of activator binding at the distant site; (**c**) blockage of transcription factor sliding to the promoter sequence from the enhancer sequence; (**d**) inhibition of binding of basal transcription factors; (**e**) a physical barrier to RNA polymerase binding; (**f**) inhibition of transcriptional initiation upon binding downstream but in contact with transcription complex; (**g**) transcriptional inhibition at the elongation step when the Py·Pu tract is located within the transcribed region of the gene. (Reproduced from Thuong and Hélène.[403])

Table 8. Inhibition of Transcription Via Intermolecular Triplex Formation

Gene	Oligomer	Conditions	Reference
Human c-*myc*	27-mer	*In vitro*	Cooney et al.[416]
Human c-*myc*	27-mer	HeLa cells (TFO uptake)	Postel et al.[555]
Mouse IL2R	28-mer	Lymphocytes (TFO uptake)	Orson et al.[554]
Mouse IL2R	15-mer-acridine	Tumor T cells HSB2 cells	Girgoriev et al.[523]
	15-mer-psoralen	(plasmid-TFO electroporation)	Girgoriev et al.[552]
E. coli bla	13-mer	*In vitro*	Duval-Valentin et al.[559]
Human dihydrofolate reductase	19-mer	*In vitro*	Gee et al.[485]
HIV-1	Various	*In vitro*	Ojwang et al.[557]
	31 and 38-mers	Human MT4 and U937 cells (TFO uptake)	McShan et al.[560]
T7 early promoter		*In vitro*	Ross et al.[486]
Maize Adh1-GUS	Various lengths	Protoplasts (DNA and TFO co-transformation)	Lu and Ferl[553]
Human platelet-derived growth factor A-chain	24-mer	*In vitro*	Wang et al.[561]
6-16 interferon-responsive element	21-mer	HeLa cells (liposome-mediated TFO delivery)	Roy[558]
Progesterone-responsive gene	38-mer	Monkey kidney CV-1 cells (cholesterol-modified TFO uptake)	Ing et al.[556]
Aldehyde dehydrogenase	21-mer	Human hepatoma cell Hep G2 (liposome-mediated TFO delivery)	Tu et al.[562]
Rat α1(I) collagen	30-mer	Rat cardiac fibroblasts (transfection) (liposome-mediated TFO delivery)	Kovacs et al.[563]
Granulocyte-macrophage colony-stimulating factor	15-mer	Jurkat T cells (TFO uptake)	Kochetkova and Shannon[564]

Table 9. Diseases that May Potentially Be Treated with
Triplex-forming Oligonucleotides and Their Analogs

Virus-associated diseases

Adenovirus, herpes simplex viruses 1 and 2, herpes zoster, cytomegalovirus, Epstein-Barr virus, human papilloma virus, influenza A and B, parainfluenza, human T cell lymphotropic virus, human immunodeficiency virus, hepatitis A and B

Oncologic diseases

Lymphoma, leukemia, melanoma, osteosarcoma, carcinoma (of colon, prostate, kidney, bladder, breast)

Other

Psoriasis, drug resistance, allergy, inflammation

agents. For many pathogenic viruses, the protein sequences and functions are relatively well understood. Many essential viral proteins whose genes should be suppressed have no human cellular analogs, which reduces the risk of undesirable suppression of normally functioning human genes. The prospects for the treatment of cancer diseases with triplex-forming oligonucleotides are less clear. Although oncogenes were suggested in the development and progression of several human tumors, the understanding of underlying processes is not clear enough. Moreover, the difference between an oncogene and its normal cellular counterpart may be in only one base pair that presents a risk of unspecific inhibition of normal cellular genes.

There are other important problems in the triplex-mediated inhibition of gene expression.[2,410,567]

1. The rate of triplex formation is slow, therefore, for oligonucleotides competing with protein factors for binding to a specific sequence on DNA, kinetic phenomena might become a limiting factor.
2. The range of limitations imposed by nonspecific oligonucleotide binding to numerous target sites (e.g., transfer RNA, small nuclear RNA, accessible single-stranded regions in ribosomal RNA, aminoacyl-tRNA-synthetases, false binding to the initiation start, etc.) remains to be elucidated.
3. Triplex-stabilizing low pH or elevated concentrations of divalent metal cations and polyamines are not available in cells. Therefore, cytosine analogs, which are not protonated at neutral pH, and stably binding purine-rich oligonucleotides containing T and G nucleotides are being developed (see ref. 2 for a review). Unfortunately, cytosine-rich pyrimidine oligonucleotides may form protonated hairpin structures, reducing the oligonucleotide amount available for triplex formation,[568] and T- and G-containing oligonucleotides may aggregate at physiological ionic strength by forming guanosine quartets.[465,569]

4. Oligonucleotides designed to form triplexes may act by interrupting other cellular processes (e.g., the cascade of events required for interferon-mediated induction of several genes, including the targeted one).[570,571] Thus an interpretation of oligonucleotide effects *in vivo* requires some caution.

5. In some cases, short oligonucleotides may not bind strongly enough to stop the enzymatic machinery. Cross-linking or intercalating agents covalently linked to oligonucleotides (Table 6) might be used to stabilize the triplexes. However, the issues of a cross-link repair and inhibition of DNA replication become important.[410]

6. Delivery of polyanionic, hydrophilic compounds to their targets through the lipid membrane must be efficient enough.[572] Since negatively charged oligonucleotides poorly penetrate hydrophobic cellular membranes, several techniques of oligonucleotide delivery into the cell via liposome-[573-575] and receptor-mediated endocytosis,[576] directly in the hydrophobized form[556] and as conjugates with polylysine,[577] have been developed. The oligonucleotides of interest can be generated at their action sites from a vector containing promoter, capping, and termination sequences of the human small nuclear U6 gene, surrounding a synthetic sequence to be synthesized.

7. When delivered into the cell, the full-length oligonucleotides may be stable enough to persist for several hours.[554] A number of other analogs have been designed to improve the triplex-forming ability and metabolic stability (see ref. 2 for a review). There is growing evidence that oligonucleotides enter cells by endocytosis into lysosomes and are inefficiently delivered to the nucleus,[410] which requires the modification of existing or development of new delivery strategies.

In summary, any disease caused by the expression of a gene can be treated at various stages of the cellular processes. Many traditional drugs are targeted against the functional proteins, and their design requires extensive information about structure–activity relationships for these drugs. The oligonucleotide-based antisense approach is directed against the formation on the mRNA template of the secondary products, proteins. However, a continuous gene expression may supply new mRNA molecules. The triplex-based antigene strategy exploits the possibility of binding oligonucleotides to DNA inside genes or regulatory regions to hamper mRNA synthesis on the DNA template. It is directed against the primary process of gene expression and, therefore, may be more efficient. Another advantage is that it avoids the necessity of determining the various specific mechanisms of drug–protein interaction. In a practical sense, several important problems must be solved before triplex-mediated gene therapy will become a reality.

VII. MISCELLANEOUS ALTERNATIVE CONFORMATIONS OF DNA

A. Slipped-Strand DNA

Slipped-strand DNA (S-DNA) structures can form in regions with direct repeat symmetry. Regions of DNA containing long tracts of repeating mono-, di-, tri-, and tetranucleotides have multiple opportunities for the formation of hydrogen bonds in an out-of-register or "slipped" fashion. To form a slipped-strand structure, a section (or all) of the repeating duplex must unwind to allow one region of the direct repeat to form a Watson–Crick base pair strand with another region of the repetitive sequence. Figure 31 shows two possible isomers of S-DNA. One isomer has loops composed of the 5' direct repeat in both strands, and the other has loops composed of the 3' direct repeats in both strands. Because this structure results in the unwinding of the DNA double helix, DNA supercoils would be lost upon the formation of S-DNA in supercoiled DNA.

Historically, S-DNA structures have been suggested to exist in eukaryotic DNA within regions of direct repeat symmetry sensitive to S1 nuclease.[578–580] However, these sequences also contained mirror repeat symmetry, and intramolecular triplex structures may form at these sites. Recent evidence for S-DNA has been presented by Pearson and Sinden.[581] By melting and reannealing DNA, they detected novel structures formed within the (CTG)·(CAG) and (CGG)·(CCG) triplet repeat tracts associated with myotonic dystrophy and fragile X syndrome, respectively. The reannealing-induced CTG- or CGG-containing DNA structures have the following properties: (1) the novel structures are formed from complementary strands; (2) the structures are formed from complementary strands of equal length; (3) the alternative DNA structure occurs within the repeat tract; (4) linear duplex DNA flanks the alternative structure; (5) formation of the alternative structure does not require superhelical tension; (6) the alternative structures are remarkably stable in linear DNA under physiological conditions; (7) the alternative structures possess single-strand character, which is expected of DNA in two loops. Moreover, the CAG strand is more susceptible to single-strand nuclease digestion than is the CTG strand, consistent with reports of a greater stability of hairpin structures in the CTG strand (see Pearson and Sinden[581] for a discussion). The results are consistent with the formation of S-DNA structures (Figure 31).

Biologically, S-DNA is very important in spontaneous frameshift mutagenesis. It is known that spontaneous deletion or addition mutations can occur within runs of a single base. In 1966, Streisinger et al. proposed a model to explain frameshift mutations within runs of a single base.[582] Since the genetic code is read as triplets, adding or deleting a single base shifts the reading frame of all bases downstream of the mutation. This will result in a mRNA that encodes amino acids that are different from those present in the wild-type protein downstream of the frameshift mutation.

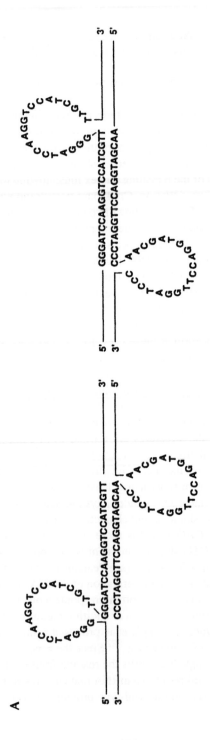

B

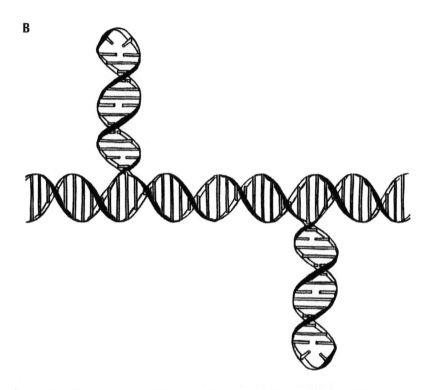

Figure 31.   Slipped-strand DNA. Slipped mispaired DNA can form within two direct repeat sequences when they pair in a misaligned fashion. (**A**) Two 20-bp direct repeats can form two different slipped mispaired isomers. In one isomer (left), the second copy of the direct repeat in the top strand is base paired with the first copy of the direct repeat on the bottom strand. The other structure (right) shows an isomer in which the first copy of direct repeat in the top strand is base paired with the second copy of the direct repeat in the bottom strand. (**B**) S-DNA formed from $(CTG)_n \cdot (CAG)_n$ and $(CGG)_n \cdot (CCG)_n$ triplet repeats. S-DNA structures are believed to form within these triplet repeat sequences following denaturation and renaturation.[581] The looped-out single strands can fold into hairpin structures stabilized by two G·C base pairs flanking an A·A or T·T mispair in the opposite strands.

B.   DNA Unwinding Elements

DNA unwinding elements or DUEs have been identified in both prokaryotic and eukaryotic DNA sequences. DUEs are A + T-rich regions of DNA that are commonly associated with replication origins and chromosomal matrix attachment sites. DUEs are A + T-rich sequences ranging in size from 30 to >100 bp in length. As a class of sequence elements, DUEs have little sequence homology in that there is no consensus sequence. The only similarity is that the sequences are A + T-rich.

In the presence of DNA supercoiling, unwinding of the double helix occurs first at A + T regions. *In vitro* DUEs unwind, and this unwound state is maintained in a stable fashion in the presence of negative DNA supercoiling. This unwound state can be detected by the sensitivity of the DUEs to digestion by single-stranded nucleases. Sheflin and Kowalski examined the pattern of cutting supercoiled DNA with mung bean nuclease.[583] The susceptibility of the DUEs to nuclease digestion is a function of the Mg^{2+} concentration. In buffer containing no Mg^{2+}, DUEs melt and are sensitive to digestion, whereas in the presence of Mg^{2+}, the A + T-rich DUE region remained double stranded. Thus the ability of these regions of DNA to unwind *in vivo* may be controlled by the level of unrestrained supercoiling and the local ionic environment in cells.

DUEs are required for the initiation of DNA replication at certain origins, as discussed above in Section IV[202,205]. A correlation exists between DNA unwinding and the proficiency of the DUE as a replication origin, as determined from certain yeast sequences. Progressive deletion of the DUE decreased function as an origin of DNA replication until, when melting no longer occurred, the region no longer functioned as a replication origin. These results suggest that unwinding at an A + T-rich DNA region near the origin of DNA replication is required for the initiation of DNA replication. DNA unwinding is also required at the *E. coli* origin of replication. DnaA proteins first bind to DnaA boxes at the *E. coli* origin of replication in supercoiled DNA and organize DNA into a tight loop. Three DnaB boxes, which are A + T-rich direct repeat sequences (effectively DUE elements) and then unwind. Following binding of DnaB and DnaC, the DnaB helicase begins unwinding the DNA double helix, forming a bubble in which DnaG, an RNA primase, binds. Subsequently, DNA polymerase holoenzyme binds and replication elongation ensues.

C. Parallel-Stranded DNA

Parallel-stranded DNA contains the two single strands in an orientation opposite that of typical Watson–Crick helix orientation, which is antiparallel.[584,585] The orientation of the two strands in parallel DNA requires that the bases be paired in a reverse Watson–Crick fashion. In the reverse Watson–Crick orientation, the glycosidic bonds and ribose sugars are extended in a *trans* position, compared with the *cis* configuration found in B-DNA (Figure 32). In a reverse Watson–Crick base pair, the thymines are hydrogen-bonded through the O2 and N3 positions, whereas the N3 and O4 positions are hydrogen bonded in a normal Watson–Crick base pair. Remarkably, the stability of parallel stranded DNA is only slightly lower than that of a corresponding parallel B-DNA helix.

There are no known examples of naturally occurring parallel DNA formed by A·T base pairs. However, G + C-rich regions in the chromosome and sequences found at telomeres can form duplex and quadruplex structures in which the strands are organized in a parallel fashion. Certain RNA sequences may also adopt a parallel helix.

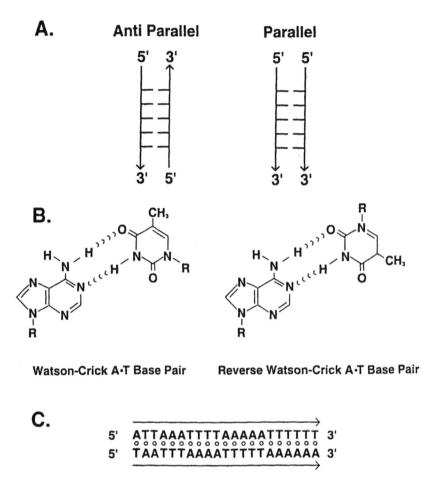

Figure 32. Parallel-stranded DNA. (**A**) Antiparallel and parallel oganization of complementary strands. Typical B-form, A-form, or Z-form DNA contains an antiparallel orientation of the complementary strands, with 5'→3' polarity in one strand and 3'→5' polarity in the opposite strand. (**B**) Parallel DNA requires the formation of reverse Watson–Crick base pairs in which one base is oriented 180° with respect to the Watson-Crick base pair orientation. (**C**) The A + T-rich sequence shown can form a parallel helix.[596] The reverse Watson–Crick base pairs are indicated by an open circle (o).

D.   Four-Stranded DNA

In recent years it has come to be realized that there are many ways four strands of DNA can be held together by various hydrogen bonding schemes forming quadruplex DNA. A G + C-rich region of DNA from the immunoglobulin heavy chain switch region was shown to exist as a four-stranded, G-quartet structure in which all strands are parallel.[586] On incubation of single-stranded oligonucleotides containing 5'-GGGGAGCTGGG-3', a higher molecular weight structure formed, as detected by electrophoresis on polyacrylamide gels. A Hoogsteen base pairing scheme involving four DNA strands organized in a parallel configuration was suggested from analysis of the chemical reactivity of the complex to dimethylsulfate. This G-quartet DNA structure may be important biologically. For example, a quadruplex might hold the four chromosomes together at meiosis.[586]

Quadruplex DNA can also form in DNA tracts found at telomeres. Telomeres are composed of purine-rich repetitive sequences (for example, $(G_4T_2)_n$ or $(G_4T_4)_n$). A multitude of alternative pairing schemes are possible with these telomere repeats (for example, see ref. 587). The *Tetrahymena* telomere DNA sequence (G_4T_2) can form a quadruplex consisting of a planar array of guanines in the *anti* configuration held together by Hoogsteen base pairs.[588] Depending on how these sequences come together, association can occur in a parallel or antiparallel orientation. A parallel orientation can occur from the interaction of four single strands (Figure 33). If a single strand first folds into a hairpin in which guanines are Hoogsteen base paired, the two strands are antiparallel. Two of these foldback structures can interact in four-stranded structures, as discussed below.

The formation of two different isomers of the *Oxytricha* telomeric DNA sequence $(G_4T_4G_4)$ have been reported.[589,590] The individual $G_4T_4G_4$ molecules can form hairpins, and two hairpins can hydrogen bond in two different ways. The organization with alternating strands arranged in an antiparallel orientation was found in an X-ray crystal structure by Kang et al.[589] Alternate nucleotides in a quartet exist in the *anti* and *syn* conformations. The pattern and direction of hydrogen bonding are reversed in adjacent quartets (Figure 34). Smith and Feigon, using NMR analysis, reported a very different quadruplex organization of the same sequence: a four-stranded structure that does not involve hairpins.[590] In this structure, the T_4 loops are on opposite ends of the quartet spanning the diagonals of the quartet at right angles to each other. In this structure, the glycosidic bond angles are *syn-syn-anti-anti* around the quartet, whereas the pattern in adjacent quartets is *anti-anti-syn-syn* (Figure 34). Like the double hairpin structure, the direction of hydrogen bonding also alternates between adjacent quartets.

The very different quartet structures identified for the same *Oxytricha* telomere sequence $(G_4T_4G_4)$ by Smith and Feigon[590] and Kang and colleagues[589] demonstrate the variability inherent in the formation of quadruplex structures from poly(dG) sequences. Presumably the variable structures represent different local ionic and buffer conditions present during crystallization or NMR analysis. The structures that form are sensitive to monovalent cations.[587,591,592] A single K^+ ion may bind tightly within the center of the quartet. If quartet structures form in living cells, the structural isomer that forms, as well as their stability, might be controlled by variations in intracellular K^+ concentration.

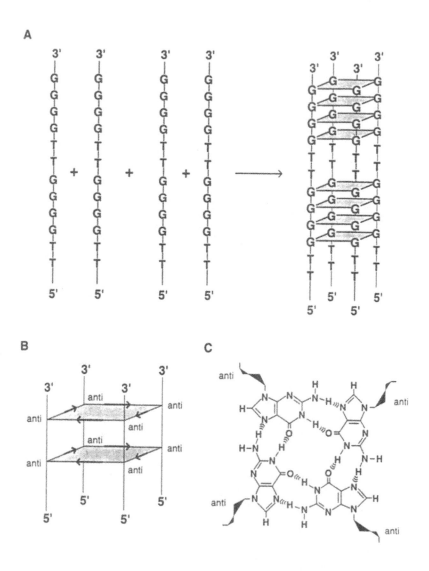

Figure 33. A G-quartet quadruplex DNA structure. (**A**) Four single strand tracts containing (G_4T_2) can form a tetraplex structure with all four strands in a parallel orientation. (**B** and **C**) The four G's, with all glycosidic bonds in the *anti* conformation, are hydrogen bonded through Hoogsteen base pairs. Moreover, the direction of hydrogen bonds are the same in each G quartet. The arrows (in **B**) indicate the direction of the hydrogen bonding from the hydrogen donor (H) to the acceptor (O or N). The direction is similar in successive base quartets.

A

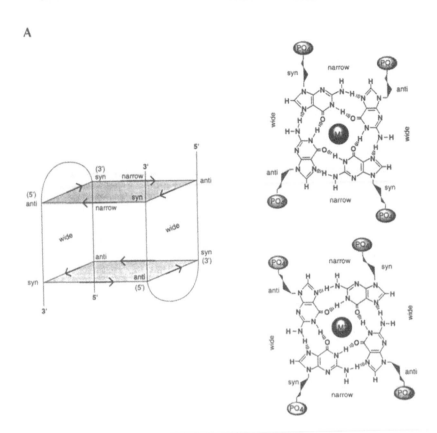

Figure 34. Different four-stranded DNA structures from a telomere sequence. Many different structural isomers can form from sequences found at telomeres. (**A**) A single strand containing $(G_4T_4)_n$ can fold into a hairpin that will be an antiparallel duplex. Two hairpins can associate in four different ways, with the hairpin loops at the same or opposite ends of the quadruplex. In each orientation the polarity of the DNA strands can be 5'-3'-5'-3' or 5'-5'-3'-3'. The *Oxytricha* telomere sequence $G_4T_4G_4$ crystallized into an antiparallel quadruplex, with the hairpin loops at opposite ends of the quartet and with 5'-3'-5'-3' polarity.[589] Within a G quartet the glycosidic bonds alternate *anti-syn-anti-syn*. Moreover, the direction of hydrogen bonding alternates in adjacent quartets. (**B**) $(G_4T_4)_n$ telomere sequences can also form a different structure formed when the two single strands fold together, forming the diagonals of a quartet with the loops at right angles to each other. There are no intrastrand hydrogen bonds, rather guanines from one strand hydrogen bond to two guanines from the second strand. There are two different isomers that can form, depending on the ways the two single strands fold. The structure shown was identified in solution for $G_4T_4G_4$ by NMR spectroscopy[590] In this orientation the glycosidic bonds are *anti-anti-syn-syn* in one quartet and *syn-syn-anti-anti* in an adjacent quartet. The direction of the hydrogen bonds is reversed between adjacent quartets.

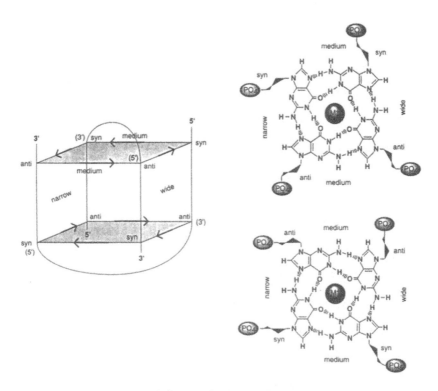

E. Higher Order Pu·Py Structures

Long regions of $(dG)_n \cdot (dC)_n$ and can form a bimolecular triplex from two smaller intramolecular triplex regions when the single strand loop from the first intramolecular triplex interacts to form the third strand of an adjacent intramolecular triplex.[429,461] When a homopurine·homopyrimidine region gets very long, there are many different opportunities for smaller triple helices to form. This is especially pronounced in very simple repetitive sequence elements such as poly(dG)·poly(dC) or polyd(G-A)·polyd(T-C). Glover and Pulleyblank[593] and Shimizu et al.[594] demonstrated considerable structural diversity in long simple polypurine·polypyrimidine mirror repeats, as is evident from multiple bands indicative of different DNA secondary transitions on two dimensional gels.

REFERENCES

1. Sinden, R.R. *DNA Structure and Function*. Academic Press, San Diego, 1994.
2. Soyfer, V.N.; Potaman, V.N. *Triple-Helical Nucleic Acids*. Springer, New York, 1995.
3. Seeman, N.C.; Rosenberg, J.M.; Suddath, F.L.; Kim, J.J.P.; Rich, A. RNA double helical fragment at atomic resolution. I. The crystal and molecular structure of sodium adenylyl-3',5'-uridine hexahydrate. *J. Mol. Biol.* **1976**, *104*, 109–144.

4. Rosenberg, J.M.; Seeman, N.C.; Day, R.O.; Rich, A. RNA double helical fragment at the atomic resolution. II. The crystal structure of sodium guanylyl-3',5'-cytidine nonahydrate. *J. Mol. Biol.* **1976**, *104*, 145–167.

5. Frank-Kamenetskii, M.D. Simplification of the empirical relationship between melting temperature of DNA, its GC content and the concentration of sodium ions in solution. *Biopolymers* **1971**, *10*, 2623–2624.

6. Hoogsteen, K. The crystal and molecular structure of a hydrogen-bonded complex between 1-methylthymine and 9-methyladenine. *Acta Crystallogr.* **1963**, *16*, 907–916.

7. Langridge, R.; Wilson, H.R.; Hooper, C.W.; Wilkins, M.H.F.; Hamilton, L.D. The molecular configuration of deoxyribonucleic acid. 1. X-ray diffraction study of a crystalline form of the lithium salt. *J. Mol. Biol.* **1960**, *2*, 19–37.

8. Langridge, R.; Marvin, D.A.; Seeds, W.E.; Wilson, H.R.; Hamilton, L.D. The molecular configuration of deoxyribonucleic acid II. Molecular models and their Fourier transforms. *J. Mol. Biol.* **1960**, *2*, 38–64.

9. Fuller, W.; Wilkins, M.H.F.; Wilson, H.R.; Hamilton, L.D.; Arnott, S. The molecular configuration of deoxyribonucleic acid. IV. X-ray diffraction study of the A form. *J. Mol. Biol.* **1965**, *12*, 60–80.

10. Fairall, L.; Martin, S.; Rhodes, D. The DNA binding site of the *Xenopus* transcription factor IIIA has a non-B-form structure. *EMBO J.* **1989**, *8*, 1809–1817.

11. Dickerson, R.E.; Drew, H.R. Structure of a B-DNA dodecamer II. Influence of base sequence on helix structure. *J. Mol. Biol.* **1981**, *149*, 761–786.

12. Dickerson, R.E. Base sequence and helix structure variation in B and A DNA. *J. Mol. Biol.* **1983**, *166*, 419–441.

13. Arnott, S.; Chandrasekaran, R.; Hall, I.H.; Puigjaner, L.C. Heteronomous DNA. *Nucleic Acids Res.* **1983**, *11*, 4141–4156.

14. Chastain, P.D.; Eichler, E.E.; Kang, S.; Nelson, D.L.; Levene, S.D.; Sinden, R.R. Anomalous rapid electrophoretic mobility of DNA containing triplet repeats associated with human disease genes. *Biochemistry* **1995**, *34*, 16125–16131.

14a. Chastain, P.D.; Sinden, R.R. CTG repeats associated with human genetic disease are inherently flexible. *J. Mol. Biol.* **1997**, in press.

15. Hagerman, P.J. Flexibility of DNA. *Annu. Rev. Biophys. Biophys. Chem.* **1988**, *17*, 265–286.

16. Bednar, J.; Furrer, P.; Katritch, V.; Stasiak, A.Z.; Dubochet, J.; Revet, B. Determination of DNA persistence length by cryoelectron microscopy—separation of the static and dynamic contributions to the apparent persistence length of DNA. *J. Mol. Biol.* **1995**, *254*, 579–591.

17. Marini, J.C.; Levene, S.D.; Crothers, D.M.; Englund, P.T. Bent helical structure in kinetoplast DNA. *Proc. Natl. Acad. Sci. USA* **1982**, *79*, 7664–7668.

18. Marini, J.C.; Levene, S.D.; Crothers, D.M.; Englund, P.T. Bent helical structure in kinetoplast DNA. *Proc. Natl. Acad. Sci. USA* **1983**, *80*, 7678.

19. Bustamante, C.; Gurrieri, S.; Smith, S.B. Towards a molecular description of pulsed-field gel-electrophoresis. *Trends Biotechnol.* **1993**, *11*, 23–30.

20. Calladine, C.R.; Drew, H.R.; McCall, M.J. The instrinsic curvature of DNA in solution. *J. Mol. Biol.* **1988**, *201*, 127–137.

21. Maroun, R.C.; Olson, W.K. Base sequence effects in double-helical DNA. II. Configurational statistis of rod-like chains. *Biopolymers* **1988**, *27*, 561

22. Travers, A.A. DNA conformation and protein binding. *Annu. Rev. Biochem.* **1995**, *58*, 427–452.

23. Wu, H.M.; Crothers, D.M. The locus of sequence-directed and protein-induced DNA bending. *Nature* **1984**, *308*, 509–513.

24. Koo, H.S.; Wu, H.M.; Crothers, D. DNA bending at adenine-thymine tracts. *Nature* **1986**, *320*, 501–506.

25. Hagerman, P.J. Sequence dependence of the curvature of DNA: A test of the phasing hypothesis. *Biochemistry* **1985**, *24*, 7033–7037.

26. Brukner, I.; Dlakic, M.; Savic, A.; Susic, S.; Pongor, S.; Suck, D. Evidence for opposite groove-directed curvature of GGGCCC and AAAAA sdequence elements. *Nucleic Acids Res.* **1993**, *21*, 1025–1029.

27. Brukner, I.; Susic, S.; Dlakic, M. Physiological concentration of magnesium ions induces a strong macroscopic curvature in GGGCCC-containing DNA. *J. Mol. Biol.* **1994**, *236*, 26–32.

28. Trifonov, E.N.; Sussman, J.L. The pitch of chromatin DNA is reflected in its nucleotide sequence. *Proc. Natl. Acad. Sci. USA* **1980**, *77*, 3816–3820.

29. Ulanovsky, L.; Bodner, M.; Trifonov, E.N.; Choder, M. Curved DNA: Design, synthesis, and circularization. *Proc. Natl. Acad. Sci. USA* **1986**, *83*, 862–866.

30. Trifonov, E.N.; Tan, R.K.Z.; Harvey, S.C. Static persistence length of DNA. In: *Structure and Expression* (Sarma, M.H.; Sarma, R.H.; eds.). Adenine Press, Albany, NY, 1988, pp. 243–254.

31. Koo, H.S.; Drak, J.; Rice, J.A.; Crothers, D.M. Determination of the extent of DNA bending by an adenine-thymine tract. *Biochemistry* **1990**, *29*, 4227–4234.

32. Selsing, E.; Wells, R.D.; Alden, C.J.; Arnott, S. Bent DNA: visualization of a base-paired and stacked A-B conformational junction. *J. Biol. Chem.* **1979**, *254*, 5417–5422.

33. Goodsell, D.S.; Kopka, M.L.; Cascio, D.; Dickerson, R.E. Crystal structure of CATGGCCATG and its implications for A-tract bending models. *Proc. Natl. Acad. Sci. USA* **1993**, *90*, 2930–2934.

34. Brahms, S.; Brahms, J.G. DNA with adenine tracts contains poly(dA)·poly(dT) conformational features in solution. *Nucleic Acids Res.* **1990**, *18*, 1559–1564.

35. Nelson, H.C.M.; Finch, J.T.; Luisi, B.F.; Klug, A. The structure of an oligo(dA)-oligo(dT) tract and its biological implications. *Nature* **1987**, *330*, 221–226.

36. Coll, M.; Frederick, C.A.; Wang, A.H.; Rich, A. A bifurcated hydrogen-bonded conformation in the d(A·T) base pairs of the DNA dodecamer d(CGCAAATTTGCG) and its complex with distamycin. *Proc. Natl. Acad. Sci. USA* **1987**, *84*, 8385–8389.

37. Peck, L.; Wang, J.C. Sequence dependence of the helical repeat of DNA in solution. *Nature* **1981**, *292*, 375

38. Rhodes, D.; Klug, A. Sequence-dependent helical periodicity of DNA. *Nature* **1981**, *292*, 378–380.

39. Burkhoff, A.M.; Tullius, T.D. Structural details of an adenine tract that does not cause DNA to bend. *Nature* **1988**, *331*, 455–457.

40. Burkhoff, A.M.; Tullius, T.D. The unusual conformation adopted by the adenine tracts in kinetoplast DNA. *Cell* **1987**, *48*, 935–943.

41. McCarthy, J.G.; Frederick, C.A.; Nicolas, A. A structural analysis of the bent kinetoplast DNA from fasciculata by high resolution chemical probing. *Nucleic Acids Res.* **1993**, *21*, 3309–3317.

42. Young, M.A.; Ravishanker, G.; Beveridge, D.L.; Berman, H.M. Analysis of local helix bending in crystal-structures of DNA oligonucleotides and DNA-protein complexes. *Biophys. J.* **1994**, *68*, 2454–2468.

43. Nadeau, J.G.; Crothers, D.M. Structural basis for DNA bending. *Proc. Natl. Acad. Sci. USA* **1989**, *86*, 2622–2626.

44. Chen, S.M.; Leupin, M.; Chazin, W.J. Conformational studies of the duplex d-(CCAAAAATTTCC)·d-(GGAAATTTTTGG) containing a (dA)₅ tract using two-dimensional ¹H-n.m.r. spectroscopy. *Int. J. Biol. Macromol.* **1992**, *14*, 57–63.

45. Leroy, J.L.; Charretier, E.; Kochoyan, M.; Gueron, M. Evidence from base-pair kinetics for two types of adenine tract structures in solution: their relation to DNA curvature. *Biochemistry* **1988**, *27*, 8894–8898.

46. Diekmann, S. Temperature and salt dependence of gel migration anomaly of curved DNA fragments. *Nucleic Acids Res.* **1987**, *15*, 247–265.

47. Chan, S.S.; Breslauer, K.J.; Hogan, M.E.; Kessler, D.J.; Austin, R.H.; Ojemann, J. et al. Physical studies of DNA premelting equilibria in duplexes with and without homo dA.dT tracts: correlations with DNA bending. *Biochemistry* **1990**, *29*, 6161–6171.

48. Chan, S.S.; Breslauer, K.J.; Austin, R.H.; Hogan, M.E. Thermodynamics and premelting conformational changes of phased (dA)₅ tracts. *Biochemistry* **1993**, *32*, 11776–11784.

49. de Souza, O.N.; Goodfellow, J.M. Molecular dynamics simulations of oligonucleotides in solution: Visualisation of intrinsic curvature. *J. Comput. Aided Mol. Des.* **1994**, *8*, 307–322.

50. Thompson, A.S.; Hurley, L.H. Solution conformation of a bizelesin A-tract duplex adduct: DNA-DNA cross-linking of an A-tract straightens out bent DNA. *J. Mol. Biol.* **1995**, *252*, 86–107.

51. Diekmann, S.; Mazzarelli, J.M.; McLaughlin, L.W.; von Kitzing, E.; Travers, A.A. DNA curvature does not require bifurcated hydrogen bonds or pyrimidine methyl groups. *J. Mol. Biol.* **1992**, *225*, 729–738.

52. Haran, T.E.; Kahn, J.D.; Crothers, D.M. Sequence elementts responsible for DNA curvature. *J. Mol. Biol.* **1994**, *244*, 135–143.

53. Bruckner, I.; Sanchez, R.; Suck, D.; Pongor, S. Sequence-dependent bending propensity of DNA as revealed by DNAse I-parameters for trinucleotides. *EMBO J.* **1995**, *14*, 1812–1818.

54. Bruckner, I.; Sanchez, R.; Suck, D.; Pongor, S. Trinucleotide models for DNA bending propensity—comparison of models based on DNase I digestion and nucleosome packaging data. *J. Biomol. Struct. Dyn.* **1995**, *13*, 309–317.

55. Drew, H.R.; Travers, A.A. DNA bending and its relation to nucleosome positioning. *J. Mol. Biol.* **1985**, *186*, 773–790.

56. Satchwell, S.C.; Drew, H.R.; Travers, A.A. Sequence periodicities in chicken nucleosome core DNA. *J. Mol. Biol.* **1986**, *191*, 659–675.

57. Wolffe, A.P.; Drew, H.R. DNA structure: implications for chromatin structure and function. In: *Chromatin Structure and Gene Expression* (Elgin, S.C.R., ed.). IRL Press, Oxford, 1995, pp. 27–48.

58. Hunter, C.A. Sequence-dependent DNA structure. The role of base stacking interactions. *J. Mol. Biol.* **1993**, *230*, 1025–1054.

59. Sprous, D.; Zacharias, W.; Wood, Z.A.; Harvey, S.C. Dehydrating agents sharply reduce curvature in DNAs containing A tracts. *Nucleic Acids Res.* **1995**, *23*, 1816–1821.

60. Dickerson, R.E.; Goodsell, D.; Kopka, M.L. MPD and DNA bending in crystals and in solution. *J. Mol. Biol.* **1996**, *256*, 108–125.

61. Calladine, C.R.; Drew, H.R. *Understanding DNA—The Molecule and How It Works.* Academic Press, San Diego, 1992.

62. Ner, S.S.; Travers, A.A.; Churchill, M.E.A. Harnessing the writhe—a role for DNA chaperones in nucleoprotein complex formation. *Trends Biochem. Sci.* **1994**, *19*, 185–187.

63. Travers, A.A.; Ner, S.S.; Churchill, M.E.A. DNA chaperones—a solution to a persistence problem. *Cell* **1994**, *77*, 167–169.

64. Bracco, L.; Kotlarz, D.; Kolb, A.; Diekmann, S.; Buc, H. Synthetic curved DNA sequences can act as transcriptional activators in *Escherichia coli*. *EMBO J.* **1989**, *8*, 4289–4296.

65. Yamada, H.; Muramatsu, S.; Mizuno, T. An *Escherichia coli* protein that preferentially binds to sharply curved DNA. *J. Biochem. (Tokyo)* **1990**, *108*, 420–425.

66. Tippner, D.; Afflerbach, H.; Bradaczek, C.; Wagner, R. Evidence for a regulatory function of the histone-like *Escherichia coli* protein H-NS in ribosomal-RNA synthesis. *Mol. Microbiol.* **1994**, *11*, 589–604.

67. Barcelo, F.; Muzard, G.; Mendoza, R.; Revet, B.; Roques, B.P.; Le Pecq, J.B. Removal of DNA curving by DNA ligands: gel electrophoresis study. *Biochemistry* **1991**, *30*, 4863–4873.

68. Owen-Hughes, T.A.; Pavitt, G.D.; Santos, D.S.; Sidebotham, J.M.; Hulton, C.; Hilton, J. et al. The chromatin-associated protein H-NS interacts with curved DNA to influence DNA topology and gene-expression. *Cell* **1992**, *71*, 255–265.

69. Dersch, P.; Schmidt, K.; Bremer, K. Synthesis of the *Escherichia coli* K-12 nucleoid-associated DNA-binding protein H-NS is subjected to growth-phase control and autoregulation. *Mol. Microbiol.* **1993**, *8*, 875–889.

70. Falconi, M.; Higgins, N.; Spurio, R.; Pon, C.; Gualerzi, C.O. Expression of the gene encoding the major bacterial nucleoid protein H-NS is subject to transcriptional auto-repression. *Mol. Microbiol.* **1993**, *10*, 273–282.

71. Ueguchi, C.; Kakeda, M.; Mizuno, T. Autoregulatory expression of the *Escherichia coli* hns gene encoding a nucleoid protein: H-NS functions as a repressor of its own transcription. *Mol. Gen. Genet.* **1993**, *236*, 171–178.

72. Zuber, F.; Kotlarz, D.; Rimsky, S.; Buc, H. Modulated expression of promoters containing upstream curved DNA sequences by the *Escherichia coli* nucleoid protein H-NS. *Mol. Microbiol.* **1994**, *12*, 231–240.

73. Yamada, H.; Yoshida, T.; Tanaka, K.; Sasakawa, C.; Mizuno, T. Molecular analysis of the *Escherichia coli* gene encoding a DNA-binding protein, which preferentially recognises curved DNA sequences. *Mol. Gen. Genet.* **1991**, *230*, 332–336.

74. Tanaka, K.; Muramatsu, S.; Yamada, H.; Mizuno, T. Systematic characterization of curved DNA segments randomly cloned from *Escherichia coli* and their functional significance. *Mol. Gen. Genet.* **1991**, *226*, 367–376.

75. Lucht, J.M.; Dersch, P.; Kempf, B.; Bremer, E. Interactions of the nucleoid-associated DNA-binding protein H-NS with the regulatory region of the osmotically controlled proU operon of *Escherichia coli. J. Biol. Chem.* **1994**, *269*, 6578–6586.

76. Mazin, A.; Timchenko, T.; Demurica, J.M.; Schreiber, V.; Angulo, J.F.; Demurcia, G. et al. Kin17, a mouse nuclear zinc-finger protein that binds preferentially to curved DNA. *Nucleic Acids Res.* **1994**, *22*, 4335–4341.

77. Gadaleta, G.; Delia, D.; Saccone, C.; Pepe, G. A 67-kDa protein-binding to the curved DNA in the main regulatory region of the rat mitochondrial genome. *Gene* **1995**, *160*, 229–234.

78. Lilley, D.M.J. HMG has DNA wrapped up. *Nature* **1992**, *357*, 282–283.

79. Schroth, G.P.; Siino, J.S.; Cooney, C.A.; Th'ng, J.P.; Ho, P.S.; Bradbury, E.M. Intrinsically bent DNA flanks both sides of an RNA polymerase I transcription start site. Both regions display novel electrophoretic mobility. *J. Biol. Chem.* **1992**, *267*, 9958–9964.

80. Schultz, S.C.; Shields, G.C.; Steitz, T.A. Crystal structure of a CAP-DNA complex: The DNA is bent by 90 degrees. *Science* **1991**, *253*, 1001–1007.

81. Ramstein, J.; Lavery, R. Energetic coupling between DNA bending and base pair opening. *Proc. Natl. Acad. Sci. USA* **1988**, *85*, 7231–7235.

82. Parvin, J.D.; McCormick, R.J.; Sharp, P.A.; Fisher, D.E. Pre-bending of a promoter sequence enhances affinity for the TATA-binding factor. *Nature* **1995**, *373*, 724–727.

83. Travers, A.A.; Klug, A. The bending of DNA in nucleosomes and its wider implications. *Philos. Trans. R. Soc. Lond. Biol.* **1987**, *317*, 537–561.

84. Collis, C.M.; Molloy, P.L.; Both, G.W.; Drew, H.R. Influence of the sequence-dependent flexure of DNA on transcription in *E. coli.. Nucleic Acids Res.* **1989**, *17*, 9447–9468.

85. Figueroa, N.; Wills, N.; Bossi, L. Common sequence determinants of the response of a prokaryotic promoter to DNA bending and supercoiling. *EMBO J.* **1991**, *10*, 941–949.

86. Perez-Martin, J.; Rojo, F.; Delorenzo, V. Promoters responsive to DNA bending—a common theme in prokaryotic gene expression. *Microbiol. Rev.* **1994**, *58*, 268–290.

87. Lamond, A.I.; Travers, A.A. Requirement for an upstream element for optimal transcription of a bacterial tRNA gene. *Nature* **1983**, *305*, 248–250.

88. Deuschle, U.; Krammerer, W.; Gentz, R.; Bujard, H. Promoters of *Escherichia coli*: a hierarchy of *in vivo* strength indicates alternate structures. *EMBO J.* **1986**, *5*, 2987–2994.

89. Knaus, R.; Bujard, H. Principles governing the activity of *E. coli* promoters. In: *Nucleic Acids and Molecular Biology* (Eckstein, F.; Lilley, D.M.J.; eds.). Springer-Verlag, Berlin, 1990, pp. 110–112.

90. Plaskon, R.R.; Wartell, R.M. Sequence distributions associated with DNA curvature are found upstream of strong *E.coli* promoters. *Nucleic Acids Res.* **1987**, *15*, 785–796.

91. Ohyama, T.; Hirota, Y. A possible function of DNA curvature in transcription. *Nucleic Acids Symp. Ser.* **1993**, *29*, 153–154.

92. Kim, J.; Klooster, S.; Shapiro, D.J. Intrinsically bent DNA in a eukaryotic transcription factor recognition sequence potentiates transcription activation. *J. Biol. Chem.* **1995**, *270*, 1282–1288.

93. Goodman, S.D.; Nash, H.A. Functional replacement of a protein-induced bend in a DNA recombination site. *Nature* **1989**, *341*, 251–254.
94. Tanaka, K.-I.; Ueguchi, C.; Mizuno, T. Importance of stereospecific positioning of the upstream cis-acting DNA element containing curved DNA structure for the functioning of the *Escherichia coli* proV promoter. *Biosci. Biotechnol. Biochem.* **1994**, *58*, 1097–1101.
95. Ross, W.; Gosink, K.; Salomon, J.; Igarashi, K.; Zhou, C.; Ishihama, A. et al. A third recognition element in bacterial promoters: DNA binding by the alpha subunit of RNA polymerase. *Science* **1993**, *262*, 1407–1413.
96. Rao, L.; Ross, W.; Appleman, J.A.; Gaal, T.; Leirmo, S.; Schlax, P.J., et al. Factor independent activation of rrnB1 P1—an "extended promoter" with an upstream element that dramatically increases promoter strength. *J. Mol. Biol.* **1994**, *235*, 1421–1435.
97. Jordi, B.J.A.M.; Owen-Hughes, T.A.; Hulton, C.S.J.; Higgins, C.F. DNA twist, flexibility and transcription of the osmoregulated proU promoter of *Salmonella typhimurium*. *EMBO J.* **1995**, *22*, 5690–5700.
98. Nadal, M.; Mirambeau, G.; Forterre, P.; Reiter, W.-D.; Duguet, M. Positively supercoiled DNA in a virus-like particle of an archaebacterium. *Nature* **1986**, *321*, 256–258.
99. Spengler, S.J.; Stasiak, A.; Cozzarelli, N.R. The stereostructure of knots and catenanes produced by phage lambda integrative recombination: Implications for mechanism and DNA structure. *Cell* **1985**, *42*, 325–334.
100. Wasserman, S.A.; Cozzarelli, N.R. Biochemical topology: applications to DNA recombination and replication. *Science* **1986**, *232*, 951–960.
101. Adams, D.E.; Shekhtman, E.M.; Zechiedrich, E.L.; Schmid, M.B.; Cozzarelli, N.R. The role of topoisomerase IV in partitioning bacterial replicons and the structure of catenated intermediates in DNA replication. *Cell* **1992**, *71*, 277–288.
102. Depew, R.E.; Wang, J.C. Conformational fluctuations of the DNA helix. *Proc. Natl. Acad. Sci. USA* **1975**, *72*, 4275–4279.
103. Pulleyblank, D.E.; Shure, M.; Tang, D.; Vinograd, J.; Vosberg, H. Action of nick-closing enzyme on supercoiled and non-supercoiled closed circular DNA: formation of a Boltzmann distribution of topological isomers. *Proc. Natl. Acad. Sci. USA* **1975**, *72*, 4280–4284.
104. Drlica, K. Biology of bacterial deoxyribonucleic acid topoisomerases. *Microbiol. Rev.* **1984**, *48*, 273–289.
105. Drlica, K. Bacterial topoisomerases and the control of DNA supercoiling. *Trends Genet.* **1990**, *6*, 433–437.
106. Gellert, M. DNA topoisomerases. *Annu. Rev. Biochem.* **1981**, *50*, 879–910.
107. Wang, J.C.; Lynch, A.S. Transcription and DNA supercoiling. *Curr. Opin. Genet. Dev.* **1993**, *3*, 764–768.
108. Chung, J.H.; Whiteley, M.; Felsenfeld, G. A 5' element of the chicken β-globin domain serves as an insulator in human erythroid cells and protects against position effect in *Drosophila*. *Cell* **1993**, *74*, 505–514.
109. Drlica, K. Control of bacterial DNA supercoiling. *Mol. Microbiol.* **1992**, *6*, 425–433.
110. Hsieh, L.S.; Burger, R.M.; Drlica, K. Bacterial DNA supercoiling and [ATP]/[ADP] changes associated with a transition to anaerobic growth. *J. Mol. Biol.* **1991**, *219*, 1–8.
111. Pruss, G.J.; Drlica, K. Topoisomerase I mutants: The gene on pBR322 that encodes resistance to tetracycline affects plasmid DNA supercoiling. *Proc. Natl. Acad. Sci. USA* **1986**, *83*, 8952–8956.
112. Lockshon, D.; Morris, D.R. Positively supercoiled plasmid DNA is produced by treatment of *Escherichia coli* with DNA gyrase inhibitors. *Nucleic Acids Res.* **1983**, *11*, 2999–3018.
113. Liu, L.F.; Wang, J.C. Supercoiling of the DNA template during transcription. *Proc. Natl. Acad. Sci. USA* **1987**, *84*, 7024–7027.
114. Esposito, F.; Sinden, R.R. DNA supercoiling and eukaryotic gene expression. *Ox. Surv. Eukaryot. Genes* **1988**, *5*, 1–50.

115. Freeman, L.A.; Garrard, W.T. DNA supercoiling in chromatin structure and gene expression. *Crit. Rev. Eukaryot Gene Expr.* **1992**, *2*, 165–209.

116. Horwitz, M.S.Z.; Loeb, L.A. An *E. coli* promoter that regulates transcription by DNA superhelix-induced cruciform extrusion. *Science* **1988**, *241*, 703–705.

117. McClure, W.P. Mechanism and control of transcription in prokaryotes. *Annu. Rev. Biochem.* **1985**, *54*, 171–204.

118. Pearson, C.E.; Zorbas, H.; Price, G.B.; Zannis-Hadjopoulos, M. Inverted repeats, stem-loops, and cruciforms: significance for initiation of DNA replication. *J. Cell. Biochem.* **1996**, *63*, 1–22.

119. Murchie, A.I.; Lilley, D.M. The mechanism of cruciform formation in supercoiled DNA: initial opening of central basepairs in salt-dependent extrusion. *Nucleic Acids Res.* **1987**, *15*, 9641–9654.

120. Zheng, G.; Sinden, R.R. Effect of base composition at the center of inverted repeated DNA sequences on cruciform transitions in DNA. *J. Biol. Chem.* **1988**, *263*, 5356–6461.

121. Courey, A.J.; Wang, J.C. Influence of DNA sequence and supercoiling on the process of cruciform formation. *J. Mol. Biol.* **1988**, *202*, 35–43.

122. Lilley, D.M. The inverted repeat as a recognizable structural feature in supercoiled DNA. *Proc. Natl. Acad. Sci. USA* **1980**, *77*, 6468–6472.

123. Panayotatos, N.; Wells, R.D. Cruciform structures in supercoiled DNA. *Nature* **1981**, *289*, 466–470.

124. Mizuuchi, K.; Mizuuchi, M.; Gellert, M. Cruciform structures in palindromic DNA are favored by DNA supercoiling. *J. Mol. Biol.* **1982**, *156*, 229–243.

125. Gellert, M.; O'Dea, M.H.; Mizuuchi, K. Slow cruciform transitions in palindromic DNA. *Proc. Natl. Acad. Sci. USA* **1983**, *80*, 5545–5549.

126. Singleton, C.K. Effects of salts, temperature, and stem length on supercoil-induced formation of cruciforms. *J. Biol. Chem.* **1983**, *258*, 7661–7668.

127. Sinden, R.R.; Pettijohn, D.E. Cruciform transitions in DNA. *J. Biol. Chem.* **1984**, *259*, 6593–6600.

128. Sullivan, K.M.; Lilley, D.M. Influence of cation size and charge on the extrusion of a salt-dependent cruciform. *J. Mol. Biol.* **1987**, *193*, 397–404.

129. Lilley, D.M. The kinetic properties of cruciform extrusion are determined by DNA. *Nucleic Acids Res.* **1985**, *13*, 1443–1465.

130. Sullivan, K.M.; Lilley, D.M. A dominant influence of flanking sequences on a local structural transition. *Cell* **1986**, *47*, 817–827.

131. Sullivan, K.M.; Murchie, A.I.; Lilley, D.M. Long range structural communication between sequences in supercoiled DNA. Sequence dependence of contextual influence on cruciform extrusion mechanism. *J. Biol. Chem.* **1988**, *263*, 13074–13082.

132. Bowater, R.; Aboul-ela, F.; Lilley, D.M.J. Large-scale stable opening of supercoiled DNA in response to temperature and supercoiling in (A + T)-rich regions that promote low-salt cruciform extrusion. *Biochemistry* **1991**, *30*, 11495–11506.

133. Morales, N.M.; Cobourn, S.D.; Muller, U.R. Effect of *in vitro* transcription on cruciform stability. *Nucleic Acids Res.* **1990**, *18*, 2777–2782.

134. Zheng, G.; Kochel, T.; Hoepfner, R.W.; Timmons, S.E.; Sinden, R.R. Torsionally tuned cruciform and Z-DNA probes for measuring unrestrained supercoiling at specific sites in DNA of living cells. Appendix: estimation of superhelical density *in vivo* from analysis of the level of cruciforms existing in living cells. *J. Mol. Biol.* **1991**, *221*, 107–129.

135. Lilley, D.M.J.; Clegg, R.M. The structure of the four-way junction in DNA. *Annu. Rev. Biophys. Biomol. Struct.* **1993**, *22*, 299–328.

136. Wemmer, D.E.; Wand, A.J.; Seeman, N.C.; Kallenbach, N.R. NMR analysis of DNA junctions: imino proton NMR studies of individual arms and intact junction. *Biochemistry* **1985**, *24*, 5745–5749.

137. Marky, L.A.; Kallenbach, N.R.; McDonough, K.A.; Seeman, N.C.; Breslauer, K.J. Melting behaviour of a DNA junction structure: a calorimetric and spectroscopic study. *Biopolymers* **1987**, *26*, 1621–1634.

138. Murchie, A.I.; Clegg, R.M.; von Kitzing, E.; Duckett, D.R.; Diekmann, S.; Lilley, D.M. Fluorescence energy transfer shows that the four-way DNA junction is a right-handed cross of antiparallel molecules. *Nature* **1989**, *341*, 763–766.

139. Churchill, M.E.A.; Tullius, T.D.; Kallenbach, N.R.; Seeman, N.C. A Holliday recombination intermediate is twofold symmetric. *Proc. Natl. Acad. Sci. USA* **1988**, *85*, 4653–4656.

140. Kimball, A.; Guo, Q.; Lu, M.; Cunningham, R.P.; Kallenbach, N.R.; Seeman, N.C., et al. Construction and analysis of parallel and antiparallel Holliday junctions. *J. Biol. Chem.* **1990**, *265*, 6544–6547.

141. Murchie, A.I.H.; Portugal, J.; Lilley, D.M.J. Cleavage of a four-way DNA junction by a restriction enzyme spanning the point of strand exchange. *EMBO J.* **1991**, *10*, 713–718.

142. Chen, J.H.; Churchill, M.E.A.; Tullius, T.D.; Kallenbach, N.R.; Seeman, N.C. Construction and analysis of monomobile DNA junctions. *Biochemistry* **1988**, *27*, 6032–6038.

143. Duckett, D.R.; Lilley, D.M.J. Effects of base mismatches on the structure of the four-way DNA junction. *J. Mol. Biol.* **1991**, *221*, 147–161.

144. Duckett, D.R.; Murchie, A.I.; Lilley, D.M. The role of metal ions in the conformation of the four-way DNA junction. *EMBO J.* **1990**, *9*, 583–590.

145. Duckett, D.R.; Murchie, A.I.H.; Diekmann, S.; von Kitzing, E.; Kemper, B.; Lilley, D.M.J. The structure of the Holliday junction, and its resolution. *Cell* **1988**, *55*, 79–89.

146. Clegg, R.M.; Murchie, A.I.H.; Zechel, A.; Carlberg, C.; Diekmann, S.; Lilley, D.M.J. Flourescence resonance energy transfer analysis of the structure of the four-way DNA junction. *Biochemistry* **1992**, *31*, 4846–4856.

147. Duckett, D.R.; Murchie, A.I.H.; Clegg, R.M.; Zechel, A.; von Kitzing, E.; Diekmann, S., et al. The structure of the Holliday junction. In: *Structure and Methods* (Sarma, M.H.; Sarma, R.H.; Eds.), Adenine Press, Albany, 1990, pp. 157–181.

148. Clegg, R.M.; Murchie, A.I.H.; Lilley, D.M.J. The solution structure of the four-way DNA junction at low salt conditions: A fluorescence resonance energy transfer analysis. *Biophys. J.* **1994**, *66*, 99–109.

149. Eis, P.S.; Millar, D.P. Conformational distributions of a four-way DNA junction revealed by time-fluorescence energy transfer. *Biochemistry* **1993**, *32*, 13852–13860.

150. Gough, G.W.; Lilley, D.M. DNA bending induced by cruciform formation. *Nature* **1985**, *313*, 154–156.

151. Pil, P.M.; Lippard, S.J. Specific binding of chromosomal protein HMG1 to DNA damaged by the anticancer drug cisplatin. *Science* **1992**, *256*, 234–237.

152. Scholten, P.M.; Nordheim, A. Diethyl pyrocarbonate: a chemical probe for DNA cruciforms. *Nucleic Acids Res.* **1986**, *14*, 3981–3993.

153. Furlong, J.C.; Lilley, D.M. Highly selective chemical modification of cruciform loops by diethyl pyrocarbonate. *Nucleic Acids Res.* **1986**, *14*, 3995–4007.

154. Blommers, M.J.J.; Van de Ven, F.J.M.; Van der Marel, G.A.; Van Boom, J.H.; Hilbers, C.W. The three-dimensional structure of a DNA hairpin in solution two dimensional NMR studies and structural analysis of d(ATCCTATTTATAGGAT). *Eur. J. Biochem.* **1991**, *201*, 33–51.

155. Haasnoot, C.A.G.; Hilbers, C.W.; Van der Marel, G.A.; Van Boom, J.H.; Singh, U.C.; Pattabiraman, N., et al. On loop-folding in nucleic acid hairpin-type structures. *J. Biomol. Struct. Dyn.* **1986**, *3*, 843–857.

156. Hirao, I.; Kawai, G.; Yoshizawa, S.; Nishimura, Y.; Ishido, Y.; Watanabe, K.; et al. Most compact hairpin-turn structure exerted by a short DNA fragment, d(GCGAAGC) in solution: an extraordinarily stable structure resistant to nucleases and heat. *Nucleic Acids Res.* **1994**, *22*, 576–285.

157. Zhou, N.; Vogel, H.J. Two dimensional NMR and restrained molecular dynamics studies of the hairpin d($T_8C_4A_8$): Detection of an extraloop cytosine. *Biochemistry* **1993**, *32*, 637–645.

158. Williamson, J.R.; Boxer, S.G. Multinuclear NMR studies of DNA hairpins. 2. Sequence-dependent structural variations. *Biochemistry* **1989**, *28*, 2831–2836.

159. Germann, M.W.; Kalisch, B.W.; Lundberg, P.; Vogel, H.J.; van de Sande, J. Perturbation of DNA hairpins containing the EcoRI recognition site by hairpin loops of varying size and composition: physical (NMR and UV) and enzymatic (EcoRI) studies. *Nucleic Acids Res.* **1990**, *18*, 1489–1498.

160. Pearson, C.E.; Zannis-Hadjopoulos, M.; Price, G.B.; Zorbas, H. A novel type of interaction between cruciform DNA and a cruciform binding protein from HeLa cells. *EMBO J.* **1995**, *14*, 1571–1580.

161. Holliday, R. A mechanism for gene conversion in fungi. *Genet. Res.* **1964**, *5*, 282–304.

162. Timsit, Y.; Moras, D. Groove-backbone interaction in B-DNA implication for DNA condensation and recombination. *J. Mol. Biol.* **1991**, *221*, 919–940.

163. Panyutin, I.G.; Hsieh, P. Formation of a single base mismatch impedes spontaneous DNA branch migration. *J. Mol. Biol.* **1993**, *230*, 413–424.

164. Panyutin, I.G.; Hsieh, P. The kinetics of spontaneous DNA branch migration. *Proc. Natl. Acad. Sci. USA* **1994**, *91*, 2021–2025.

165. Panyutin, I.G.; Biswas, I.; Hsieh, P. A pivotal role for the structure of the Holliday junction in DNA branch migration. *EMBO J.* **1995**, *14*, 1819–1826.

166. Mulrooney, S.B.; Fishel, R.A.; Hejna, J.A.; Warner, R.C. Preparation of figure 8 and cruciform DNAs and their use in studies of the kinetics of branch migration. *J. Biol. Chem.* **1996**, *271*, 9648–9659.

167. Evans, D.H.; Kolodner, R. Effect of DNA structure and nucleotide sequence on Holliday junction resolution by a *Saccharomyces cerevisiae* endonuclease. *J. Mol. Biol.* **1987**, *262*, 969–980.

168. Evans, D.H.; Kolodner, R. Construction of a synthetic Holliday junction analog and characterization of its interaction with a *Saccharomyces cerevisiae* endonuclease that cleaves Holliday junctions. *J. Biol. Chem.* **1987**, *262*, 9160–9165.

169. Passanti, C.; Davies, B.; Ford, M.; Fried, M. Structure of an inverted duplication formed as a first step in a gene amplification event: implications for a model of gene amplification. *EMBO J.* **1987**, *6*, 1697–1703.

170. Mueller, J.E.; Newton, C.J. Resolution of Holliday junction analogs by T4 endonuclease VII can be directed by substrate structure. *J. Biol. Chem.* **1990**, *265*, 13918–13924.

171. Dunderdale, H.J.; Benson, F.E.; Parsons, C.A.; Sharples, G.J.; Lloyd, R.G.; West, S.C. Formation and resolution of recombination intermediates by *E. coli* RecA and RuvC proteins. *Nature* **1991**, *354*, 5506–5510.

172. Bennett, R.J.; Dunderale, H.J.; West, S.C. Resolution of Holliday junctions by RuvC resolvase: cleavage specificity and DNA distortion. *Cell* **1993**, *74*, 1021–1031.

173. Waga, S.; Mizuno, S.; Yoshida, M. Chromosomal protein HMG1 removes the transcriptional block caused by the cruciform in supercoiled DNA. *J. Biol. Chem.* **1990**, *265*, 19424–19428.

174. Shiba, T.; Iwasaki, H.; Nakata, A.; Shinagawa, H. SOS-inducible DNA repair proteins, RuvA and RuvB, of *Escherichia coli*: functional interactions between RuvA and RuvB for ATP hydrolysis and renaturation of the cruciform structure in supercoiled DNA. *Proc. Natl. Acad. Sci. USA* **1991**, *88*, 8445–8449.

175. Adams, D.E.; West, S.C. Relaxing and unwinding on Holliday: DNA helicase mediated branch migration. *Mutat. Res.* **1995**, *337*, 149–159.

176. Lilley, D.M. *In vivo* consequences of plasmid topology. *Nature* **1981**, *292*, 380–382.

177. Palecek, E. Local supercoil-stabilized DNA structures. *CRC Crit. Rev. Biochem. Mol. Biol.* **1991**, *26*, 151–226.

178. Furlong, J.C.; Sullivan, K.M.; Murchie, A.I.; Gough, G.W.; Lilley, D.M. Localized chemical hyperreactivity in supercoiled DNA: evidence for base unpairing in sequences that induce low-salt cruciform extrusion. *Biochemistry* **1989**, *28*, 2009–2017.

179. Lilley, D.M.; Kemper, B. Cruciform-resolvase interactions in supercoiled DNA. *Cell* **1984**, *36*, 413–422.

180. Sinden, R.R.; Broyles, S.S.; Pettijohn, D.E. Perfect palindromic lac operator DNA sequence exists as a stable cruciform structure in supercoiled DNA *in vitro* but not *in vivo*. *Proc. Natl. Acad. Sci. USA* **1983**, *80*, 1797–1801.

181. Frappier, L.; Price, G.B.; Martin, R.G.; Zannis-Hadjopoulos, M. Monoclonal antibodies to cruciform DNA structures. *J. Mol. Biol.* **1987**, *193*, 751–758.

182. Cimino, G.D.; Gamper, H.B.; Isaacs, S.T.; Hearst, J.E. Psoralens as photoactive probes of nucleic acid structure and function: organic chemistry, photochemistry, and biochemistry. *Annu. Rev. Biochem.* **1985**, *54*, 1151–1193.

183. Ussery, D.W.; Hoepfner, R.W.; Sinden, R.R. Probing DNA structure with psoralen *in vitro*. *Methods Enzymol.* **1992**, *212*, 242–262.

184. Sinden, R.R.; Ussery, D.W. Analysis of DNA structure *in vivo* using psoralen photobinding: measurement of supercoiling, topological domains, and DNA-protein interactions. *Methods Enzymol.* **1992**, *212*, 319–335.

185. Lis, J.T.; Neckameyer, W.; Mirault, M.-E.; Artavanis-Tsakonas, S.; Lall, P.; Martin, P.; et al. DNA sequences flanking the starts of the hsp70 and alpha beta heat shock genes are homologous. *Dev. Biol.* **1981**, *83*, 291–300.

186. Panayotatos, N.; Fontaine, A. A native cruciform DNA structure probed in bacteria by recombinant T7 endonuclease. *J. Biol. Chem.* **1987**, *262*, 11364–11368.

187. McClellan, J.A.; Boublikova, P.; Palecek, E.; Lilley, D.M. Superhelical torsion in cellular DNA responds directly to environmental and genetic factors. *Proc. Natl. Acad. Sci. USA* **1990**, *87*, 8373–8377.

188. Haniford, D.B.; Pulleyblank, D.E. Transition of a cloned $d(AT)_n$-$d(AT)_n$ tract to a cruciform *in vivo*. *Nucleic Acids Res.* **1985**, *13*, 4343–4361.

189. Ward, G.K.; McKenzie, R.; Zannis-Hadjopoulos, M.; Price, G.B. The dynamic distribution and quantification of DNA cruciforms in eukaryotic nuclei. *Exp. Cell. Res.* **1990**, *188*, 235–246.

190. Ward, G.K.; Shihab-el-Deen, A.; Zannis-Hadjopoulos, M.;. Price, G.B. DNA cruciforms and the nuclear supporting structure. *Exp. Cell. Res.* **1991**, *195*, 92–98.

191. Jackson, D.A.; Dickinson, P.; Cook, P.R. The size of chromatin loops in HeLa cells. *EMBO J.* **1990**, *9*, 567–571.

192. Burkholder, G.D.; Latimer, L.J.P.; Lee, J.S. Immunofluorescent staining of mammalian nuclei and chromosomes with a monoclonal antibody to triplex DNA. *Chromosoma* **1988**, *97*, 185–192.

193. Wittig, B.; Dorbic, T.; Rich, A. Transcription is associated with Z-DNA formation in metabolically active permeabilized mammalian cell nuclei. *Proc. Natl. Acad. Sci. USA* **1991**, *88*, 2259–2263.

194. Wittig, B.; Dorbic, T.; Rich, A. The level of Z-DNA in metabolically active, permeabilized mammalian cell nuclei is regulated by torsional strain. *J. Cell. Biol.* **1989**, *108*, 755–764.

195. Klein, W.J.; Beiser, S.M.; Erlanger, B.F. Nuclear fluoresence employing antinucleoside immunoglobins. *J. Exp. Med.* **1967**, *125*, 61–70.

196. Tan, E.M.; Lerner, R.A. An immunological study of the fates of nuclear and nucleolar macromolecules during the cell cycle. *J. Mol. Biol.* **1972**, *68*, 107–114.

197. Collins, J.M. Deoxyribonucleic acid structure in human diploid fibroblasts stimulated to proliferate. *J. Biol. Chem.* **1977**, *252*, 141–147.

198. Conrad, M.N.; Newlon, C.S. Stably denatured regions of chromosomal DNA from the cdc 2 *Saccharomyces cerevisiae* cell cycle mutant. *Mol. Cell. Biol.* **1983**, *3*, 1665–1669.

199. Stillman, B. Smart machines at the replication fork. *Cell* **1994**, *78*, 725–728.

200. DePamphilis ML. Origins of DNA replication in metazoan chromosomes. *J. Biol. Chem.* **1993**, *268*, 1–4.

201. Hagerman, P.J. Sequence-directed curvature of DNA. *Annu. Rev. Biochem.* **1990**, *59*, 755–781.

202. Umek, R.M.; Kowalski, D. The ease of DNA unwinding as a determinant of initiation at yeast replication origins. *Cell* **1988**, *52*, 559–567.

203. Natale, D.A.; Umek, R.M.; Kowalski, D. Ease of unwinding is a conserved property of yeast replication origins. *Nucleic Acids Res.* **1993**, *21*, 555–560.

204. Kowalski, D.; Eddy, M.J. The DNA unwinding element: a novel, cis-acting component that facilitates opening of the Escherichia coli replication orgin. *EMBO J.* **1989**, *8*, 4335–4344.

205. Umek, R.M.; Kowalski, D. Yeast regulatory sequences preferentially adopt a non-B conformation in supercoiled DNA. *Nucleic Acids Res.* **1987**, *15*, 4467–4480.

206. Lin, S.; Kowalski, D. DNA helical instability facilitates initiation at the SV40 replication origin. *J. Mol. Biol.* **1994**, *235*, 496–507.

207. Hay, R.T.; DePamphilis, M.L. Initiation of SV40 DNA replication in vivo: location and structure of 5' ends of DNA synthesized in the ori region. *Cell* **1982**, *28*, 767–779.

208. Hay, R.T.; Hendrickson, E.A.; DePamphilis, M.L. Sequence specificity for the initiation of RNA-primed simian virus 40 DNA synthesis *in vivo*. *J. Mol. Biol.* **1984**, *175*, 131–157.

209. Yagil, G. Paranemic structures of DNA and their role in DNA unwinding. *CRC Crit. Rev. Biochem. Mol. Biol.* **1991**, *26*, 475–559.

210. Murchie, A.I.H.; Bowater, R.; Aboul-ela, F.; Lilley, D.M.J. Helix opening transitions in supercoiled DNA. *Biochim. Biophys. Acta* **1992**, *1131*, 1–15.

211. Williams, L.; Kowalski, D. Easily unwound DNA sequences and hairpin structures in the Epstein-Barr virus origin of plasmid replication. *J. Virol.* **1993**, *67*, 2707–2715.

212. Weller, S.K.; Spadaro, A.; Schaffer, J.E.; Murray, A.W.; Maxam, A.M.; Schaffer, P.A. Cloning, sequencing, and functional analysis of oriL, a herpes simplex virus type 1 origin of DNA synthesis. *Mol. Cell Biol.* **1985**, *5*, 930–942.

213. Lockshon, D.; Galloway, D.A. Cloning and characterization of oriL2, a large palindromic DNA replication origin of herpes simplex virus type 2. *J. Virol.* **1986**, *58*, 513–521.

214. Noirot, P.; Bargonetti, J.; Novick, R.P. Initiation of rolling-circle replication in pT181 plasmid: initiator protein enhances cruciform extrusion at the origin. *Proc. Natl. Acad. Sci. USA* **1990**, *87*, 8560–8564.

215. Frappier, L.; Price, G.B.; Martin, R.G.; Zannis-Hadjopoulos, M. Characterization of the binding specificity of two anticruciform DNA monoclonal antibodies. *J. Biol. Chem.* **1989**, *264*, 334–341.

216. Steinmetzer, K.; Zannis-Hadjopoulos, M.; Price, G.B. Anti-cruciform monoclonal antibody and cruciform DNA interaction. *J. Mol. Biol.* **1995**, *254*, 29–37.

217. Zannis-Hadjopoulos, M.; Frappier, L.; Khoury, M.; Price, G.B. Effect of anti-cruciform DNA monoclonal antibodies on DNA replication. *EMBO J.* **1988**, *7*, 1837–1844.

218. Fried, M.; Feo, S.; Heard, E. The role of inverted duplication in the generation of gene amplification in mammalian cells. *Biochim. Biophys. Acta* **1991**, *1090*, 143–155.

219. Windle, B.E.; Wahl, G.M. Molecular disection of mammalian gene amplification: new mechanistic insights revealed by analysis of very early events. *Mutat. Res.* **1992**, *276*, 199–224.

220. Legouy, E.; Fossar, N.; Lhomond, G.; Brison, O. Structure of four amplified DNA novel joints. *Somat. Cell Mol. Genet.* **1989**, *15*, 309–320.

221. Hyrien, O. Large inverted duplications in amplified DNA of mammalian cells form hairpins *in vitro* upon DNA extraction but not *in vivo*. *Nucleic Acids Res.* **1989**, *17*, 9557–9569.

222. Esposito, F.; Sinden, R.R. Supercoiling in eukaryotic and prokaryotic DNA: changes in response to topological pertubation of SV40 in CV-1 cells and of plasmids in *E. coli*. *Nucleic Acids Res.* **1987**, *15*, 5105–5124.

223. Cohen, S.; Hassin, D.; Karby, S.; Lavi, S. Hairpin structures are the primary amplification products: a novel mechanism for generation of inverted repeats during gene amplification. *Mol. Cell Biol.* **1994**, *14*, 7782–7791.

224. Berko-Flint, Y.; Karby, S.; Hassin, D.; Lavi, S. Carcinogen-induced DNA amplification *in vitro*, overreplication of simian virus 40 origin region in extracts from carcinogen-treated CO60 cells. *Mol. Cell Biol.* **1990**, *10*, 75–83.

225. Aladjem, M.I.; Lavi, S. The mechanism of carcinogen-induced DNA amplification *in vivo* and *in vitro*. *Mutat. Res.* **1992**, *276*, 339–344.

226. Workman, J.L.; Taylor, C.A.; Kingston, R.E. Activation domains of stably bound GAL4 derivatives alleviate repression of promoters by nucleosomes. *Cell* **1991**, *64*, 533–544.

227. Cheng, L.; Kelly, T.J. The transcriptional activator nuclear factor 1 stimulates the replication of SV40 minichromosomes *in vivo* and *in vitro*. *Cell* **1989**, *59*, 541–551.

228. Simpson, R.T. Nucleosome positioning: Occurrence, mechanisms, and functional consequences. *Prog. Nucleic Acids Res. Mol. Biol.* **1990**, *40*, 143–184.

229. Weintraub, H. A dominant role for DNA secondary structure in forming hypersensitive structures in chromatin. *Cell* **1983**, *32*, 1191–1203.

230. Weintraub, H.; Cheng, P.F.; Conrad, K. Expression of transfected DNA depends on DNA topology. *Cell* **1986**, *46*, 115–122.

231. Nickol, J.; Martin, R.G. DNA stem-loop structures bind poorly to histone octamer cores. *Proc. Natl. Acad. Sci. USA* **1983**, *80*, 4669–4673.

232. Nobile, C.; Nickol, J.; Martin, R.G. Nucleosome phasing on a DNA fragment from the replication origin of simian virus 40 and rephasing upon cruciform formation of the DNA. *Mol. Cell Biol.* **1986**, *6*, 2916–2922.

233. Battistoni, A.; Leoni, L.; Sampaolese, B.; Savino, M. Kinetic persistence of cruciform structures in reconstituted minichromosomes. *Biochim. Biophys. Acta* **1988**, *950*, 161–171.

234. Kotani, H.; Kmiec, E.B. DNA cruciforms facilitate *in vitro* strand transfer on nucleosomal templates. *Mol. Gen. Genet.* **1994**, *243*, 681–690.

235. Nickol, J.; Behe, M.; Felsenfeld, G. Effect of the B-Z transition in poly(dG-m^{5}dC)poly(dG-m^{5}dC) on nucleosome formation. *Biochemistry* **1982**, *79*, 1771–1775.

236. Miller, F.D.; Dixon, G.H.; Rattner, J.B.; van de Sande, J.H. Assembly and characterization of nucleosomal cores on B- vs. Z-form DNA. *Biochemistry* **1985**, *24*, 102–109.

237. Garner, M.M.; Felsenfeld, G. Effect of Z-DNA on nucleosome placement. *J. Mol. Biol.* **1987**, *196*, 581–590.

238. Ausio, J.; Zhou, G.; van Holde, K. A re-examination of the reported B-Z DNA transition in nucleosomes reconstituted with poly(dG-m^{5}dC)poly(dG-m^{5}dC). *Biochemistry* **1987**, *26*, 5595–5599.

239. Gross, D.S.; Huang, S.Y.; Garrard, W.T. Chromatin structure of the potential Z-forming sequence $(dT-dG)_n \times (dC-dA)_n$. Evidence for an "alternating-B" conformation. *J. Mol. Biol.* **1985**, *183*, 251–265.

240. Lee-Chen, G.; Woodworth-Gutai, M. Simian virus 40 DNA replication: functional organization of regulatory elements. *Mol. Cell Biol.* **1986**, *6*, 3086–3093.

241. Chandrasekharappa, S.C.; Subramanian, K.N. Effects of position and orientation of the 72-base-pair-repeat transcriptional enhancer on replication from the simian virus 40 core origin. *J. Virol.* **1987**, *61*, 2973–2980.

242. Jakobovits, E.B.; Bratosin, S.; Aloni, Y. A nucleosome-free region in SV40 minichromosomes. *Nature* **1980**, *285*, 263–265.

243. Saragosti, S.; Moyne, G.; Yanif, M. Absence of nucleosomes in a fraction of SV40 chromatin between the origin of replication and the region coding for the late reader RNA. *Cell* **1980**, *20*, 65–73.

244. Solomon, M.J.; Varshavsky, A. A nuclease-hypersensitive region from *de novo* after chromosome replication. *Mol. Cell Biol.* **1987**, *7*, 3822–3825.

245. Nordheim, A.; Rich, A. Negatively supercoiled simian virus 40 DNA contains Z-DNA segments within transcriptional enhancer sequences. *Nature* **1983**, *303*, 674–679.

246. Azorin, F.; Rich, A. Isolation of Z-DNA binding proteins from SV40 minichromosomes: evidence for binding to the viral control region. *Cell* **1985**, *41*, 365–374.

247. Sage, E.; Leng, M. Conformational changes of poly(dG-dC)·poly(dG-dC) modified by the carcinogen N-acetoxy-N-acetyl-2-aminofluorene. *Nucleic Acids Res.* **1981**, *9*, 1241

248. Nordheim, A.; Herrera, R.E.; Rich, A. Binding of anti-Z-DNA antibodies to negatively super-coiled SV40 DNA. *Nucleic Acids Res.* **1987**, *15*, 1661–1677.

249. Muller, R.C.; Raphael, A.L.; Barton, J.K. Evidence for altered DNA conformations in the sim-ian virus 40 genome: site-specific DNA cleavage by the chiral complex Λ-tris (4,7-dihenyl-1,10-phenanthroline) cobalt (III). *Proc. Natl. Acad. Sci. USA* **1987**, *84*, 1764–1768.

250. Sheflin, L.G.; Kowalski, D. Altered DNA conformations detected by mung bean nuclease occur in promoter and terminator regions of supercoiled pBR322 DNA. *Nucleic Acids Res.* **1985**, *13*, 6137–6155.

251. Iacono-Conners, L.; Kowalski, D. Altered conformations in gene regulatory regions of torsion-ally stressed SV40 DNA. *Nucleic Acids Res.* **1986**, *14*, 8949–8962.

252. Muller, U.R.; Fitch, W.M. Evolutionary selection for perfect hairpin structures in viral DNAs. *Nature* **1982**, *298*, 582–585.

253. Greenberg, M.E.; Siegfried, Z.; Ziff, E.B. Mutation of the c-*fos* dyad symmetry element inhibits serum inducibility *in vivo* and nuclear regulatory factor binding *in vitro*. *Mol. Cell Biol.* **1987**, *7*, 1217–1225.

254. Martinez-Arias, A.; Yost, H.J.; Casadaban, M.J. Role of an upstream regulatory element in leu-cine repression of the *Saccharomyces cerevisiae* leu2 gene. *Nature* **1984**, *307*, 740–742.

255. Shuster, J.; Yu, J.; Cox, D.; Chan, R.V.L.; Smith, M.; Young, E. ADR1-mediated regulation of ADH2 requires an inverted repeat sequence. *Mol. Cell Biol.* **1986**, *6*, 1894–1902.

256. Schon, E.; Evans, T.; Welsh, J.; Efstratiadis, A. Conformation of promoter DNA: fine mapping of S1-hypersensitive sites. *Cell* **1983**, *35*, 837–848.

257. Glucksmann, M.A.; Markiewicz, P.; Malone, C.; Rothman-Denes, L.B. Specific sequences and a hairpin structure in the template strand are required for N4 virion RNA polymerase promoter recognition. *Cell* **1992**, *70*, 491–500.

258. Horwitz, M.S.Z.; Loeb, L.A. Structure function relationships in *Escherichia coli* promoter DNA. *Prog. Nucleic Acids Res. Mol. Biol.* **1990**, *38*, 137–164.

259. Markiewicz, P.; Malone, C.; Chase, J.W.; Rothman-Denes, L.B. *Escherichia coli* single-stranded DNA-binding protein is a supercoiled template-dependent transcriptional activator of N4 virion RNA polymerase. *Genes Dev.* **1992**, *6*, 2010–2019.

260. Wu, H.Y.; Shyy, S.H.; Wang, J.C.; Liu, L.F. Transcription generates positively and negatively supercoiled domains in the template. Cell 1988, 53, 433–440.

261. Wang, J.C.; Liu, L.F. DNA replication: topological aspects and the roles of DNA topoi-somerases. In: *DNA Topology and Its Biological Effects* (Cozzarelli, N.R.; Wang, J.C.; Eds.). Cold Spring Harbor, NY, Cold Spring Harbor Laboratory, 1990, pp. 321–340.

262. Dayn, A.; Malkhosyan, S.; Mirkin, S. Transcriptionally driven cruciform formation *in vivo*. *Nucleic Acids Res.* **1992**, *20*, 5991–5997.

263. Wang, J.C. DNA topoisomerases. *Annu. Rev. Biochem.* **1985**, *54*, 665–697.

264. Horwitz, M.S. Transcription regulation *in vitro* by an *E. coli* promoter containing a DNA cruci-form in the 3'-5' region. *Nucleic Acids Res.* **1989**, *17*, 5537–5545.

265. Bagga, R.; Ramesh, N.; Brahmachari, S.K. Supercoil-induced unusual structures as transcrip-tional block. *Nucleic Acids Res.* **1990**, *18*, 3363–3369.

266. Hamada, H.; Bustin, M. Hierarchy of binding sites for chromosomal proteins HMG1 and HMG2 in supercoiled deoxyribonucleic acid. *Biochemistry* **1985**, *24*, 1428–1433.

267. Waga, S.; Mizuno, S.; Yoshida, M. Nonhistone protein HMG1 removes the transcriptional block caused by left-handed Z-form segment in a supercoiled DNA. *Biochem. Biophys. Res. Commun.* **1988**, *153*, 334–339.

268. Sastry, S.S.; Kun, E. The interaction of adenosine diphosphoribosyl transferase (ADPRT) with a cruciform DNA. *Biochem. Biophys. Res. Commun.* **1990**, *167*, 842–847.

269. Oei, S.L.; Herzog, H.; Hirsch-Kauffmann, M.; Schneider, R.; Auer, B.; Schweiger, M. Tran-scriptional regulation and autoregulation of the human gene for ADP-ribosyltransferase. *Mol. Cell Biochem.* **1994**, *138*, 99–104.

270. Cozzarelli, N.R.; Wang, J.C. *DNA Topology and Its Biolocical Effects.* Cold Spring Harbor, NY, Cold Spring Harbor Laboratory, 1990.

271. White, J.H.; Bauer, W.R. Superhelical DNA with local substructures. A generalization of the topological constraint in terms of the intersection number and the ladder-like correspondence surface. *J. Mol. Biol.* **1987**, *195*, 205–213.

272. Pearson, C.E.; Ruiz, M.T.; Price, G.B.; Zannis-Hadjopoulos, M. Cruciform DNA binding protein in HeLa cell extracts. *Biochemistry* **1994**, *33*, 14185–14196.

273. Diekmann, S.; Lilley, D.M. The anomalous gel migration of a stable cruciform: temperature and salt dependence, and some comparisons with curved DNA. *Nucleic Acids Res.* **1987**, *15*, 5765–5774.

274. McLean, M.J.; Wells, R.D. The role of DNA sequence in the formation of Z-DNA versus cruciforms in plasmids. *J. Biol. Chem.* **1988**, *263*, 7370–7377.

275. Klysik, J.; Stirdivant, S.; Wells, R. Left-handed DNA. *J. Biol. Chem.* **1982**, *257*, 10152–10158.

276. Burd, J.F.; Wartell, R.M.; Dodgson, J.B.; Wells, R.D. Transmission of stability (telestability) in deoxyribonucleic acid. Physical and enzymatic studies on the duplex block polymer d($C_{15}A_{15}$)-d($T_{15}G_{15}$). *J. Biol. Chem.* **1975**, *250*, 5109–5113.

277. Klein, R.; Wells, R. Effects of neighboring DNA homopolymers on the biochemical and physical properties of the *Escherichia coli* lactose promoter. *J. Biol. Chem.* **1982**, *257*, 12962–12969.

278. Gierer, A. Model for DNA and protein interactions and the function of the operator. *Nature* **1966**, *212*, 1480–1481.

279. deMassey, B.; Studier, F.W.; Dorgai, L.; Appelbaum, F.; Weisberg, R.A. Enzymes and the sites of genetic recombination: studies with gene-3 endonuclease of phage T7 and with site-affinity mutants of lambda phage. *Cold Spring Harb. Symp. Quant. Biol.* **1984**, *49*, 715–726.

280. West, S.C.; Korner, A. Cleavage of cruciform DNA structures by an activity from *Saccharomyces cerevisiae*. *Proc. Natl. Acad. Sci. USA* **1985**, *82*, 6445–6449.

281. Symington, L.; Kolodner, R. Partial purification of an endonuclease from *Saccharomyces cerevisiae* that cleave Holliday junctions. *Proc. Natl. Acad. Sci. USA* **1985**, *82*, 7247–7251.

282. West, S.C.; Parsons, C.A.; Picksley, S.M. Purification and properties of a nuclease from *Saccharomyces cerevisiae* that cleaves DNA at cruciform junctions. *J. Biol. Chem.* **1987**, *262*, 12752–12758.

283. Taylor, A.F.; Smith, G.R. Action of RecBCD enzyme on cruciform DNA. *J. Mol. Biol.* **1990**, *211*, 117–134.

284. Connolly, B.; Parsons, C.A.; Benson, F.E.; Dunderdale, H.J.; Sharples, G.J.; Lloyd, R.G.; et al. Resolution of Holliday junctions *in vitro* requires the *Escherichia coli* ruvC gene product. *Proc. Natl. Acad. Sci. USA* **1991**, *88*, 6063–6067.

285. Parsons, C.A.; Kemper, B.; West, S.C. Interaction of a four-way junction in DNA with T4 endonuclease VII. *J. Biol. Chem.* **1990**, *265*, 9285–9289.

286. Parsons, C.A.; West, S.C. Specificity of binding to four-way junctions in DNA by bacteriophage T7 endonuclease I. *Nucleic Acids Res.* **1990**, *18*, 4377–4384.

287. Bhattacharyya, A.; Murchie, A.I.; von Kitzing, E.; Diekmann, S.; Kemper, B.; Lilley, D.M.J. A model for the interaction of DNA junctions and resolving enzymes. *J. Mol. Biol.* **1991**, *221*, 1191–1207.

288. Duckett, D.R.; Murchie, A.I.H.; Bhattacharyya, A.; Clegg, R.M.; Diekmann, S.; von Kitzing, E.; et al. The structure of DNA junctions and their interaction with enzymes. *Eur. J. Biochem.* **1992**, *207*, 285–295.

289. Bianchi, M.E. Interaction of a protein from rat liver nuclei with cruciform DNA. *EMBO J.* **1988**, *7*, 843–849.

290. Bonne-Andrea, C.; Harper, F.; Puvion, E.; Delpech, M. Nuclear accumulation of HMG1 is correlated to DNA synthesis. *Biol. Cell* **1986**, *58*, 185–194.

291. Bustin, M.; Lehn, D.A.; Landsman, D. Structural features of the HMG chromosomal proteins and their genes. *Biochim. Biophys. Acta* **1990**, *1049*, 231–243.

292. Einck, L.; Bustin, M. The intracellular distribution and function of the high mobility group chromosomal proteins. *Exp. Cell Res.* **1985**, *156*, 295–310.

293. Bottger, M.; Vogel, F.; Platzer, U.; Kiessling, U.; Grade, K.; Strauss, M. Condensation of vector DNA by the chromosomal protein HMG1 results in efficient transfection. *Biochim. Biophys. Acta* **1988**, *950*, 221–228.

294. Mathis, D.J.; Kindelis, A.; Spadafora, C. HMG proteins (1+2) from beaded structures when complexed with closed circular DNA. *Nucleic Acids Res.* **1980**, *8*, 2577–2590.

295. Javaherian, K.; Sadeghi, M.; Liu, L.F. Nonhistone proteins HMG1 and HMG2 unwind DNA double helix. *Nucleic Acids Res.* **1979**, *6*, 3569–3580.

296. Javaherian, K.; Liu, L.F.; Wang, J.C. Nonhistone proteins HMG1 and HMG2 change the DNA helical structure. *Science* **1978**, *199*, 1345–1346.

297. Yoshida, M. High glutamic and aspartic regions in nonhistone protein HMB(1+2) unwinds DNA double helical structure. *J. Biochem.* **1987**, *101*, 175–180.

298. Yoshida, M.; Shimura, K. Unwinding of DNA by nonhistone chromosomal protein HMG(1+2) from pig thymus as determined with endonuclease. *J. Biochem.* **1984**, *95*, 117–124.

299. Paull, T.T.; Haykinson, M.; Johnson, R.C. The nonspecific DNA-binding and bending proteins HMG1 and HMG2 promote the assembly of complex nucleoprotein structures. *Genes Dev.* **1993**, *7*, 1521–1534.

300. Pil, P.M.; Chow, C.S.; Lippard, S.J. High-mobility-group 1 protein mediates DNA bending determined by ring closures. *Proc. Natl. Acad. Sci. USA* **1993**, *90*, 9465–9469.

301. Hodges-Garcia, Y.; Hagerman, P.J.; Pettijohn, D.E. DNA ring closure mediated by protein HU. *J. Biol. Chem.* **1989**, *264*, 14621–14623.

302. Bianchi, M.E. Prokaryotic HU and eukaryotic HMG1: a kinked relationship. *Mol. Microbiol.* **1994**, *14*, 1–5.

303. Jackson, J.B.; Rill, R.L. Circular dichroism, thermal denaturation, and deoxyribonuclease I digestion studies of nucleosomes highly enriched in high mobility group proteins HMG1 and HMG2. *Biochemistry* **1981**, *20*, 1042–1046.

304. Singh, J.; Dixon, G.H. High mobility group proteins 1 and 2 function as general class II transcription factors. *Biochemistry* **1990**, *29*, 6295–6302.

305. Tremethick, D.J.; Molloy, P.L. Effects of high mobility group proteins 1 and 2 on initiation and elongation of specific transcription by RNA polymerase II *in vitro*. *Nucleic Acids Res.* **1988**, *16*, 11107–11123.

306. Tremethick, D.J.; Molloy, P.L. High mobility group proteins 1 and 2 stimulate transcription *in vitro* by RNA polymerase II and III. *J. Biol. Chem.* **1986**, *261*, 6986–6992.

307. Watt, F.; Molloy, P.L. High mobility group proteins 1 and 2 stimulate binding of a specific transcription factor in the adenovirus major late promoter. *Nucleic Acids Res.* **1988**, *16*, 1471–1486.

308. Bonne, C.; Sautiere, P.; Duguet, M.; De Recondo, A.-M. Identification of a single-stranded DNA binding protein from rat liver with high mobility group protein 1. *J. Biol. Chem.* **1982**, *257*, 2722–2725.

309. Alexandrova, E.A.; Marekov, L.N.; Beltchev, B.G. Involvement of protein HMG1 in DNA replication. *FEBS Lett.* **1984**, *178*, 153–156.

310. Alexandrova, E.A.; Beltchev, B.G. Acetylated HMG1 protein interacts specifically with homologous DNA polymerase alpha *in vitro*. *Biochem. Biophys. Res. Commun.* **1988**, *154*, 918–927.

311. Bianchi, M.E.; Falciola, L.; Ferrari, S.; Lilley, D.M.J. The DNA binding site of HMG1 protein is composed of two similar segments (HMG boxes), both of which have counterparts in other eukaryotic regulatory proteins. *EMBO J.* **1992**, *11*, 1055–1063.

312. Ner, S.S. HMGs everywhere. *Curr. Biol.* **1992**, *2*, 208–210.

313. Isackson, P.J.; Fishback, J.L.; Bidney, D.L.; Reeck, G.R. Preferential affinity of high molecular weight high mobility group non-histone chromatin proteins for single-stranded DNA. *J. Biol. Chem.* **1979**, *254*, 5569–5572.

314. Ferrari, S.; Harley, V.R.; Pontigga, A.; Goodfellow, P.N.; Lovell-Badge, R.; Bianchi, M.E. SRY, like HMG1, recognizes sharp angles in DNA. *EMBO J.* **1992**, *11*, 4497–4506.

315. Harley, V.R.; Goodfellow, P.N. The biochemical role of SRY in sex determination. *Mol. Reprod. Dev.* **1994**, *39*, 184–193.

316. Bianchi, M.E.; Beltrame, M.; Paonessa, G. Specific recognition of cruciform DNA by nuclear protein HMG1. *Science* **1989**, *243*, 1056–1059.

317. Gut, S.H.; Bischoff, M.; Hobi, R.; Kuenzle, C.C. Z-DNA binding proteins from bull testis. *Nucleic Acids Res.* **1987**, *15*, 9691–9705.

318. Christen, T.; Bischoff, M.; Hobi, R.; Kuenzle, C.C. High mobility group proteins 1 and 2 bind preferentially to brominated poly(dG-dC)·poly(dG-dC) in the Z-DNA conformation but not to other types of Z-DNA. *FEBS Lett.* **1990**, *267*, 139–141.

319. Rohner, K.J.; Hobi, R.; Kuenzle, C.C. Z-DNA-binding proteins. Identification critically depends on the proper choice of ligands. *J. Biol. Chem.* **1990**, *265*, 19112–19115.

320. Bruhn, S.L.; Pil, P.M.; Essigman, J.M.; Houseman, D.E.; Lippard, S.J. Isolation and characterization of human cDNA clones encoding a high mobility group box protein that recognizes structural distortions to DNA caused by binding of the anticancer agent cisplatin. *Proc. Natl. Acad. Sci. USA* **1992**, *89*, 2307–2311.

321. Bianchi, M.E.; Beltrame, M.; Falciola, L. The HMG box motif. In: *Nucleic Acids and Molecular Biology* (Eckstein, F.; Lilley, D.M.J.; Eds.). Springer-Verlag, Berlin, 1992, 112–128.

322. Jantzen, H.-M.; Admon, A.; Bell, S.P.; Tijan, R. Nucleolar transcription factor hUBF contains a DNA-binding motif with homology to HMG proteins. *Nature* **1990**, *344*, 830–836.

323. Travis, A.; Amsterdam, A.; Berlanger, C.; Grosschedl, R. LEF-1 gene encoding a lymphoid-specific protein with an HMG domain, regulates T-cell receptor a enhancer function. *Genes Dev.* **1991**, *5*, 880–894.

324. Diffley, J.F.X.; Stillman, B. Purification of a yeast protein that binds to origins of DNA replication and a trascriptional silencer. *Proc. Natl. Acad. Sci. USA* **1988**, *85*, 2120–2124.

325. Shirakata, M.; Huppi, K.; Usuda, S.; Okazaki, K.; Yoshida, K.; Sakano, H. HMG1-related DNA-binding protein isolated with V-(D)-J recombination signal probes. *Mol. Cell Biol.* **1991**, *11*, 4528–4536.

326. Sinclair, A.H.; Berta, P.; Palmer, M.S.; Hawkins, J.R.; Griffiths, B.L.; Smith, M.J.; et al. A gene from the human sex-determining region encodes a protein with homology to a conserved DNA-binding motif. *Nature* **1990**, *346*, 240–244.

327. Copenhaver, G.P.; Putnam, C.D.; Denton, M.L.; Pikaard, C.S. The RNA polymerase 1 transcription factor UBF is a sequence-tolerant HMG-box protein that can recognize structured nucleic acids. *Nucleic Acids Res.* **1994**, *22*, 2651–2657.

328. Laudet, V.; Stehelin, D.; Clevers, H. Ancestry and diversity of the HMG box superfamily. *Nucleic Acids Res.* **1993**, *21*, 2493–2501.

329. Varga-Weisz, P.; Zlatanova, J.; Leuba, S.H.; Schroth, G.P.; van Holde, K. Binding of histones H1 and H5 and their globular domains to four-way junction DNA. *Proc. Natl. Acad. Sci. USA* **1994**, *91*, 3525–3529.

330. Carballo, M.; Puigdomenech, P.; Palau, J. DNA and histone H1 interact with different domains of HMG1. *EMBO J.* **1983**, *2*, 1759–1764.

331. Kohlstaedt, L.A.; Cole, R.D. Specific interaction between H1 histone and high mobility group protein, HMG1. *Biochemistry* **1994**, *33*, 570–575.

332. Kohlstaedt, L.A.; Sung, E.C.; Fujishige, A.; Cole, R.D. Non-histone chromosomal protein HMG1 modulates the histone H1-induced condensation of DNA. *J. Biol. Chem.* **1987**, *262*, 524–526.

333. Bonne, C.; Harper, C.F.; Sobczak, J.; De Recondo, A.-M. Rat liver HMG1: a physiological nucleosome assembly factor. *EMBO J.* **1984**, *3*, 1193–1199.

334. Zechiedrich, E.L.; Osheroff, N. Eukaryotic topoisomerases recognize nucleic acid topology by preferentially interacting with DNA crossovers. *EMBO J.* **1990**, *9*, 4555–4562.

335. Pognan, F.; Paoletti, C. Does cruciform DNA provide a recognition signal for DNA-topoisomerase II? *Biochimie* **1992**, *74*, 1019–1023.

336. Sekiguchi, J.; Seeman, N.C.; Shuman, S. Resolution of Holliday junctions by eukaryotic DNA topoisomerase. *Proc. Natl. Acad. Sci. USA* **1996**, *93*, 785–789.

337. Froelich-Ammon, S.J.; Gale, K.C.; Osheroff, N. Site-specific cleavage of a DNA hairpin by topoisomerase II. *J. Biol. Chem.* **1994**, *269*, 7719–7725.

338. Arndt-Jovin, D.J.; Uduardy, A.; Garner, M.M.; Ritter, S.; Jovin, T.M. Z-DNA binding and inhibition by GTP of *Drosophila* topoisomerase II. *Biochemistry* **1993**, *32*, 4862–4872.

339. McMurray, C.T.; Wilson, W.D.; Douglass, J.O. Hairpin formation within the enhancer region of the human enkephalin gene. *Proc. Natl. Acad. Sci. USA* **1991**, *88*, 666–670.

340. Spiro, C.; Richards, J.P.; Chandrasekaran, S.; Brennan, R.G.; McMurray, C.T. Secondary structure creates mismatched base pairs required for high-affinity binding of cAMP response element-binding protein to the human enkephalin enhancer. *Proc. Natl. Acad. Sci. USA* **1993**, *90*, 4606–4610.

341. Carmichael, E.P.; Roome, J.M.; Wahl, A.F. Binding of a sequence-specific single-stranded DNA-binding factor to the simian virus 40 core origin inverted repeat domain is cell cycle regulated. *Mol. Cell Biol.* **1993**, *13*, 408–420.

342. Quinn, J.P.; McAllister, J. A preprotachykinin promoter interacts with a sequence specific single stranded DNA binding protein. *Nucleic Acids Res.* **1996**, *21*, 1637–1641.

343. Kim, C.; Snyder, R.O.; Wold, M.S. Binding properties of replication protein A from human and yeast cells. *Mol. Cell Biol.* **1992**, *12*, 3050–3059.

344. Mitsui, Y.; Langridge, R.; Grant, R.C.; Kodama, M.; Wells, R.D.; Shortle, B.E.; et al. Physical and enzymatic studies on poly(dI-dC)·poly(dI-dC), an unusual double-helical DNA. *Nature* **1970**, *228*, 1166–1169.

345. Sutherland, J.C.; Griffen, K.P. Vacuum ultraviolet circular dichroism of poly(dI-dC)·poly(dI-dC): no evidence for a left-handed double helix. *Biopolymers* **1983**, *22*, 1445–1448.

346. Pohl, F.M.; Jovin, T.M. Salt-induced co-operative conformational change of a synthetic DNA: equilibrium and kinetic studies with poly (dG-dC). *J. Mol. Biol.* **1972**, *67*, 375–396.

347. Wang, A.H.J.; Quigley, G.J.; Kolpak, F.J.; Crawford, J.L.; Van Boom, J.H.; Van der Marel, G.; et al. Molecular structure of a left-handed double helical DNA fragment at atomic resolution. *Nature* **1979**, *282*, 680–686.

348. Rich, A.; Nordheim, A.; Wang, A.-J. The chemistry and biology of left-handed Z-DNA. *Annu. Rev. Biochem.* **1984**, *53*, 791–846.

349. Klysik, J.; Zacharias, W.; Galazka, G.; Kwinkowski, M.; Uznanski, B.; Okruszek, A. Structural interconversion of alternating purine-pyrimidine inverted repeats cloned in supercoiled plasmids. *Nucleic Acids Res.* **1988**, *16*, 6915–6933.

350. Nejedly, K.; Klysik, J.; Palecek, E. Supercoil-stabilized left handed DNA in the plasmid (dA-dT)$_{16}$ insert formed in the presence of Ni^{2+}. *FEBS Lett.* **1989**, *243*, 313–317.

351. Singleton, C.K.; Klysik, J.; Stirdivant, S.M.; Wells, R.D. Left-handed Z-DNA is induced by supercoiling in physiological ionic conditions. *Nature* **1982**, *299*, 312–316.

352. Peck, L.J.; Nordheim, A.; Rich, A.; Wang, J.C. Flipping of cloned d(pCpG)$_n$·d(pGpC)$_n$ DNA sequences from right-to-left-handed helical structure by salt, Co(III), or negative supercoiling. *Proc. Natl. Acad. Sci. USA* **1982**, *79*, 4560–4564.

353. Klysik, J.; Stirdivant, S.M.; Larson, J.E.; Hart, P.A.; Wells, R.D. Left-handed DNA in restriction fragments and a recombinant plasmid. *Nature* **1981**, *290*, 672–677.

354. Stirdivant, S.M.; Klysik, J.; Wells, R.D. Energetic and structural inter-relationship between DNA supercoiling and the right- to left-handed Z helix transitions in recombinant plasmids. *J. Biol. Chem.* **1982**, *257*, 10159–10165.

355. Haniford, D.B.; Pulleyblank, D.E. Facile transition of poly[d(TG)·d(CA)] into a left-handed helix in physiological conditions. *Nature* **1983**, *302*, 632–634.

356. Nordheim, A.; Rich, A. The sequence (dC-dA)$_n$·(dG-dT)$_n$ forms left-handed Z-DNA in negatively supercoiled plasmids. *Proc. Natl. Acad. Sci. USA* **1983**, *80*, 1821–1825.

357. Ellison, M.J.; Kelleher, R.J., III; Wang, A.H.-J.; Habener, J.F.; Rich, A. Sequence-dependent energetics of the B-Z transition in supercoiled DNA containing nonalternating purine-pyrimidine sequences. *Proc. Natl. Acad. Sci. USA* **1985**, *82*, 8320–8324.

358. Blaho, J.A.; Larson, J.E.; McLean, M.J.; Wells, R.D. Multiple DNA secondary structures in perfect inverted repeat inserts in plasmids. Right-handed B-DNA, cruciforms, and left-handed Z-DNA. *J. Biol. Chem.* **1988**, *263*, 14446–14455.

359. Sinden, R.R.; Kochel, T.J. Reduced 4,5',8-trimethylpsoralen cross-linking of left-handed Z-DNA stabilized by DNA supercoiling. *Biochemistry* **1987**, *26*, 1343–1350.

360. Kochel, T.J.; Sinden, R.R. Analysis of trimethylpsoralen photoreactivity to Z-DNA provides a general *in vivo* assay for Z-DNA: analysis of the hypersensitivity of $(GT)_n$ B-Z junctions. *Biotechniques* **1988**, *6*, 532–543.

361. Lafer, E.M.; Sousa, R.; Rich, A. Anti-Z-DNA antibody binding can stabilize Z-DNA in relaxed and linear plasmids under physiological conditions. *EMBO J.* **1985**, *4*, 3655–3660.

362. Lafer, E.; Moller, A.; Nordheim, A.; Stollar, B.D.; Rich, A. Antibodies specific for left-handed Z-DNA. *Proc. Natl. Acad. Sci. USA* **1981**, *78*, 3546–3550.

363. Moller, A.; Gabriels, J.E.; Lafer, E.M.; Nordheim, A.; Rich, A.; Stollar, D.B. Monoclonal antibodies recognize diffrent parts of Z-DNA*. *J. Biol. Chem.* **1982**, *257*, 12081–12085.

364. Peck, L.J.; Wang, J.C. Energetics of B-to-Z transition in DNA. *Proc. Natl. Acad. Sci. USA* **1983**, *80*, 6206–6210.

365. Nordheim, A.; Peck, L.J.; Lafer, E.M.; Stollar, B.D.; Wang, J.C.; Rich, A. Supercoiling and left-handed Z-DNA. *Cold Spring Harb. Symp. Quant. Biol.* **1983**, *47 Pt 1*, 93–100.

366. Johnston, B.H.; Rich, A. Chemical probes of DNA conformation: detection of Z-DNA at nucleotide resolution. *Cell* **1985**, *42*, 713–724.

367. McLean, M.J.; Lee, J.W.; Wells, R.D. Characteristics of Z-DNA helices formed by imperfect (purine-pyrimidine) sequences in plasmids. *J. Biol. Chem.* **1988**, *263*, 7378–7385.

368. Rio, P.; Leng, M. N-Hydroxyaminofluorene: a chemical probe for DNA conformation. *J. Mol. Biol.* **1986**, *191*, 569–572.

369. Kochel, T.J.; Sinden, R.R. Hyperreactivity of B-Z junctions to 4,5',8-trimethylpsoralen photobinding assayed by an exonuclease III/photoreversal mapping procedure. *J. Mol. Biol.* **1989**, *205*, 91–102.

370. Hoepfner, R.W.; Sinden, R.R. Amplified primer extension assay for psoralen photoproducts provides a sensitive assay for a $(CG)_6TA(CG)_2(TG)_8$Z-DNA torsionally tuned probe: preferential psoralen photobinding to one strand of a B-Z junction. *Biochemistry* **1993**, *32*, 7542–7548.

371. Azorin, F.; Hahn, R.; Rich, A. Restriction endonucleases can be used to study B-Z junction in supercoiled DNA. *Biochemistry* **1984**, *81*, 5714–5718.

372. Singleton, C.K.; Klysik, J.; Wells, R.D. Conformational flexibility of junctions between contiguous B- and Z-DNAs in supercoiled plasmids. *Proc. Natl. Acad. Sci. USA* **1983**, *80*, 2447–2451.

373. Zacharias, W.; Larson, J.E.; Kilpatrick, M.W.; Wells, R.D. HhaI methylase and restriction endonuclease as probes for B to Z DNA conformational changes in d(GCGC) sequences. *Nucleic Acids Res.* **1984**, *12*, 7677–7692.

374. Vardimon, L.; Rich, A. In Z-DNA the sequence G-C-G-C is neither methylated by Hha I methyltransferase nor cleaved by Hha I restriction endonuclease. *Proc. Natl. Acad. Sci. USA* **1984**, *81*, 3268–3272.

375. Zhang, S.; Lockshin, C.; Herbert, A.; Winter, E.; Rich, A. Zuotin, a putative Z-DNA binding protein in *Saccharomyces cerevisiae*. *EMBO J.* **1992**, *11*, 3787–3796.

376. Jaworski, A.; Hsieh, W.T.; Blaho, J.A.; Larson, J.E.; Wells, R.D. Left-handed DNA *in vivo*. *Science* **1987**, *238*, 773–777.

376a. Haniford, D.B.; Pulleyblank, D.E. The *in vivo* occurrence of Z DNA. *J. Biomol. Struct. Dyn.* **1983**, *1*, 593–609.

377. Rahmouni, A.R.; Wells, R.D. Stabilization of Z DNA *in vivo* by localized supercoiling. *Science* **1989**, *246*, 358–363.

378. Nordheim, A.; Pardue, M.L.; Lafer, E.M.; Moller, A.; Stollar, B.D.; Rich, A. Antibodies to left-handed Z-DNA bind to interband regions of *Drosophila* polytene chromosomes. *Nature* **1981**, *294*, 417–422.

379. Hill, R.J.; Stollar, B.D. Dependence of Z-DNA antibody binding to polytene chromosomes on acid fixation and DNA torsional strain. *Nature* **1983**, *305*, 338–340.

380. Jackson, D.A.; Cook, P.R. A general method for preparing chromatin containing intact DNA. *EMBO J.* **1985**, *4*, 913–918.

381. Santoro, C.; Costanzo, F.; Ciliberto, G. Inhibition of eukaryotic tRNA transcription by potential Z-DNA sequences. *EMBO J.* **1984**, *3*, 1553–1559.

382. Banerjee, R.; Carothers, A.M.; Grunberger, D. Inhibition of the herpes simplex virus thymidine kinase gene transfection in Ltk-cells by potential Z-DNA forming polymers. *Nucleic Acids Res.* **1985**, *13*, 5111–5126.

383. Banerjee, R.; Grunberger, D. Enhanced expression of the bacterial chloramphenicol acetyltransferase gene in mouse cells cotransfected with synthetic polynucleotides able to form Z-DNA. *Proc. Natl. Acad. Sci. USA* **1986**, *83*, 4988–4992.

384. Naylor, L.H.; Clark, E.M. d(TG)$_n$·d(CA)$_n$ sequences upstream of the rat prolactin gene form Z-DNA and inhibit gene transcription. *Nucleic Acids Res.* **1990**, *18*, 1595–1601.

385. Blaho, J.A.; Wells, R.D. Left-handed Z-DNA and genetic recombination. *Prog. Nucleic Acids Res. Mol. Biol.* **1989**, *37*, 107–126.

386. Molineaux, S.M.; Engh, H.; de-Ferra, F.; Hudson, L.; Lazzarini, R.A. Recombination within the myelin basic protein gene created the dysmyelinating shiverer mouse mutation. *Proc. Natl. Acad. Sci. USA* **1986**, *83*, 7542–7546.

387. Weinreb, A.; Collier, D.A.; Birshtein, B.K.; Wells, R.D. Left-handed Z-DNA and intramolecular triplex formation at the site of an unequal sister chromatid exchange. *J. Biol. Chem.* **1990**, *265*, 1352–1359.

388. Weinreb, A.; Katzenberg, D.R.; Gilmore, G.L.; Birshtein, B.K. Site of unequal sister chromatid exchange contains a potential Z-DNA-forming tract. *Proc. Natl. Acad. Sci. USA* **1988**, *85*, 529–533.

389. Boehm, T.; Mengle-Gaw, L.; Kees, U.R.; Spurr, N.; Lavenir, I.; Forster, A.; et al. Alternating purine-pyrimidine tracts may promote chromosomal translocations seen in a variety of human lymphoid tumours. *EMBO J.* **1989**, *8*, 2621–2631.

390. Wahls, W.P.; Wallace, L.J.; Moore, P.D. The Z-DNA motif d(TG)$_{30}$ promotes reception of information during gene conversion events while stimulating homologous recombination in human cells in culture. *Mol. Cell Biol.* **1990**, *10*, 785–793.

391. Choo, K.B.; Lee, H.H.; Liew, L.N.; Chong, K.Y.; Chou, H.F. Analysis of the unoccupied site of an integrated human papillomavirus 16 sequence in a cervical carcinoma. *Virology* **1990**, *178*, 621–625.

392. Kmiec, E.B.; Angelides, K.J.; Holloman, W.K. Left-handed DNA and the synaptic pairing reaction promoted by Ustilago rec1 protein. *Cell* **1985**, *40*, 139–145.

393. Blaho, J.A.; Wells, R.D. Left-handed Z-DNA binding by the recA protein of *Escherichia coli*. *J. Biol. Chem.* **1987**, *262*, 6082–6088 [published erratum appears in *J. Biol. Chem.* **1988**, *263*, 11015].

394. Fishel, R.A.; Detmer, K.; Rich, A. Identification of homologous pairing and strand-exchange activity from a human tumor cell line based on Z-DNA affinity chromatography. *Proc. Natl. Acad. Sci. USA* **1988**, *85*, 36–40.

395. Herbert, A.G.; Spitzner, J.R.; Lowenhaupt, K.; Rich, A. Z-DNA binding protein from chicken blood nuclei. *Proc. Natl. Acad. Sci. USA* **1993**, *90*, 3339–3342.

396. Gut, S.H.; Bischoff, M.; Hobi, R.; Kuenzle, C.C. Z-DNA-binding proteins from bull testis. *Nucleic Acids Res.* **1987**, *15*, 9691–9705.

397. Krishna, P.; Kennedy, B.P.; van de Sande, J.H.; McGhee, J.D. Yolk proteins from nematodes, chickens, and frogs bind strongly and preferentially to left-handed Z-DNA. *J. Biol. Chem.* **1988**, *263*, 19066–19070.

398. Lafer, E.M.; Sousa, R.J.; Rich, A. Z-DNA-binding proteins in *Escherichia coli* purification, generation of monoclonal antibodies and gene isolation. *J. Mol. Biol.* **1988**, *203*, 511–516.

399. Lafer, E.M.; Sousa, R.; Rosen, B.; Hsu, A.; Rich, A. Isolation and characterization of Z-DNA binding proteins from wheat germ. *Biochemistry* **1985**, *24*, 5070–5076.
400. Nordheim, A.; Lafer, E.M.; Peck, L.J.; Wang, J.C.; Stollar, B.D.; Rich, A. Negatively supercoiled plasmids contain left-handed Z-DNA segments as detected by specific antibody binding. *Cell* **1982**, *31*, 309–318.
401. Felsenfeld, G.; Davies, D.R.; Rich, A. Formation of a three-stranded polynucleotide molecule. *J. Am. Chem. Soc.* **1957**, *79*, 2023–2024.
402. Wells, R.D.; Collier, D.A.; Hanvey, J.C.; Shimizu, M.; Wohlrab, F. The chemistry and biology of unusual DNA structures adopted by oligopurine·oligopyrimidine sequences. *FASEB J.* **1988**, *2*, 2939–2949.
403. Thuong, N.T.; Hélène, C. Sequence-specific recognition and modification of double-helical DNA by oligonucleotides. *Angew. Chem. Int. Ed. Engl.* **1993**, *32*, 666–690.
404. Mirkin, S.M.; Frank-Kamenetskii, M.D. H-DNA and related structures. *Annu. Rev. Biophys. Biomol. Struct.* **1994**, *23*, 541–576.
405. Radhakrishnan, I.; Patel, D.J. DNA triplexes: solution structures, hydration sites, energetics, interactions, and function. *Biochemistry* **1994**, *33*, 11405–11416.
406. Frank-Kamenetskii, M.D.; Mirkin, S.M. Triplex DNA structures. *Annu. Rev. Biochem.* **1995**, *64*, 65–95.
407. Chubb, J.M.; Hogan, M.E. Human therapeutics based on triple helix technology. *Trends Biotechnol.* **1992**, *10*, 132–136.
408. Helene, C.; Thuong, N.T.; Harrel-Bellan, A. Control of gene expression by triple helix-forming oligonucleotides. The antigene strategy. *Ann. N.Y. Acad. Sci.* **1992**, *660*, 27–36.
409. Strobel, S.A.; Dervan, P.B. Triple helix-mediated single-site enzymatic cleavage of megabase genomic DNA. *Methods Enzymol.* **1992**, *216*, 309–321.
410. Maher, L.J., III. Prospects for the therapeutic use of antigene oligonucleotides. *Cancer Invest.* **1996**, *14*, 66–82.
411. Felsenfeld, G.; Miles, H.T. The physical and chemical properties of nucleic acids. *Annu. Rev. Biochem.* **1967**, *26*, 407–468.
412. Michelson, A.M.; Massoulie, J.; Guschlbauer, W. Synthetic polynucleotides. *Prog. Nucleic Acids Res. Mol. Biol.* **1967**, *6*, 83–141.
413. Cheng, Y.K.; Pettitt, B.M. Stabilities of double- and triple-strand helical nucleic acids. *Prog. Biophys. Mol. Biol.* **1992**, *58*, 225–257.
414. Moser, H.E.; Dervan, P.B. Sequence-specific cleavage of double helical DNA by triple helix formation. *Science* **1987**, *238*, 645–650.
415. Le Doan, T.; Perroualt, L.; Praseuth, D.; Habhoub, N.; Decoult, J.L.; Thuong, N.T.; et al. Sequence-specific recognition, photocrosslinking and cleavage of the DNA double-helix by an oligo-[α]-thymidilate covalently linked to an azidoproflavine derivative. *Nucleic Acids Res.* **1987**, *15*, 7749–7760.
416. Cooney, M.; Czernuszewicz, G.; Postel, E.H.; Flint, S.J.; Hogan, M.E. Site-specific oligonucleotide binding represses transcription of the human c-*myc* gene *in vitro*. *Science* **1988**, *241*, 456–459.
417. Lyamichev, V.I.; Mirkin, S.M.; Frank-Kamenetskii, M.D. Structures of homopurine-homopyrimidine tract in superhelical DNA. *J. Biomol. Struct. Dyn.* **1986**, *3*, 667–669.
418. Kohwi, Y.; Kohwi-Shigematsu, T. Magnesium ion-dependent triple-helix structure formed by homopurine-homopyrimidine sequences in supercoiled plasmid DNA. *Proc. Natl. Acad. Sci. USA* **1988**, *85*, 3781–3785.
419. Piriou, J.M.; Ketterle, C.; Gabarro-Arpa, J.; Cognet, J.A.H.; Le Bret, M. A database of 32 DNA triplets to study triple helices by molecular mechanics and dynamics. *Biophys. Chem.* **1994**, *50*, 323–343.
420. van Vlijmen, H.W.T.; Ramé, G.L.; Pettitt, B.M. A study of model energetics and conformational properties of polynucleotide triplexes. *Biopolymers* **1990**, *30*, 517–532.

421. Helene, C.; Lancelot, G. Interactions between functional groups in protein-nucleic acid associations. *Prog. Biophys. Mol. Biol.* **1982**, *39*, 1–68.

422. Laughton, C.A.; Neidle, S. Prediction of the structure of the Y+.R-.R(+)-type DNA triple helix by molecular modelling. *Nucleic Acids Res.* **1992**, *20*, 6535–6541.

423. Arnott, S.; Selsing, E. Structures for the polynucleotide complexes poly(dA)∗poly(dT) and poly(dT)∗poly(dA)∗poly(dT). *J. Mol. Biol.* **1974**, *88*, 509–521.

424. Raghunathan, G.; Miles, H.T.; Sasisekharan, V. Symmetry and molecular structure of a DNA triple helix: d(T)$_n$·d(A)$_n$·d(T)$_n$. *Biochemistry* **1993**, *32*, 455–462.

425. Macaya, R.; Wang, E.; Schultze, P.; Sklenar, V.; Feigon, J. Proton nuclear magnetic resonance assignments and structural characterization of an intramolecular DNA triplex. *J. Mol. Biol.* **1992**, *225*, 755–773.

426. Shin, C.; Koo, H.S. Helical periodicity of GA-alternating triple-stranded DNA. *Biochemistry* **1996**, *35*, 968–972.

427. Htun, H.; Dahlberg, J.E. Topology and formation of triple-stranded H-DNA. *Science* **1989**, *243*, 1571–1576.

428. Hanvey, J.C.; Shimizu, M.; Wells, R.D. Intramolecular DNA triplexes in supercoiled plasmids. II. Effect of base composition and noncentral interruptions on formation and stability. *J. Biol. Chem.* **1989**, *264*, 5950–5956.

429. Panyutin, I.G.; Wells, R.D. Nodule DNA in the (GA)$_{37}$·(CT)$_{37}$ insert in superhelical plasmids. *J. Biol. Chem.* **1992**, *267*, 5495–5501.

430. Maher, L.J.; Dervan, P.B.; Wold, B.J. Kinetic analysis of oligodeoxyribonucleotide-directed triple-helix formation on DNA. *Biochemistry* **1990**, *29*, 8820–8826.

431. Rougee, M.; Faucon, B.; Mergny, J.L.; Barcelo, F.; Giovannangeli, C.; Garestier, T.; et al. Kinetics and thermodynamics of triple-helix formation: effects of ionic strength and mismatches. *Biochemistry* **1992**, *31*, 9269–9278.

432. Durland, R.H.; Kessler, D.J.; Gunnel, S.; Duvic, M.; Pettitt, B.M.; Hogan, M.E. Binding of triple helix forming oligonucleotides to sites in gene promoters. *Biochemistry* **1991**, *30*, 9246–9255.

433. Krakauer, H.; Sturtevant, J.M. Heats of the helix-coil transitions of the poly A–poly U complexes. *Biopolymers* **1968**, *6*, 491–512.

434. Glaser, R.; Gabbay, E.J. Topography of nucleic acid helices in solutions. III. Interactions of spermine and spermidine derivatives with polyadenylic-polyuridylic and polyinosinic-polycytidylic acid helices. *Biopolymers* **1968**, *6*, 243–254.

435. Hampel, K.J.; Crosson, P.; Lee, J.S. Polyamines favor DNA triplex formation at neutral pH. *Biochemistry* **1991**, *30*, 4455–4459.

436. Hanvey, J.C.; Williams, E.M.; Besterman, J.M. DNA triple-helix formation at physiologic pH and temperature. *Antisense Res. Dev.* **1991**, *1*, 307–317.

437. Singleton, S.F.; Dervan, P.B. Equilibrium association constants for oligonucleotide-directed triple helix formation at single DNA sites: linkage to cation valence and concentration. *Biochemistry* **1993**, *32*, 13171–13179.

438. Thomas, T.; Thomas, T.J. Selectivity of polyamines in triplex DNA stabilization. *Biochemistry* **1993**, *32*, 14068–14074.

439. Potaman, V.N.; Sinden, R.R. Stabilization of triple-helical nucleic acids by basic oligopeptides. *Biochemistry* **1995**, *34*, 14885–14992.

440. Kiessling, L.L.; Griffin, L.C.; Dervan, P.B. Flanking sequence effects within the pyrimidine triple-helix motif characterized by affinity cleaving. *Biochemistry* **1992**, *31*, 2829–2834.

441. Völker, J.; Klump, H.H. Electrostatic effects in DNA triple helices. *Biochemistry* **1994**, *33*, 13502–13508.

442. Darnell, J.; Lodish, H.; Baltimore, D. *Molecular Cell Biology.* Scientific American Books, New York, 1990.

442a. Sarhan, S.; Seiler, N. On the subcellular localization of polyamines. *Biol. Chem. Hoppe-Seyler* **1989**, *370*, 1279–1284.

443. Davis, R.W.; Morris, D.R.; Coffino, P. Sequestered end products and enzyme regulation: the case of ornitine decarboxylase. *Microbiol. Rev.* **1992**, *56*, 280–290.

444. Frank-Kamenetskii, M.D. Protonated DNA structures. *Methods Enzymol.* **1992**, *211*, 180–191.

445. Bernues, J.; Beltran, R.; Casasnovas, J.M.; Azorin, F. DNA-sequence and metal-ion specificity of the formation of *H-DNA. *Nucleic Acids Res.* **1990**, *18*, 4067–4073.

446. Roberts, R.W.; Crothers, D.M. Specificity and stringency in DNA triplex formation. *Proc. Natl. Acad. Sci. USA* **1991**, *88*, 9397–9401.

447. Weerasinghe, S.; Smith, P.E.; Mohan, V.; Cheng, Y.K.; Pettitt, B.M. Nanosecond dynamics and structure of a model DNA triple helix in saltwater solution. *J. Am. Chem. Soc.* **1995**, *117*, 2147–2158.

448. Lyamichev, V.I.; Mirkin, S.M.; Frank-Kamenetskii, M.D. A pH-dependent structural transition in the homopurine-homopyrimidine tract in superhelical DNA. *J. Biomol. Struct. Dyn.* **1985**, *3*, 327–338.

449. Lyamichev, V.I.; Mirkin, S.M.; Kumarev, V.P.; Baranova, L.V.; Vologodskii, A.V.; Frank-Kamenetskii, M.D. Energetics of the B-H transition in supercoiled DNA carrying $d(CT)_x \cdot d(AG)_x$ and $d(C)_n \cdot d(G)_n$ inserts. *Nucleic Acids Res.* **1989**, *17*, 9417

450. Schroth, G.P.; Ho, P.S. Occurrence of potential cruciform and H-DNA forming sequences in genomic DNA. *Nucleic Acids Res.* **1995**, *23*, 1977–1983.

451. Horne, D.A.; Dervan, P.B. Recognition of mixed-sequence duplex DNA by alternate-strand triple-helix formation. *J. Am. Chem. Soc.* **1990**, *112*, 2435–2437.

451a. Behe, M.J. An overabundance of long oligopurine tracts occurs in the genome of simple and complex eukaryotes. *Nucleic Acids Res.* **1995**, *23*, 507–511.

452. Pilch, D.S.; Breslauer, K.J. Ligand-induced formation of nucleic acid triple helices. *Proc. Natl. Acad. Sci. USA* **1994**, *91*, 9332–9336.

453. Duval-Valentin, G.; de Bizemont, T.; Takasugi, M.; Mergny, J.L.; Bisagni, E.; Helene, C. Triple-helix specific ligands stabilize H-DNA conformation. *J. Mol. Biol.* **1995**, *247*, 847–858.

454. Wilson, W.D.; Mizan, S.; Tanious, F.A.; Yao, S.; Zon, G. The interaction of intercalators and groove-binding agents with DNA triple-helical structures: the influence of ligand structure, DNA backbone modifications and sequence. *J. Mol. Recogn.* **1994**, *7*, 89–98.

455. Nielsen, P.E.; Egholm, M.; Buchardt, O. Peptide nucleic acid (PNA). A DNA mimic with a peptide backbone. *Bioconjug. Chem.* **1994**, *5*, 3–7.

456. Malkov, V.A.; Voloshin, O.N.; Soyfer, V.N.; Frank-Kamenetskii, M.D. Cation and sequence effects on stability of intermolecular pyrimidine-purine-purine triplex. *Nucleic Acids Res.* **1993**, *21*, 585–591.

457. Belotserkovskii, B.P.; Veselkov, A.G.; Filippov, S.A.; Dobrynin, V.N.; Mirkin, S.M.; Frank-Kamenetskii, M.D. Formation of intramolecular triplex in homopurine-homopyrimidine mirror repeats with point substitutions. *Nucleic Acids Res.* **1990**, *18*, 6621–6624.

458. Xodo, L.E.; Alunni-Fabbroni, M.; Manzini, G.; Quadrifoglio, F. Sequence-specific DNA-triplex formation at imperfect homopurine-homopyrimidine sequences within a DNA plasmid. *Eur. J. Biochem.* **1993**, *212*, 395–401.

459. Wells, R.D.; Amirhaeri, S.; Blaho, J.A.; Collier, D.A.; Dohrman, A.; Griffin, J.A.; et al. Biology and chemistry of Z-DNA and triplexes. In: *The Bacterial Chromosome* (Drlica, K.; Riley, M.; Eds.). American Society for Microbiology, Washington, DC, 1990, pp. 187–194.

460. Collier, D.A.; Wells, R.D. Effect of length, supercoiling, and pH on intramolecular triplex formation. Multiple conformers at pur·pyr mirror repeats. *J. Biol. Chem.* **1990**, *265*, 10652–10658.

461. Kohwi-Shigematsu, T.; Kohwi, Y. Detection of triple-helix related structures adopted by poly(dG)·poly(dC) sequences in supercoiled plasmid DNA. *Nucleic Acids Res.* **1991**, *19*, 4267–4271.

462. Naylor, R.; Gilham, P.T. Studies on some interactions and reactions of oligonucleotides in aqueous solution. *Biochemistry* **1966**, *5*, 2722–2728.

463. Cassani, G.R.; Bollum, F.J. Oligodeoxythymidilate:polydeoxyadenylate and oligodeoxyadenylate: polydeoxythymidilate interactions. *Biochemistry* **1969**, *8*, 3928–3936.

464. Broitman, S.L.; Im, D.D.; Fresco, J.R. Formation of the triple-stranded polynucleotide helix, poly(A·U·U). *Proc. Natl. Acad. Sci. USA* **1987**, *84*, 5120–5124.

465. Cheng, A.J.; Van Dyke, M.W. Oligodeoxyribonucleotide length and sequence effects on intermolecular purine-purine-pyrimidine triple-helix formation. *Nucleic Acids Res.* **1994**, *22*, 4742–4747.

466. Johnston, B.H. The S1-sensitive form of d(C-T)$_n$·d(A-G)$_n$: chemical evidence for a three-stranded structure in plasmids. *Science* **1988**, *241*, 1800–1804.

467. Hanvey, J.C.; Klysik, J.; Wells, R.D. Influence of DNA sequence on the formation of non-B right-handed helices in oligopurine·oligopyrimidine inserts in plasmids. *J. Biol. Chem.* **1988**, *263*, 7386–7396.

468. Lyamichev, V.I.; Frank-Kamenetskii, M.D.; Soyfer, V.N. Protection against UV-induced pyrimidine dimerization in DNA by triplex formation. *Nature* **1990**, *344*, 568–570.

469. Ussery, D.W.; Sinden, R.R. Environmental influences on the *in vivo* level of intramolecular triplex DNA in *Escherichia coli*. *Biochemistry* **1993**, *32*, 6206–6213.

470. Lee, J.S.; Burkholder, G.D.; Latimer, L.J.P.; Haug, B.L.; Braun, R.P. A monoclonal antibody to triplex DNA binds to eucaryotic chromosomes. *Nucleic Acids Res.* **1987**, *15*, 1047–1061.

471. Agazie, Y.M.; Lee, J.S.; Burkholder, G.D. Characterization of new monoclonal antibody to triplex DNA and immunofluorescent staining of mammalian chromosomes. *J. Biol. Chem.* **1994**, *269*, 7019–7023.

472. Lee, J.S.; Latimer, L.J.; Haug, B.L.; Pulleyblank, D.E.; Skinner, D.M.; Burkholder, G.D. Triplex DNA in plasmids and chromosomes. *Gene* **1989**, *82*, 191–199.

473. Karlovsky, P.; Pecinka, P.; Vojtiskova, M.; Makaturova, E.; Palecek, E. Protonated triplex DNA in *E. coli* cells as detected by chemical probing. *FEBS Lett.* **1990**, *274*, 39–42.

474. Kohwi, Y.; Malkhosyan, S.R.; Kohwi-Shigematsu, T. Intramolecular dG·dG·dC triplex detected in *Escherichia coli* cells. *J. Mol. Biol.* **1992**, *223*, 817–822.

475. Glaser, R.L.; Thomas, G.H.; Siegfried, E.; Elgin, S.C.; Lis, J.T. Optimal heat-induced expression of the Drosophila *hsp26* gene requires a promoter sequence containing (CT)$_n$·(GA)$_n$ repeats. *J. Mol. Biol.* **1990**, *211*, 751–761.

476. Parniewski, P.; Kwinkowski, M.; Wilk, A.; Klysik, J. Dam methyltransferase sites located within the loop region of the oligopurine-oligopyrimidine sequences capable of forming H-DNA are undermethylated *in vivo*. *Nucleic Acids Res.* **1990**, *18*, 605–611.

477. Jaworski, A.; Blaho, J.A.; Larson, J.E.; Shimizu, M.; Wells, R.D. Tetracycline promoter mutations decrease non-B DNA structural transitions, negative linking differences and deletions in recombinant plasmids in *Escherichia coli*. *J. Mol. Biol.* **1989**, *207*, 513–526.

478. Sarkar, P.S.; Brahmachari, S.K. Intramolecular triplex potential sequence within a gene down regulates its expression *in vivo*. *Nucleic Acids Res.* **1992**, *20*, 5713–5718.

479. Rao, B.S. Pausing of simian virus 40 DNA replication fork movement *in vivo* by (dG-dA)$_n$·(dT-dC)$_n$ tracts. *Gene* **1994**, *140*, 233–237.

480. Kinniburgh, A.J.; Firulli, A.B.; Kolluri, R. DNA triplexes and regulation of the c-*myc* gene. *Gene* **1994**, *149*, 93–100.

481. Droge, P. Transcription-driven site-specific DNA recombination *in vitro*. *Proc. Natl. Acad. Sci. USA* **1993**, *90*, 2759–2763.

482. Bowater, R.P.; Chen, D.; Lilley, D.M.J. Elevated unconstrained supercoiling of plasmid DNA generated by transcription and translation of the tetracycline resistance gene in eubacteria. *Biochemistry* **1994**, *33*, 9266–9275.

483. van Holde, K.; Zlatanova, J. Unusual DNA structures, chromatin and transcription. *Bioessays* **1994**, *16*, 59–68.

484. Lee, J.S.; Woodsworth, M.L.; Latimer, J.P.; Morgan, A.R. Poly(pyrimidine)·poly(purine) synthetic DNAs containing 5-methylcytosine form stable triplexes at neutral pH. *Nucleic Acids Res.* **1984**, *12*, 6603–6614.

485. Gee, J.E.; Blume, S.; Snyder, R.C.; Ray, R.; Miller, D.M. Triplex formation prevents Sp1 binding to the dihydrofolate reductase promoter. *J. Biol. Chem.* **1992**, *267*, 11163–11167.

486. Postel, E.H.; Berberich, S.J.; Flint, S.J.; Ferrone, C.A. Human c-*myc* transcription factor PuF identified as nm23-H2 nucleoside diphosphate kinase, a candidate suppressor of tumor metastasis. *Science* **1993**, *261*, 478–480.

487. Kolluri, R.; Torey, T.A.; Kinniburgh, A.J. A CT promoter element binding protein: definition of a double-strand and a novel single-strand DNA-binding motif. *Nucleic Acids Res.* **1992**, *20*, 111–116.

488. Yee, H.A.; Wong, A.K.C.; van de Sande, J.H.; Rattner, J.B. Identification of novel single-stranded $d(TC)_n$ binding proteins in several mammalian species. *Nucleic Acids Res.* **1991**, *19*, 949–953.

489. Muraiso, T.; Nomoto, S.; Yamazaki, H.; Mishima, Y.; Kominami, R. A single-stranded DNA binding protein from mouse tumor cells specifically recognizes the C-rich strand of the $(AGG:CCT)_n$ repeats that can alter DNA conformation. *Nucleic Acids Res.* **1992**, *20*, 6631–6635.

490. Goller, M.; Funke, B.; Gehe-Becker, C.; Kroger, B.; Lottspeich, F.; Horak, I. Murine protein which binds preferentially to oligo-C-rich single-stranded nucleic acids. *Nucleic Acids Res.* **1994**, *22*, 1885–1889.

491. Aharoni, A.; Baran, N.; Manor, H. Characterization of a multisubunit human protein which selectively binds single stranded $d(GA)_n$ and $d(TC)_n$ sequence repeats in DNA. *Nucleic Acids Res.* **1993**, *21*, 5221–5228.

492. Hollingworth, M.A.; Closken, C.; Harris, A.; McDonald, C.D.; Pahwa, G.S.; Maher, L.J., III. A nuclear factor that binds purine-rich, single-stranded oligonucleotides derived from S1-sensitive elements upstream of the CFTR gene and the MUC1 gene. *Nucleic Acids Res.* **1994**, *22*, 1138–1146.

493. Kiyama, R.; Camerini-Otero, R.D. A triplex DNA-binding protein from human cells: Purification and characterization. *Proc. Natl. Acad. Sci. USA* **1991**, *88*, 10450–10454.

494. Chen, A.; Reyes, A.; Akeson, R. A homopurine:homopyrimidine sequence derived from the rat neuronal cell adhesion molecule-encoding gene alters expression in transient transfections. *Gene* **1993**, *128*, 211–218.

495. Raghu, G.; Tevosian, S.; Anant, S.; Subramanian, K.N.; George, D.L.; Mirkin, S.M. Transcriptional activity of the homopurine-homopyrimidine repeat of the c-Ki-*ras* promoter is independent of its H-forming potential. *Nucleic Acids Res.* **1994**, *22*, 3271–3279.

496. Bucher, P.; Yagil, G. Occurrence of oligopurine-oligopyrimidine tracts in eukaryotic and prokaryotic genes. *DNA Seq.* **1991**, *1*, 157–172.

497. Tripathi, J.; Brahmachari, S.K. Distribution of simple repetitive$(TG/CA)_n$ and $(CT/AG)_n$ sequences in human and rodent genomes. *J. Biomol. Struct. Dyn.* **1991**, *9*, 387–397.

498. Usdin, K.; Furano, A.V. Insertion of L1 elements into sites that can form non-B DNA. Interactions of non-B DNA-forming sequences. *J. Biol. Chem.* **1989**, *264*, 20736–20743.

499. Pestov, D.G.; Dayn, A.; Siyanova, E.Y.; George, D.L.; Mirkin, S.M. H-DNA and Z-DNA in the mouse c-Ki-*ras* promoter. *Nucleic Acids Res.* **1991**, *19*, 6527–6532.

500. Nelson, K.L.; Becker, N.A.; Pahwa, G.S.; Hollingworth, M.A.; Maher, L.J., III. Potential for H-DNA in the human MUC1 mucin gene promoter. *J. Biol. Chem.* **1996**, *271*, 18061–18067.

501. Potaman, V.N.; Ussery, D.W.; Sinden, R.R. The formation of a combined H-DNA/open TATA box structure in the promoter sequence of the human Na, K-ATPase $\alpha 2$ gene. *J. Biol. Chem.* **1996**, *271*, 13441–13447.

502. Menzel, R.; Gellert, M. Regulation of the genes for *E. coli* DNA gyrase: homeostatic control of DNA supercoiling. *Cell* **1983**, *34*, 105–113.

503. Borowiec, J.A.; Gralla, J.D. Supercoiling response of the lac ps promoter *in vitro*. *J. Mol. Biol.* **1985**, *184*, 587–598.

504. Brahms, J.G.; Dargouge, O.; Brahms, S.; Ohara, Y.; Vagner, V. Activation and inhibition of transcription by supercoiling. *J. Mol. Biol.* **1985**, *181*, 455–465.

505. Dorman, C.J.; Barr, G.C.; Bhriain, N.N.; Higgins, C.F. DNA supercoiling and the anaerobic and growth phase regulation of tonB gene expression. *J. Bacteriol.* **1988**, *170*, 2816–2826.

506. Parvin, J.D.; Sharp, P.A. DNA topology and a minimal set of basal factors for transcription by RNA polymerase II. *Cell* **1993**, *73*, 533–540.

507. Kato, M.; Shimizu, N. Effect of the potential triplex DNA region on the *in vivo* expression of bacterial β-lactamase gene in superhelical recombinant plasmids. *J. Biochem.* **1992**, *112*, 492–494.

508. Kohwi, Y.; Kohwi-Shigematsu, T. Altered gene expression correlates with DNA structure. *Genes Dev.* **1991**, *5*, 2547–2554.

509. Clark, S.P.; Lewis, C.D.; Felsenfeld, G. Properties of BPG1, a poly(dG)-binding protein from chicken erythrocytes. *Nucleic Acids Res.* **1990**, *18*, 5119–5126.

510. Frank-Kamenetskii, M.D.; Malkov, V.A.; Voloshin, O.N.; Soyfer, V.N. Stabilization of PyPuPu triplexes with bivalent cations. *Nucleic Acids Symp. Ser.* **1991**, *24*, 159–162.

511. Johnson, C.A.; Jinno, Y.; Merlino, G.T. Modulation of epidermal growth factor receptor proto-oncogene transcription by a promoter site sensitive to S1 nuclease. *Mol. Cell. Biol.* **1988**, *8*, 4174–4184.

512. Davis, T.L.; Firulli, A.B.; Kinniburgh, A.J. Ribonucleoprotein and protein factors bind to an H-DNA-forming c-myc DNA element: possible regulators of the c-*myc* gene. *Proc. Natl. Acad. Sci. USA* **1989**, *86*, 9682–9686.

513. Postel, E.H.; Mango, S.E.; Flint, S.J. A nuclease-hypersensitive element of the human c-*myc* promoter interacts with a transcription initiation factor. *Mol. Cell. Biol.* **1989**, *9*, 5123–5133.

514. Mavrothalassitis, G.J.; Watson, D.K.; Papas, T.S. Molecular and functional characterization of the promoter of ETS2, the human c-*ets*-2 gene. *Proc. Natl. Acad. Sci. USA* **1990**, *87*, 1047–1051.

515. Santra, M.; Danielson, K.G.; Iozzo, R.V. Structural and functional characterization of the human decorin gene promoter. A homopurine-homopyrimidine S1 nuclease-sensitive region is involved in transcriptional control. *J. Biol. Chem.* **1994**, *269*, 579–587.

516. Hoffman, E.K.; Trusko, S.P.; Murphy, M.; George, D.L. An S1 nuclease-sensitive homopurine/homopyrimidine domain in the c-Ki-*ras* promoter interacts with a nuclear factor. *Proc. Natl. Acad. Sci. USA* **1990**, *87*, 2705–2709.

517. Lafyatis, R.; Denhez, F.; Williams, T.; Sporn, M.; Roberts, A. Sequence-specific protein binding to and activation of the TGF-β3 promoter through a repeated TCCC motif. *Nucleic Acids Res.* **1991**, *19*, 6419–6425.

518. Chung, Y.T.; Keller, B. Regulatory elements mediating transcription from the *Drosophila melanogaster* actin 5C proximal promoter. *Mol. Cell. Biol.* **1990**, *10*, 206–216.

519. Firulli, A.B.; Maibenco, D.C.; Kinniburgh, A.J. Triplex forming ability of a c-*myc* promoter element predicts promoter strength. *Arch. Biochem. Biophys.* **1994**, *310*, 236–242.

520. Mollegaard, N.E.; Buchardt, O.; Egholm, M.; Nielsen, P.E. Peptide nucleic acid-DNA strand displacement loops as artificial transcription promoters. *Proc. Natl. Acad. Sci. USA* **1994**, *91*, 3892–3895.

521. Daube, S.S.; von Hippel, P.H. Functional transcription elongation complexes from synthetic RNA-DNA bubble duplexes. *Science* **1992**, *258*, 1320–1324.

522. Biggin, M.D.; Tjian, R. Transcription factors that activate the Ultrabithorax promoter in developmentally staged extracts. *Cell* **1988**, *53*, 699

523. Grigoriev, M.; Praseuth, D.; Robin, P.; Hemar, A.; Saison-Behmoaras, T.; Dautry-Varsat, A.; et al. A triple helix-forming oligonucleotide-intercalator conjugate acts as a transcriptional repressor via inhibition of NF kB binding to interleukin-2 receptor α-regulatory sequence. *J. Biol. Chem.* **1992**, *267*, 3389–3395.

524. Spencer, C.A.; Groudine, M. Transcription elongation and eukaryotic gene regulation. *Oncogene* **1990**, *5*, 777–785.

525. Ulrich, M.J.; Gray, W.J.; Ley, T.J. An intramolecular DNA triplex is disrupted by point mutations associated with hereditary persistence of fetal hemoglobin. *J. Biol. Chem.* **1992**, *267*, 18649–18658.

526. Rao, B.S.; Manor, H.; Martin, R.G. Pausing in simian virus 40 DNA replication by a sequence containing $(dG-dA)_{27} \cdot (dT-dC)_{27}$. *Nucleic Acids Res.* **1988**, *16*, 8077–8094.

527. Baran, N.; Lapidot, A.; Manor, H. Formation of DNA triplexes accounts for arrests of DNA synthesis at $d(TC)_n$ and $d(GA)_n$ tracts. *Proc. Natl. Acad. Sci. USA* **1991**, *88*, 507–511.

528. Lapidot, A.; Baran, N.; Manor, H. $(dT-dC)_n$ and $(dG-dA)_n$ tracts arrest single stranded DNA replication *in vitro*. *Nucleic Acids Res.* **1989**, *17*, 883–900.

529. Samadashwily, G.M.; Dayn, A.; Mirkin, S. Suicidal nucleotide sequences for DNA polymerization. *EMBO J.* **1993**, *12*, 4975–4983.

530. Samadashwily, G.M.; Mirkin, S.M. Trapping DNA polymerases using triplex-forming oligodeoxyribonucleotides. *Gene* **1994**, *149*, 127–136.

531. Maine, I.P.; Kodadek, T. Efficient unwinding of triplex DNA by a DNA helicase. *Biochem. Biophys. Res. Commun.* **1994**, *204*, 1119–1124.

532. Kopel, V.; Pozner, A.; Baran, N.; Manor, H. Unwinding of the third strand of a DNA triple helix, a novel activity of the SV40 large T-antigen helicase. *Nucleic Acids Res.* **1996**, *24*, 330–335.

533. Baran, N.; Neer, A.; Manor, H. "Onion skin" replication of integrated polyoma virus DNA and flanking sequences in polyoma-transformed rat cells: termination within a specific cellular DNA segment. *Proc. Natl. Acad. Sci. USA* **1983**, *80*, 105–109.

534. Brinton, B.T.; Caddle, M.S.; Heintz, N.H. Position and orientation-dependent effects of a eukaryotic Z-triplex DNA motif on episomal DNA replication in COS-7 cells. *J. Biol. Chem.* **1991**, *266*, 5153–5161.

535. Birnboim, H.C. Spacing of polypyrimidine regions in mouse DNA as determined by poly(adenylate,guanylate) binding. *J. Mol. Biol.* **1978**, *121*, 541–559.

536. Manor, H.; Rao, B.S.; Martin, R.G. Abundance and degree of dispersion of genomic $d(CA)_n \cdot d(TC)_n$ sequences. *J. Mol. Evol.* **1988**, *27*, 96–101.

537. Lee, J.S.; Morgan, A.R. Novel aspects of the structure of the *Escherichia coli* nucleoid investigated by a rapid sedimentation assay. *Can. J. Biochem.* **1982**, *60*, 952–961.

538. Hampel, K.J.; Burkholder, G.D.; Lee, J.S. Plasmid dimerization mediated by triplex formation between polypyrimidine-polypurine repeats. *Biochemistry* **1993**, *32*, 1072–1077.

539. Blackburn, E.H. Telomeres. *Trends Biol. Sci.* **1991**, *16*, 378–381.

540. Sen, D.; Gilbert, W. Guanine quartet structures. *Methods Enzymol.* **1992**, *211*, 191–199.

541. Veselkov, A.G.; Malkov, V.A.; Frank-Kamenetskii, M.D.; Dobrynin, V.N. Triplex model of chromosome ends. *Nature* **1993**, *364*, 496

542. Radding, C.M. Helical interactions in homologous pairing and strand exchange driven by RecA protein. *J. Biol. Chem.* **1991**, *266*, 5355–5358.

543. Stasiak, A. Three-stranded DNA structure: is this the secret of DNA homologous recognition? *Mol. Microbiol.* **1992**, *6*, 3267–3276.

544. Camerini-Otero, R.D.; Hsieh, P. Parallel DNA triplexes, homologous recombination, and other homology-dependent DNA interactions. *Cell* **1993**, *73*, 217–223.

545. Zhurkin, V.B.; Raghunathan, G.; Ulyanov, N.B.; Camerini-Otero, R.D.; Jernigan, R.L. A parallel DNA triplex as a model for the intermediate in homologous recombination. *J. Mol. Biol.* **1994**, *239*, 181–200.

546. Collier, D.A.; Griffin, J.A.; Wells, R.D. Non-B right-handed DNA conformations of homopurine.homopyrimidine sequences in the murine immunoglobulin C alpha switch region. *J. Biol. Chem.* **1988**, *263*, 7397–7405.

547. Kato, M. Polypyrimidine/polypurine sequence in plasmid DNA enhances formation of dimer molecules in *Escherichia coli*. *Mol. Biol. Rep.* **1993**, *18*, 183–187.

548. Rooney, S.M.; Moore, P.D. Antiparallel, intramolecular triplex DNA stimulates homologous recombination in human cells. *Proc. Natl. Acad. Sci. USA* **1995**, *92*, 2141–2144.

549. Kohwi, Y.; Panchenko, Y. Transcription-dependent recombination induced by triple-helix formation. *Genes Dev.* **1993**, *7*, 1766–1778.

550. Morgan, A.R.; Wells, R.D. Specificity of the three-stranded complex formation between double-stranded DNA and single-stranded RNA containing repeating nucleotide sequences. *J. Mol. Biol.* **1968**, *37*, 63–80.

551. Maher, L.J.I.; Dervan, P.B.; Wold, B. Analysis of promoter-specific repression by triple-helical DNA complexes in a eukaryotic cell-free transcription system. *Biochemistry* **1992**, *31*, 70–81.

552. Grigoriev, M.; Praseuth, D.; Guyesse, A.L.; Robin, P.; Thuong, N.T.; Helene, C.; et al. Inhibition of gene expression by triple helix-directed DNA cross-linking at specific genes. *Proc. Natl. Acad. Sci. USA* **1993**, *90*, 3501–3505.

553. Lu, G.; Ferl, R.J. Site-specific oligodeoxynucleotide binding to maize Adh1 gene promoter represses Adh1-GUS gene expression *in vivo*. *Plant Mol. Biol.* **1992**, *19*, 715–723.

554. Orson, F.M.; Thomas, D.W.; McShan, W.M.; Kessler, D.J.; Hogan, M.E. Oligonucleotide inhibition of IL2Rα mRNA transcription by promoter region collinear triplex formation in lymphocytes. *Nucleic Acids Res.* **1991**, *19*, 3435–3441.

555. Postel, E.H.; Flint, S.J.; Kessler, D.J.; Hogan, M.E. Evidence that a triplex-forming oligonucleotide binds to the c-*myc* promotor in HeLa cells, thereby reducing c-myc mRNA levels. *Proc. Natl. Acad. Sci. USA* **1991**, *88*, 8227–8231.

556. Ing, N.H.; Beekman, J.M.; Kessler, D.J.; Murphy, M.; Jayaraman, K.; Zendegui, J.G.; et al. *In vivo* transcription of a progesterone-responsive gene is specifically inhibited by a triplex-forming oligonucleotide. *Nucleic Acids Res.* **1993**, *21*, 2789–2796.

557. Ojwang, J.; Elbaggari, A.; Marshall, H.B.; Jayaraman, K.; McGrath, M.S.; Rando, R.F. Inhibition of human immunodeficiency virus type 1 activity *in vitro* by oligonucleotides composed entirely of guanosine and thymidine. *J. Acquired Immune Def. Synd.* **1994**, *7*, 560–570.

558. Roy, C. Triple-helix formation interferes with the transcription and hinged DNA structure of the interferon-inducible 6-16 gene promoter. *Eur. J. Biochem.* **1994**, *220*, 493–503.

559. Duval-Valentin, G.; Thuong, N.T.; Helene, C. Specific inhibition of transcription by triple helix-forming oligonucleotides. *Proc. Natl. Acad. Sci. USA* **1992**, *89*, 504–508.

560. McShan, W.M.; Rossen, R.D.; Laughter, A.H.; Trial, J.; Kessler, D.J.; Zendegui, J.G.; et al. Inhibition of transcription of HIV-1 in infected human cells by oligodeoxynucleotides designed to form DNA triple helices. *J. Biol. Chem.* **1992**, *267*, 5712–5721.

561. Wang, Z.Y.; Lin, X.H.; Nobuyoshi, M.; Qui, Q.Q.; Deuel, T.F. Binding of single-stranded oligonucleotides to a non-B-form DNA structure results in loss of promoter activity of the platelet-derived growth factor A-chain gene. *J. Biol. Chem.* **1992**, *267*, 13669–13674.

562. Tu, G.C.; Cao, Q.N.; Israel, Y. Inhibition of gene expression by triple helix formation in hepatoma cells. *J. Biol. Chem.* **1995**, *270*, 28402–28407.

563. Kovacs, A.; Kandala, J.C.; Weber, K.T.; Guntaka, R.V. Triple helix-forming oligonucleotide corresponding to the polypyrimidine sequence in the rat α1(I) collagen promoter specifically inhibits factor binding and transcription. *J. Biol. Chem.* **1996**, *271*, 1805–1812.

564. Kochetkova, M.; Shannon, M.F. DNA triplex formation selectively inhibits granulocyte-macrophage colony-stimulating factor gene expression in human T cells. *J. Biol. Chem.* **1996**, *271*, 14438–14444.

565. Xodo, L.E.; Alunni-Fabbroni, M.; Manzini, G.; Quadrifoglio, F. Pyrimidine phosphorothioate oligonucleotides form triple-stranded helices and promote transcription inhibition. *Nucleic Acids Res.* **1994**, *22*, 3322–3330.

566. Akhtar, S.; Kole, R.; Juliano, R.L. Stability of antisense oligonucleotide analogs in cellular extracts and sera. *Life Sci.* **1991**, *49*, 1793–1801.

567. Crooke, S.T. The future of sequence-specific transcriptional inhibition. *Cancer Invest.* **1996**, *14*, 89–90.

568. Lacroix, L.; Mergny, J.L.; Leroy, J.L.; Helene, C. Inability of RNA to form the i-motif: implications for triplex formation. *Biochemistry* **1996**, *35*, 8715–8722.

569. Olivas, W.M.; Maher, L.J., III. Competitive triplex/quadruplex equilibria involving guanine-rich oligonucleotides. *Biochemistry* **1995**, *34*, 278–284.

570. Fedoseyeva, E.V.; Li, Y.; Huey, B.; Tam, S.; Hunt, C.A.; Benichou, G.; et al. Inhibition of inter-feron-gamma-mediated immune functions by oligonucleotides. Suppression of human T cell proliferation by downregulation of IFN-gamma-induced ICAM-1 and Fc-receptor on accessory cells. *Transplantation* **1994**, *7*, 606–612.

571. Ramanathan, M.; Lantz, M.; MacGregor, R.D.; Garovoy, M.R.; Hunt, C.A. Characterization of the oligodeoxynucleotide-mediated inhibition of interferon-gamma-induced major histocom-patibility complex Class I and intercellular adhesion molecule-1. *J. Biol. Chem.* **1994**, *269*, 24564–24574.

572. Gewirtz, A.M.; Stein, C.A.; Glazer, P.M. Facilitating oligonucleotide delivery: helping anti-sense deliver on its promise. *Proc. Natl. Acad. Sci. USA* **1996**, *93*, 3161–3163.

573. Leonetti, J.P.; Machy, P.; Degols, G.; Lebleu, B.; Leserman, L. Antibody-targeted liposomes containing oligodeoxyribonucleotides complementary to viral RNA selectively inhibit viral replication. *Proc. Natl. Acad. Sci. USA* **1990**, *87*, 2448–2451.

574. Bennett, C.F.; Chiang, M.Y.; Chan, H.; Shoemaker, J.E.; Mirabelli, C.K. Cationic lipids enhance cellular uptake and activity of phosphorothioate antisense oligonucleotides. *Mol. Phar-macol.* **1992**, *41*, 1023–1033.

575. Lewis, J.G.; Lin, K.Y.; Kothavale, A.; Flanagan, W.M.; Matteucci, M.D.; DePrince, R.B.; et al. A serum-resistant cytofectin for cellular delivery of antisense oligodeoxynucleotides and plas-mid DNA. *Proc. Natl. Acad. Sci. USA* **1996**, *93*, 3176–3181.

576. Loke, S.L.; Stein, C.A.; Zhang, X.A.; Mori, K.; Nakanishi, M.; Subasinghe, C.; et al. Charac-terization of oligonucleotide transport into living cells. *Proc. Natl. Acad. Sci. USA* **1989**, *86*, 3874–3878.

577. Clarenc, J.P.; Degols, G.; Leonetti, J.P.; Michaud, P.; Lebleu, B. Delivery of antisense oligonu-cleotides by poly(L-lysine) conjugation and liposome encapsulation. *Anticancer Drug Des.* **1993**, *8*, 81–94.

578. Hentschel, C.C. Homocopolymer sequences in the spacer of a sea urchin histone gene repeat are sensitive to S1 nuclease. *Nature* **1982**, *295*, 714–716.

579. Mace, H.A.F.; Pelham, H.R.B.; Travers, A.A. Association of an S1 nuclease-sensitive structure with short direct repeats 5' of *Drosophila* heat shock genes. *Nature* **1983**, *304*, 555–557.

580. McKeon, C.; Schmidt, A.; de Crombrugghe, B. A sequence conserved in both the chicken and mouse α2(I) collagen promoter contains sites sensitive to S1 nucleases. *J. Biol. Chem.* **1984**, *259*, 6636–6640.

581. Pearson, C.E.; Sinden, R.R. Alternative DNA structures within the trinucleotide repeats of the myotonic dystrophy and fragile X locus. *Biochemistry* **1996**, *35*, 5041–5053.

582. Streisinger, G.; Okada, Y.; Emrich, J.; Newton, J.; Tsugita, A.; Terzaghi, E.; et al. Frameshift mutations and the genetic code. *Cold Spring Harb. Symp. Quant. Biol.* **1966**, *31*, 77–84.

583. Sheflin, L.G.; Kowalski, D. Mung bean nuclease cleavage of dA + dT-rich sequence or an inverted repeat sequence in supercoiled PM2 DNA depends on ionic environment. *Nucleic Acids Res.* **1984**, *12*, 7087–7104.

584. van de Sande, J.H.; Ramsing, N.B.; Germann, M.W.; Elhorst, W.; Kalisch, B.W.; von Kitzing, E.; et al. Parallel stranded DNA. *Science* **1988**, *241*, 551–557.

585. Jovin, T.M.; Rippe, K.; Ramsing, N.B.; Klement, R.; Elhorst, W.; Vojtiskova, M. Parallel stranded DNA. In: *Structure and Methods* (Sarma, M.H.; Sarma, R.H.; eds.). Adenine Press, Albany, NY, 1990, pp. 155–174.

586. Sen, D.; Gilbert, W. Formation of parallel four-stranded complexes by guanine-rich motifs in DNA and its implications for meiosis. *Nature* **1988**, *334*, 364–366.

587. Hardin, C.C.; Henderson, E.; Watson, T.; Prosser, J.K. Monovalent cation induced structural transitions in telomeric DNAs: G-DNA folding intermediates. *Biochemistry* **1991**, *30*, 4460–4472.

588. Williamson, J.R.; Raghuraman, M.K.; Cech, T.R. Monovalent cation-induced structure of telo-meric DNA: the G-quartet model. *Cell* **1989**, *59*, 871–880.

589. Kang, C.K.; Zhang, X.; Ratliff, R.; Moyzis, R.; Rich, A. Crystal structure of four-stranded *Oxytricha* telomeric DNA. *Nature* **1992**, *356*, 126–131.

590. Smith, F.W.; Feigon, J. Quadruplex structure of *Oxtricha* telomeric oligonucleotides. *Nature* **1992**, *356*, 164–168.

591. Sen, D.; Gilbert, W. A sodium-potassium switch in the formation of four-stranded G4-DNA. *Nature* **1990**, *344*, 410–414.

592. Scaria, P.V.; Shire, S.J.; Shafer, R.H. Quadruplex structure of $d(G_3T_4G_3)$ stabilized by K^+ or Na^+ is an asymmetric hairpin dimer. *Proc. Natl. Acad. Sci. USA* **1992**, *89*, 10336–10340.

593. Glover, J.N.; Pulleyblank, D.E. Protonated polypurine/polypyrimidine DNA tracts that appear to lack the single-stranded pyrimidine loop predicted by the "H" model. *J. Mol. Biol.* **1990**, *215*, 653–663.

594. Shimizu, M.; Hanvey, J.C.; Wells, R.D. Multiple non-B-DNA conformations of polypurine·polypyrimidine sequences in plasmids. *Biochemistry* **1990**, *29*, 4704–4713.

595. Johnson, D.; Morgan, A.R. Unique structures formed by pyrimidine-purine DNAs which may be four-stranded. *Proc. Natl. Acad. Sci. USA* **1978**, *75*, 1637–1641.

596. Ramsing, N.B.; Jovin, T.M. Parallel stranded duplex DNA. *Nucleic Acids Res.* **1988**, *16*, 6659–676.

INTRON-EXON STRUCTURES:
FROM MOLECULAR TO POPULATION BIOLOGY

Manyuan Long and Sandro J. de Souza

Advances in Genome Biology
Volume 5A, pages 143-178.
Copyright © 1998 by JAI Press Inc.
All rights of reproduction in any form reserved.
ISBN: 0-7623-0079-5

I. INTRODUCTION

The discovery of introns in the 1970s[1,2] marked the end of the traditional view that there was a colinearity between proteins and genes. The coding regions of most eukaryotic genes are not continuous, but are interrupted by introns that are spliced from primary RNA transcripts before the formation of mature RNA. The discovery of introns also immediately raised more questions: What is the molecular process of RNA splicing? What intron-exon structures do various eukaryotic organisms have? How did introns originate and what roles do introns play in the evolution of genomes? Are introns just a pile of useless DNA junk in eukaryotic genes?

The significant developments in the field over the past 10 years are essentially the progress made in molecular techniques and the creation of enormous data bases of DNAs and proteins. Today, we have a more sophisticated arsenal of molecular techniques by which we can easily check the intron-exon structure of a gene. We have a data set currently containing more than 20,000 intron-exon structures, which is rapidly expanding as a consequence of genome projects. We have computers connected to the internet in many molecular biological laboratories, so once the sequence of a gene is determined, we can compare it with hundreds of thousands of genes from the data base maintained by NCBI. There is no doubt that today we have a better understanding of some features of the intron-exon structures, such as the distribution of introns, but final answers to some questions, for example, the origin of introns, are still lacking and the issues have become more challenging than ever.

Until now, the study of intron-exon structure has been comparative in terms of methodology. One could use structure to infer whether two genes in question belong to the same gene family; one could also compare a group of exon-intron sequences to derive common biological rules; one could even find evidence of the ancestral relationship of two introns in distantly related organisms. Mount[3] was among the first to investigate the splicing signal from a comparison of a group of intron sequences. As more sequence data accumulated, more statistical approaches were used to address various problems pertinent to the intron-exon structure. Dorit, Schoenbach and Gilbert,[4] Fichant,[5] and Fedorov et al.[6] represent some more recent examples. However, the systematic description of the intron-exon structures across various organisms has perhaps just begun, and much work remains to be done.

Current gene structures are the products of evolutionary processes. To understand the current states of intron-exon structures, it will be necessary to examine the process itself. In fact, our understanding of intron-exon organization has been associated with investigations of the origin and evolution of these structures. Soon after the intron was discovered, speculation as to its origin and significance in evolution began[7] and eventually developed into an active field of molecular evolution (e.g., a query of *Biological Abstracts* for 1995 only using the search words "intron and evolution" found more than 100 research articles). Although significant findings have been made in the field of intron-exon evolution, there is disagreement among researchers concerning the origin of introns.[8]

This chapter will present an overview of the major progress made in the past, especially in the last 5 years. First we will describe general issues of the topic, how intron-exon structures are organized in eukaryotic genes, and how introns are spliced out. Then we will discuss whether introns have any biological function, a question of paramount interest not only to the cognoscenti of the field, but to newcomers as well. We will also describe how intron-exon structures change via movement of exons and introns. Thereafter, we will address the controversial issue of the origin of introns. Finally, we will discuss various studies concerning the evolution of intron-exon structures, covering the topic from the exon universe, the evolutionary relationship among three major types of introns, to molecular population genetics that pinpointed the forces that control the evolution of intron sequences.

II. CLASSIFICATION AND STRUCTURAL FEATURES

In principle, any intervening sequence that is spliced out at the RNA level can be classified as an intron. In the last decade, it became clear that introns fall into several groups. We can use structural as well as functional features to classify an intron into a specific group. On the basis of functional features, it is possible to make a distinction between self-splicing and spliceosomal introns. On the other hand, we can clearly see structural differences between group I and group II introns. In this section we will give a brief description of each major group of introns, their structures, and how they are excised from RNA to produce a mature message.

A. Spliceosomal Introns

We do not intend to give a detailed description of spliceosomal introns and how they are excised from mRNA precursors. For a detailed view, we suggest a number of reviews (refs. 9–11), which together cover all of the basic aspects involving spliceosomal introns.

Spliceosomal introns have been characterized in almost all eukaryotes, with the exception of a few early protists. A major feature is their dependence for splicing on a ribonucleoprotein complex called the spliceosome. Spliceosome structure seems to be very conserved in eukaryotes, being very well described in fungi and humans, for example.

Introns in mRNA precursors are believed to be excised via a two-step mechanism. The first step comprises a cleavage at the 5' splice site immediately followed by the joining of the 5' end of the intron and an internal adenosine through a 2',5'-phosphodiester bond. The region where the adenosine is located is called the branch site. The intron forms a lariat structure that is similar to that observed in the splicing process of group II introns, thus supporting the argument that spliceosomal introns and group II introns may have evolved from a common ancestor. The second step involves a nucleophilic attack of the 3' end of the upstream exon toward the 3' splice site, which leads to the joining of the two exons and the release of the intron. Although several details of this complex two-step splicing process are known, the crucial question of whether the catalytic activity of the spliceosome comes from RNA or protein remains unanswered. Most researchers in this field believe that RNA is the catalytic agent in the process.

It has recently been proposed that there are two different active sites in the spliceosome, one directed to the first transesterification reaction, and the second one responsible for the last step of the splicing reaction.[9] The strongest evidence for this proposal came from chemical and stereochemical studies by Moore and Sharp,[12] who showed that the two reactions are inhibited by the same phosphorothioate diastereomer (the splicing substrate has a chiral phosphorothioate), which would not be expected if both steps were opposite reactions catalyzed by the same catalytic center. Moreover, mutations affecting RNA components of the spliceosome have different effects in each transesterification reaction.[13,14]

How is the spliceosome assembled? It is quite clear that the spliceosome assembles in situ following a specific pathway, involving at least six snRNPs. The first step of the pathway is the formation of a commitment complex, which involves the binding of U1 snRNP (small nuclear ribonucleoprotein particle). The term "commitment" is applied because mRNA precursors are processed preferentially by spliceosomes in vitro if they are preincubated with yeast extract.[15] U1 snRNP binds to the 5' splice site, and this binding is dependent on a base-pairing interaction between both RNAs. Additional U1 snRNP binding to the branch site may also be involved in the formation of the commitment complex. The next step involves the binding of U2 snRNP to form complex A, a process that is ATP-dependent (in contrast to the formation of the commitment complex). The sequence in the intron involved in the U2 snRNP binding is the branch site, especially an adjacent polypirimidine stretch. A growing body of evidence has implicated several proteins in the formation of complex A. Interestingly, most of them, at least in mammals, contain a serine- and arginine-rich structural domain[16] that is probably involved in RNA-binding activity.

Subsequent to the formation of complex A, complex B1 is formed by the addition of three different snRNPs. It seems that complex B1 is formed by the joining of complex A with a complex containing U4/U6, U5 snRNPs, and several proteins. U4/U6 snRNPs are tightly bound when they enter the spliceosome, but this binding is weakened and probably disrupted after entry. This rearrangement characterizes stage B2 in the spliceosome complex. An additional modification is the binding of U6 snRNP to U2 snRNP. Complex C1 is characterized by the formation of the lariat structure. Complex C2 is formed after the two flanking exons are joined, which are then released from the spliceosome. The excised intron is then degraded, and the spliceosome components are probably recycled, since they have a long half-life.

Is there any structural feature in the splice site that drives the assembly of the spliceosome? It was generally believed that the sequence AG...GT present in both 5' and 3' splice sites in exons was a recognition signal for splicing. Statistical analysis using a larger set of genes has indicated, however, that the exonic part of this putative signal is not well conserved.[17] Highly significant conservation exists for the intronic sequences, which usually use GT...AG in intron-exon junctions as a splicing signal. Moreover, the recognition site must be more complex, involving more than one site, because unspliced AGGT sequences exist within large introns. Alternatively, Robberson et al.[18] suggested that the assembly sites reside in the exons.

Recently it was found that a minor class of nuclear introns lack GT...AG consensus sequences and use AT...AC signals for splicing.[19–21] This class of introns represents less than 0.1% of known eukaryotic nuclear introns.[22] The examples for the intron of this kind include introns in the *CMP* gene encoding cartilage matrix protein in human and chicken, the *Rep-3* gene encoding mismatch repair protein, the *P120* gene for nucleolar protein in mammal, and the *prospero* gene of *Drosophila* for homeobox protein.[23] These introns are spliced by a novel form of spliceosome that contains the minor snRNPs U11 and U12.[20,21]

Trans-Splicing

The process that we just described occurs in most of the mRNA precursors in most organisms. In some cases, however, a different kind of splicing leads to the joining of exons from two different mRNA precursors. This process has been called *trans*-splicing, as opposed to the classical *cis*-splicing reaction. The most well characterized example of *trans*-splicing occurs in trypanosomatides, where a splice leader RNA is linked to the 5' end of all mRNA precursors.[24] *Trans*-splicing is also observed in some other species, such as the nematodes *C. elegans* and *A. lumbricoides*,[25] in plant organelles,[26] and in mammalian cells in culture.[27]

Comparative studies between *trans*- and *cis*-splicing have shown that the two processes are identical in terms of catalytic activity. This is important because *trans*-splicing reactions do not seem to be dependent on U1 and U5 snRNAs, suggesting that these snRNAs are not directly involved with catalysis in the splicing reaction.

In vitro models of *trans*-splicing have been established.[28,29] Findings obtained from these models suggest that *trans*-splicing is a slower and less efficient process than *cis*-splicing. Establishment of *in vitro* models like this will certainly become a rich source of information concerning the mechanism of *trans*-splicing.

Alternative Splicing

Alternative splicing produces several mRNA messages from a single gene. It can affect gene expression in a significant way. Detailed reviews can be found in this topic.[30,31] A key question in this field is how the selection of the specific 5' and 3' splice sites occur. The most interesting model of alternative splicing is sex determination in *Drosophila*.[32] The sex determination genes, sex-lethal (*sxl*), transformer (*tra*) and doublesex (*dsx*), are regulated by specific splicing pathways. *sxl* and *tra* produce functional proteins only in females, and *dsx* produces different proteins in each sex. In the case of *tra* mRNA, a default splicing pathway occurs in both sexes. In females, however, *sxl* proteins bind to *tra* pre-mRNA, inhibiting the default pathway and inducing the expression of its own splicing pathway. Finally, female-specific splicing of *dsx* is promoted by tra and tra-2 proteins.

B.   Group I Introns

The discovery made by Cech and his colleagues that a group I intron was able to promote a self-splicing reaction has had a great impact in several areas of research in chemistry and biology. [33,34] The self-splicing activity of group I introns is due to a series of transesterification reactions, the first reaction being initiated by a free guanosine, which promotes an attack in the 5' end of the group I intron. The second transesterification reaction occurs when the 3'-OH of the upstream exon attacks the 3' end of the intron joining the flanking exons, releasing the intron, which after an additional transesterification reaction can then be cyclized.

Catalytic activity is a consequence of the secondary structure of the intron. This secondary structure consists of several pairing elements, designated P1 to P10. Also important in the context of secondary structure are the internal guide sequences (IGSs) that pair with exon sequences flanking the 3' and 5' ends of the group I intron. On the basis of structural as well as sequence features, group I intron can be classified in different subgroups: IA, IB, IC, and ID. One of the most impressive features of the group I introns is the variety of reactions they can catalyze. Those include cleavage of RNA and DNA, RNA polymerization, RNA ligation, and amino ester cleavage, among others.[34] This broad catalytic power reinforces the notion that group I introns and ribozymes in general had an important role in early phases of gene evolution, serving as both the store of genetic information and the target of natural selection. Several proteins have been shown to assist some group I introns in the process of splicing.[35,36] Since the catalysis is still due to RNA, these proteins probably stabilize the core structure of the intron, as has been shown for a tyrosyl-tRNA synthetase.[37]

Group I introns are widely distributed. They are present in eubacteria, eukaryotes, bacteriophages and several organelles. They can interrupt rRNAs as well as protein-encoding genes. It is accepted that their broad distribution is a product of their mobility. Group I introns can jump into another locus, a process named transposition, or into an intronless allele, a process called homing. Intron-encoded proteins seem to be essential in these processes. Group I intron can contain an open reading frame (ORF) that in most cases codes for a maturase. Maturases in group I introns have a specific LAGLI-DADG motif, which is also found in a family of endonucleases. Interestingly, some of these endonucleases, not encoded within an intron, are mobile elements. This suggests that the mobility of group I introns is due entirely to the presence of the ORF, as has already been shown.[38] This leads to a hypothesis suggesting that the ORF is an independent genetic element that invaded introns during evolution. Recent findings seem to support this idea.[34]

C. Group II Introns

Group II introns interrupting tRNAs, rRNAs, and protein-encoding genes are found in bacterial and organelle genomes. Eubacterial species that contain group II introns are either cyanobacterias or proteobacterias, believed to be the ancestors of chloroplasts and mitochondria, respectively.

Group II introns have ribozyme activity, but only *in vitro* and under special and nonphysiological conditions. *In vivo*, splicing activity is only achieved with the help of some proteins. The splicing of group II introns shares some similarities with the splicing of nuclear introns, which led to the suggestion that the group II introns share a common ancestor with nuclear introns. The reaction starts with the formation of a lariat, the result of a 2'-5' phosphodiester bond between the 5' end of the intron and a nucleotide, usually A, in the middle of the intron. The transesterification reaction then occurs, joining the 3' end of the upstream exon with the 5' end of the downstream exon. Like group I introns, the different classes of group II share a common secondary structure and contain an ORF that encodes maturases that are involved in the mobility activity of those introns, including homing. Unlike group I introns, homing in group II introns is RNA-mediated and depends of a reverse transcriptase activity of the encoded maturase. Recently the homing process of yeast mitochondrial group II intron al2 was dissected by Zimmerly et al.[40] Surprisingly, they found that the homing process is mediated by a ribonucleoprotein complex containing both the al2 protein and al2 RNA. In the process, both components show endonuclease activity, the RNA cutting one DNA strand and the protein cutting the remaining one. More recently it was shown that the same process was also responsible for the transposition of this group II intron.[41] This was the first report showing a complete reverse splicing reaction for a group

Table 1. Comparison of Three Intron Groups

	Group I Introns	*Group II Introns*	*Nuclear Introns*
Distribution	Eubacteria, eukaryotes, bacteriophage, and organelles	Organelles, eubacteria	Eukaryotes
Splicing *in vivo*	Self	Protein-independent	Protein-independent
Lariat formation	No	Yes	Yes
Ribozyme activity	Yes	Yes	No
Maturase	Yes	Yes	No

II intron. These and other observations led to the suggestion that group II introns are related to retrotransposons.

D.   Group III and Twintrons

The chloroplast in *Euglena gracilis* has a very intriguing genome. The most amazing feature is the presence of more than 150 introns that make up more than a third of the genome. Most of the introns are group II introns, which are, however, slightly different from the group II introns of other organisms.[42] *Euglena* also contains a new type of intron, the group III intron, which is very small and shares some similarities with group II, such as the presence of a consensus 5' splicing site. It differs from group II in that the famous domain V is absent from group III.[43] It seems that at least some group III introns can be excised as a lariat.

Euglena chloroplasts also have introns within introns (twintrons). The formation of twintrons can involve both group II and group III introns, sometimes in the same twintron. The number of nested introns is usually two, but more complex twintrons can be found. A remarkable example is the twintron in the rps 18 gene that contains four group III introns.[44] Usually a twintron is formed by an intron insertion event occurring within a preexisting intron. This insertion normally takes place in a functional domain of the external intron. How are twintrons excised? First the internal intron splices out, followed by the splicing of the external intron.

Recently a huge twintron containing several alternative splicing pathways was described by Hong and Hallick.[42] It is fascinating that in one of the splicing pathways, a group III intron is formed from pieces of two group II introns, which strongly suggests that group III arose from group II introns. In addition, this finding also led to the proposal of a model for the evolution of nuclear introns from group II introns.

A comparison of the major properties of three groups of introns is presented in Table 1.

III.   FEATURES OF THE INTRON-EXON STRUCTURE

What is the general picture of intron-exon structures in eukaryotic genomes? Since the first intron-exons structures reported in the early 1980s, it has been known that intron-exon structures vary from gene to gene, from species to species, and in number and length. An early description of the structure can be found in Hawkins[45] and Dorit et al.[4] Recent investigations motivated by the search of the origin of introns made it possible to characterize more completely the distribution of introns and exons across genes and species.[46]

A.   Exon Length

Long et al.[46] presented a distribution of exon length in an exon data base with 1925 independent genes and 13,042 exons derived from a primary sequence data base, GenBank release 84. Although exon sizes range from a few amino acid residues to more than 3000 residues, most exons are distributed around a peak of 35–40 amino acids, as described in an early study.[4] The longest exons known to date include exon 10 in the *unc-22* gene for twitchin in *Caenorhabditis elegans,* which has 3205 amino acid residues[47] and the central exon of human mucin gene, which encodes a 3570-amino-acid-long peptide.[177]

B.   Intron Length

The length of introns is much more variable than that of exons. Unfortunately, a complete distribution of intron lengths has not yet been reported. From the available literature, we know that the smallest intron could be smaller than 10 bp, most of which were thought to be derived from sequencing errors.

There may be a limit for the minimum size of introns because of the splicing requirement.[48–50] The actual limit may be species-specific and gene-specific. For example, for chicken type II procollagen gene, 80 bp may be the minimum; but for many genes in *Paramecium tetraurelia,* a ciliated protozoan, 20- to 33-nucleotide-long introns are common.[51] The introns in nucleomorph protein genes of the endosymbiont in chlorarachniophyte algae are even smaller. Twelve introns in SnRNP E, *clpP, S4,* and *S13* ribosomal protein genes in this small nucleomorph genome of 380 kb are 18–20 nucleotides long.[52]

The largest introns were reported in dystrophin genes, whose 79 exons of which are scattered over a genomic region of 2.4 Mb.[53] Eight introns in this gene are more than 100 kb long, the largest of which is 400 kb long.[54] A difficulty in studying intron size distribution is that many large introns reported are not completely sequenced.

The statistical distribution of intron size has been studied in a number of species. Mount et al.[49] surveyed intron size distribution in *Drosophila* using a data set that contains 209 completely sequenced introns. They found that the smallest introns are 51 nucleotides, more than half of the introns are shorter than 80 nucleotides in length, and the size for most of these introns ranges from 59–67 nucleotides, with a median length of 79. The *C. elegans* introns have a tighter distribution and a smaller median size.[55] The distribution of human introns (Long, unpublished data) has a higher median than that of *Drosophila.* Hughes and Hughes[56] reported

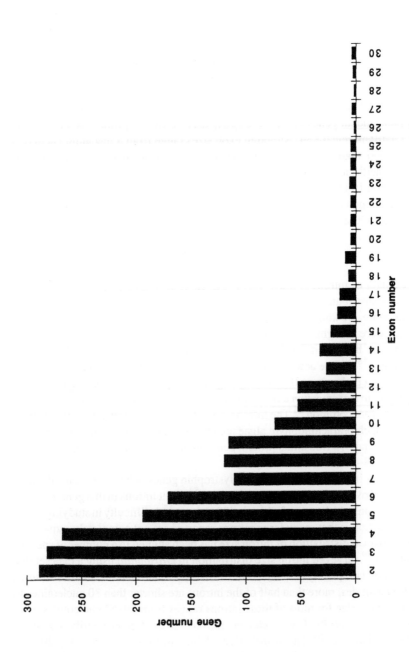

Figure 1. The distribution of exon number per gene in the exon data base, which was developed from Gen Bank release 84 (1994).[46]

a small intron size in birds that have small genomes. These results showed that intron size appears to be correlated with genome size.

C. Intron-Exon Number per Gene

The number of exons (introns) per gene in eukaryotic genomes has a smooth distribution of reverse J shape, as our analysis of the exon data base shows (Figure 1). Forty-four percent of genes in the data base consist of two to four exons (i.e., one to three introns) and 36% have five to nine exons (or four to eight introns). The remaining 20% of genes contain more than 10 exons. The largest number is 94 exons, in the human perlecan gene (HSPG2).[57]

D. Variation of Intron Number among Organisms

Palmer and Logsdon[58] made perhaps the first large-scale investigation of the distribution of intron number. They observed that intron numbers are different among various organisms. Plants, animals, and fungi have two to six introns per kilobase of coding region, whereas protists have fewer or even no introns. It appears that the distribution of introns is phylogenetically specific. Although recent additions to the data base showing some new introns in protists have modified this picture to some extent,[59] it appears to be true that introns are unevenly distributed in different eukaryotic organisms.

E. Conservation of Intron-Exon Structures

In general, intron positions evolve more slowly than DNA sequences. This is because any change that affects the splicing site would possibly bring deleterious effects to the gene, either interrupting splicing or causing a shift in the reading frame. Brown et al.[60] used this property of intron-exon structure to develop a computing method to classify gene families by comparing the similarity of positions and phases of introns, which is dubbed "exon fingerprints."

IV. ARE INTRONS FUNCTIONAL?

Soon after intron splicing was discovered, many researchers wondered whether introns have any functions. In many eukaryotic genes, intron sequences are much larger than exon sequences. Because a large portion of genes consist of introns, one can easily speculate that these introns perhaps carry out some functions, including (1) regulation of splicing functions; (2) regulation of transcription; (3) evolutionary function[61]; (4) coding capacity (more strictly, this is beyond the defined question of intron function, since this is a question of gene structures). Indeed, although there are some cases among the hundreds of thousands of introns

for classes 2 and 4, there are no clear examples for classes 1 and 3. We will summarize available examples of functional introns and discuss their significance in a
general perspective.

A. Examples of Functional Introns

Group I Introns as Ribozyme

It is clear that this class of introns has some functions, the discovery of which
led to the revolutionary concept of a nonproteic enzyme[33] and to a solution to the
chicken-egg paradox of the origin of life. Initially it was found that group I introns
in *Tetrahymena* can splice themselves from RNA. Later, the intron was found to
have nucleotidyltransferase, RNA ligase, and phosphatase activities.[34] However,
even though Schaefer et al.[62] observed that a group I intron in fission yeast
encodes an endonuclease, most intron activity seems to be related to the splicing
and mobility activity of the intron itself.

Two comments perhaps are necessary when we discuss the obvious function of
the group I intron. First, the distribution of the intron is sporadic in both prokaryotes and eukaryotes.[63,64] Second, group I introns appeared in the very early stage
of life, perhaps the beginning of the genetic systems. Its functional features are
perhaps only related to its early origin; and as introns evolved, their function
would become weak or completely lost, as the situation of spliceosomal introns
tells us.

Group II Introns Encoding Proteins and snRNAs

Structurally, group II introns have a lariat intermediate similar to nuclear
introns, whereas their internal structures are more specific. These internal structures were believed to have evolved into snRNAs, the fundamental components of
spliceosomes.[65,66] Although the group II introns are self-splicing, an efficient
splicing process is only achieved by the presence RNA folding proteins,[67] which
are encoded in nuclear genes or in the group II introns.

Maturase

This activity was first observed in the first intron of subunit I of cytochrome oxidase in yeast in an analysis of splicing-defective mutants by Carignani et al.,[68]
who found that the maturase activity facilitates the splicing of the intron where the
protein is encoded. Similar intron-encoded proteins have also been found in some
group I introns, including the fourth intron of the cytochrome *b* gene in yeast,[69] the
intron of cytochrome *c* oxidase subunit 3 of the alkane yeast *Yarrowia lipolytica*,[70]
and the intron of the *trnK* gene, which codes for a tRNA-lys in potato.[71]

Reverse Transcriptase

It was observed that the yeast aI1 and aI2 introns and the same group II introns from other organisms encode proteins similar to reverse transcriptase.[72,73] Kennell et al.[74] demonstrated biochemically that the maturase-encoding introns aI1 and aI2 of the yeast *coxI* gene also encode reverse transcriptase-like proteins. They suggest that these group I and II introns are mobile retro-elements, which can copy the intron.

Endonuclease

Goguel et al.[69] analyzed the relationship between the two intron-encoded proteins in yeast, bI4 RNA maturase (encoded by the fourth intron of the cytochrome *b* gene) and aI4 DNA endonuclease (encoded by the fourth intron of the cytochrome oxidase subunit I gene), and proposed that there was an evolutionary switch from DNA endonuclease to RNA maturase.

The enzymatic activities of reverse transcriptase, maturase, and endonuclease are also involved in intron homing,[75] and reverse transcriptase may also play some role in transposition (for example, aI5b intron of yeast *COXI* gene, Mueller et al.[76]; *Podospora* mtDNA intron transposition, Sellem et al.[77]; *Schizosaccharomyces pombe* mtDNA intron, Schmidt et al.[78]) and loss of the group II introns.[79]

Nuclear Pre-mRNA Introns

We have discussed the fact that group I and group II introns have some important biological functions. In contrast to these ancient introns, most nuclear pre-mRNA introns do not appear to have any obvious biochemical function. However, some of them can serve as regulatory elements, harbor unrelated genes,or encode snRNAs.

i) Regulatory elements. There are a number of cases in which regulatory elements, such as promoters, enhancers, and attenuators, are located in introns.

Promoter. The promoter for larval mRNAs in the *Adh* gene of *D. melanogaster* is in the first intron.[80] More examples of intron promoters can be found in the rat phosphodiesterase genes[81] and in the actin gene of *Saccharomyces cerevisiae*.[82]

Enhancer. Enhancers were found in the J-C intron of the immunoglobulin genes,[83] the first intron of the human purine nucleoside phosphorylase-encoding gene,[84] and the second intron of the human apolipoprotein B gene.[85]

Attenuator. Xie et al.[86] found an attenuator sequence in intron 2 of the c-*fms* proto-oncogene, which controls c-*fms* expression in macrophages by down-regulating transcription elongation. This is the first example of the attenuator property of the intron.

Intron transcripts binding to promoters. To regulate gene expression, the RNA sequence of introns can pair with the promoters. One such an example is the human globin genes,[87,88] where a significant sequence similarity was detected between the promoter and introns of the same gene for each of the six genes in the human globin complex. Pairing of the intron transcript with the antisense strand (in the case of ε-globin) was believed to be associated with up-regulation, whereas pairing with the sense strand (in the cases of γ-, δ-, and β-globins) may lead to down-regulation.

ii) Nested genes. In the first intron of *Gart* locus in *Drosophila*, a gene coding three purine pathway enzymes, a pupal cuticle protein gene is located on the opposite strand.[89] A more striking example for nested genes was reported in *C. elegans*.[90] In this organism, it was shown that the five genes are located within the large introns of gene *F10F2.2,* which encodes FGAM synthase (four of them are present within one intron). Like the pupal cuticle gene within the *Gart* gene, all of these genes are encoded on the strand opposite that of F10F2.

iii) Small nuclear RNAs (snRNAs). snRNAs include many classes of RNAs, among which U14-U21, U23-U24, and E3 are encoded by introns of protein-coding genes. Most of these snRNAs are functionally related to translation.[91,92] Some snRNAs (e.g., U20 encoded in intron 11 of the nucleolin gene in mammals), may participate in the formation of the small ribosome.[93] In an extreme case, only introns are functional and the associated exons are not, as was reported recently in the U22 host gene (UHG) in mammalian genomes.[94] In this gene, eight snRNAs, U22 and U25-U31, are encoded within different introns. U22 snRNA plays an essential role for the maturation of 18S rRNA, whereas all other seven snRNAs, U25–U31, have short pieces of nucleotides with extensive complementarity to 18S or 28S ribosome RNAs. These snRNAs may function in the folding of the pre-rRNA for correct processing and ribosomal protein assembly in the nucleus. More surprisingly, the mature RNA after splicing does not seem to encode any protein because of the poor conservation in the exons, short open reading frames, and rapid degradation of the UHG mRNA in cytoplasm. This is the first example in which only introns, rather than exons, carry out some functions.

B.   Junkyards Are Still Full of Junk

Having reviewed the above examples of functional introns, can we paint a new picture of the role of introns in genomes? Yes and no. The analysis of group I and group II introns suggests a vital role of such introns in the early stages of evolution. The available functions for these introns, especially for the group I introns, seem to depict an important role in the early stages of life. However, in eukaryotic genomes, group I and II introns are quantitatively minor compared to the dominant nuclear pre-mRNA introns. Although the examples discussed above clearly show that not all that was originally dismissed as genomic junk is without function, the vast majority of intron-containing nuclear genes, which number in hundreds of thousands, still seem to contain intron sequences of no known value. Therefore, the long-held image in the eye of the molecular biologist, that introns in general are nonfunctional, needs no change, at least as long as it appears that examples described above can be viewed as exceptions or vestiges inherited from the evolutionary history of genomes.

Duplicated Genes

As previously discussed, introns might play some role in gene regulation. There are many gene families that include both members with and without introns. However, expression levels do not appear to differ between the two groups, suggesting that the presence or absence of introns is not important in regulating expression levels. Rather, evolutionary analysis provides more promising clues to the function of introns.

Evolutionary Rates

Li and Graur[95] summarized a large-scale survey of the substitution rates in introns, exons, intergenic regions, and pseudogenes. They found that the substitution rates of introns are very close to those of pseudogenes (a rate of 3×10^{-9}/ nucleotide site/year, which is four times higher than the nonsynonymous rate). This implies that like pseudogenes, these introns are free from functional constraints. Furthermore, insertions and deletions occur frequently within introns. As a general practice, introns are normally taken as selectively neutral and become a calibration for the phylogenetic analysis.

V. MOVEMENTS OF INTRONS AND EXONS

Both exons and introns move within eukaryotic genes. Exon insertion and exon exchange lead to changes in protein structures, which may be subject to natural selection. On the contrary, movements of introns, intron sliding, and intron slippage do not promote radical changes in the protein structure or function and therefore are expected to be evolutionarily neutral or nearly neutral.

A. Exon Movement: Exon Shuffling

Soon after the discovery that eukaryotic genes were discontinuous and interrupted by introns, Gilbert[7] proposed the concept of exon shuffling. According to this theory, introns might represent hot spots for recombination that would promote exon shuffling. As we will discuss later in this review, one strength of this proposal is that it provides a mechanism for the creation of gene diversity. Prior to this argument, the creation of new genes was believed to be dependent on gene duplication, as suggested by Ohno.[96]

Almost 20 years after Gilbert's proposal, the accumulation of sequence data in the data bases shows that exon shuffling has been used during evolution to promote gene diversity. This is evident in multicellular animals. Structures related to multicellularity were constructed by exon shuffling. It seems clear, for example, that the origin of multicellular organisms is correlated with the origin of an extracellular matrix (ECM). Most ECM components are mosaic proteins that share several protein modules, which in turn correspond generally to exons or sets of exons. The same is true of several cell surface receptors. Indeed, the first examples of exon shuffling were the genes encoding low density lipoprotein (LDL) receptor and epidermal growth factor (EGF)[97] and the genes involved in blood coagulation.[98]

There are at least two possible mechanisms for exon shuffling. The first one is seen at the DNA level, where recombination within the introns is believed to bring together exons from unrelated genes. The second way occurs at the RNA level. A mature message can be copied to DNA by a reverse transcriptase and inserted back into the genome. If this insertion occurs in a nontranscribed region, the cDNA becomes a pseudogene. If, however, the cDNA is inserted into an intron of a pre-existing gene, it will become a complex 3' exon. A clear example of this process is the *Drosophila* gene *jingwei*.[99] In this gene, the mature message for alcohol dehydrogenase was inserted into the genome exactly in the third intron of the gene *yande*. This new chimerical gene now contained three 5' exons from *yande* and a complex 5' exon from the ADH cDNA. In addition, another RNA-mediated mechanism for exon shuffling is *trans*-splicing followed by reverse transcription and insertion into the genome.

The Example of Cytochrome c1

Many examples of exon shuffling have been reported.[8,100–103] Although the statistical significance of some of these cases has often been questioned, the recent finding of exon shuffling between cytochrome c_1 and GapC in plant genes represents a clear example and demonstrates that the shuffled exons confer new functions on the acceptor gene.[104]

Many genes in the nuclear genome have an organellar origin. After the endosymbiotic event that gave rise to mitochondria and chloroplasts, a number of genes in these organelles were transferred to the nucleus, where most retained their organelle function. For these genes, acquisition of an organelle-targeting activity was fundamental. Organelle-targeting activity is usually due to the presence of a presequence (mitochondria) or transit peptide (chloroplasts) at the N-terminus of the protein. Palmer and his colleagues have speculated about the role of exon shuffling in the acquisition of these extra sequences.[105,106] This speculation was confirmed in the origin of the presequence of cytochrome c_1 proteins. A significant similarity between the first three exons of pea cytosolic GAPDH and potato cytochrome c_1 was observed (44% identity and 64% similarity over 41 amino acids without insertion of any gap in the alignment). Besides the significant sequence similarity, moreover, intron positions and phases were very well conserved. Since the sequence is conserved among all GAPDHs characterized so far, it was possible to establish a donor–recipient relationship in the shuffling event (from GAPDH to cytochrome c_1). The shuffling was recent since this presequence is not even present in *Arabidopsis*.

Quantitative Assessment of Exon Shuffling

What is the contribution of exon shuffling to gene evolution? Long et al.[46] made a quantitative estimation by estimating the excess of symmetric intron phases in an exon data base. Assuming the biased phase distribution as a consequence of certain unknown selection, they made a most conservative estimation that at least 19% of exons in the data base have been involved in exon shuffling. If we assume a random distribution of intron phase, the contribution of exon shuffling would be more than 28%. Thus we have a first quantitative estimate of the contribution of exon shuffling, one of the two major mechanisms for the origin of new genes. This fact suggests that the gene duplication is perhaps not the only important mechanism in the origin of new genes. Exon shuffling also plays a very important role as well.

In light of the excess of symmetric exons in the data base and its implication as reported,[46] Hurst and McVean[107] argued that the excess of (1,1) exons over (0,0) would seem to favor the model of nonrandom intron insertion associated with selection. To them, the existence of exon shuffling of (1,1) exons is not understandable, because, for one thing, "the accumulation of such exons may lead to the alteration of amino acids flanking the exon." Apparently, they neglected two important papers of Patthy,[100,101] who extensively collected many actual cases of exon shuffling and found that almost all of these examples are (1,1) exons. Therefore, they failed to see the agreement between the empirical facts and the statistical conclusion.

B. Intron Movement: Intron Slippage and Sliding

It is important to make a distinction between intron mobility and intron movement. Intron mobility refers to the capacity of group I and group II introns to insert themselves in the same locus (homing) or in a different locus (transposition). The term "movement" that we are using now is related to a putative capacity of introns, including nuclear introns, to shift their positions by just a few base pairs in a gene. When one compares the position of the introns in a specific gene among different species, the generated phylogenetic pattern sometimes suggests that introns could move laterally in the gene, since these introns are located very close to each other. The difference of a few base pairs in intron positions among different species is extremely frequent in intron data bases. But how can introns move? Cerff[108] recently discussed two types of mechanisms that can possibly account for the movement of introns. The first one, slippage, is the result of a reverse splicing reaction in which an intron is reinserted immediately in the vicinity of the original position after its splicing. Obviously this mechanism assumes a process of reverse transcription and gene conversion to add the new gene to the genome. Interestingly, spliceosomal introns have been found in the genes encoding U2 and U6 snRNAs in several yeasts.[109–111]

The second mechanism is DNA-dependent and is called sliding. Sliding can occur if a mutation affects the normal splice site, which is substituted by a cryptic one. This leads to an insertion or deletion in the amino acid sequence. Another possibility that leads to sliding is an insertion or deletion that is compensated for by a deletion or insertion in the opposite side of the exon/intron junction. Clear examples of intron sliding were reported in the histone gene of *Volox*[112] and the globin genes.[113]

VI.   ORIGIN OF INTRONS

As is well known today, a major difference between prokaryotic and eukaryotic genes is that only eukaryotic genes have spliceosomal introns. An intuitive way of thinking about the time of origin of this major type of intron is to assume that these introns originated only after prokaryotes and eukaryotes diverged, because we only observe them in eukaryotic genomes. This is often called the introns-late theory. According to this line of thinking, transposable elements would be a candidate for the source of the introns. However, intuition may not provide the final answer to the question, although this is the first step toward a solution of the puzzle. The weakness of this attitude has been demonstrated in the analysis of questions from a simple process and system, e.g., a renewal paradox in the renewal process.[114] One needs to go further in both the logical derivation and interpretation of available observations. The alternative explanation of the origin of introns is to put the time of origin back to the progenote, the ancestor of eukaryotes and prokaryotes. This view, the introns-early theory, is still not fully accepted, although the proposal is supported by several lines of evidence. The controversy between the two views, which have few points of agreement, has not ceased since the discovery of introns in the late 1970s.

It is important to understand the origin of the intron-exon structure, one of the most general structures of genomes. In the evolutionary processes of genes, one can see that many types of introns are involved; one can see many different phenomena; one can see introns in many, many genes. Unfortunately, this complexity in the field caused much confusion, either over the evolutionary processes themselves or over the arguments proposed to explain these processes. We list here several common points of confusion. (1) Group I, group II, and nuclear introns are distinct types of introns, both structurally and in splicing mechanisms, so the properties unique to one type of intron may not be applicable to others. (2) The concept of exon shuffling deals with a general molecular mechanism of exon movements to generate new genes but does not address the question of how introns originated.

(3) Exon shuffling within genes and partial duplication look like a semantic problem of definitions, but the two concepts may represent very different ways of viewing gene and genome evolution. A clear understanding of these issues is important for a correct assessment of the studies in the field.

A. Introns—Early Theories

There are several versions of the introns-early theory. The first one, with a complete form and greater impact, is the exon theory of genes, advocated by Gilbert.[115] The other two related versions include the minigenes theory proposed by Knowles' group,[116,117] and the split-gene theory proposed by Senapathy.[117–119] A recent version of the introns-early hypothesis was proposed by Forsdyke,[120,121] whose considerations completely differ from those of the previous authors.

Exon Theory of Genes

As a direct consequence of the discovery of introns one year before, 1978 was a celebrated year for the field of gene and genome evolution. Gilbert began to examine the general role of introns and concluded that introns can speed up evolution by increasing recombination between exons.[7] Because recombination can take place at any point within introns, the probability of a successful recombination is increased a million times when compared with an intronless gene. Years later, intron-mediated recombination of exons, i.e., exon shuffling, was observed and more and more examples documented, as summarized in Section V. In the same year, realizing that insertions of transposable elements in genomes could be deleterious, it was suggested that introns could have existed before the divergence of prokaryotes and eukaryotes.[122,123] The explanation for the lack of introns in prokaryotes is that the genomes of these organisms were subjected to a "streamlining" process in which the pressure for rapid replication led to intron loss. Meanwhile, Blake inferred a corresponding relationship between exons and protein structures.[124] Two years later, Go[125] observed a correlation between intron positions and a protein structural element (which she called a "module") in hemoglobin.

Gilbert synthesized and extended these theories and observations into a cohesive whole, dubbed the "exon theory of genes."[115] The three major components in this theory are the early existence of introns, exon shuffling as a major force to create new genes, and the exon-coded modules as minimum protein functional units. In addition, this theory also discussed the evolutionary relationship among the self-splicing and spliceosomal introns. The general picture painted from the theory is that ancient introns existed as a form of today's group I or group II introns in an RNA world,[113] followed by the appearance of increased catalytic efficiency in the form of proteins, followed by the origin of DNAs. After a DNA world came into existence, the spliceosomal machinery evolved in eukaryotic genomes to splice the introns inherited directly from ancestral forms of group II introns. In the early stage of this process, exons were small, but by retroposition mechanisms, exons were fused to create more complex exons. A clear example of this process was found in the *jingwei* gene in *Drosophila*.[99] Many introns were gradually lost in various lineages.

From this discussion we can see that exon shuffling is a mechanism for recombining exons to form new genes, and it is not a synonym for the introns-early hypothesis, which explores the origin of introns. Moreover, we can see that the exon theory of genes not only deals with the origin of nuclear introns, but also accounts for the whole evolutionary process of genes.

Minigenes and the Split-Gene Hypotheses

The two hypotheses have some points in common, such as the evolution of the splicing signal from termination codons. In the split-gene hypothesis, the origin of introns is correlated with noncoding sequences that split ancestral genes. In the microgene hypothesis, however, introns are seen as "failed exons," originally a minigene that failed to bring any functional protein. One interesting point about the minigene is that it assumes that functional proteins were assembled by complementation of the exon products. Recently, experimental evidence for this step has been provided in the genes for triosephosphate isomerase and isoleucyl-tRNA synthetase.[126,127]

Stem Loop "Kissing" Model and Origin of Introns

This argument, proposed by Forsdyke, is based on a recent conceptual change in how the initiation of meiotic recombination might occur. It is now believed that there is a process of sequence-based homology search before chromosome pairing, instead of the prior formation of a synaptonemal complex.[128] It is believed that stem loop structures in DNA play an important role in the initiation of homologous recombination by the "kissing" interaction between the tips of step loop structures.[129,130] What is the realtionship between the stem loop structures and origin of introns?

Forsdyke analyzed the distribution of the minimum free energy for folding of natural sequences along a 68-kb human chromosome 19 segment and other long DNA fragments from *Drosophila, E coli.*, and bacteriophage λ. Forsdyke made two observations.[120,121] First, he found that the difference between the mean minimum free energy for the random sequences and the minimum free energy value for folding (designated as FORS-D value) are always positive. This difference is a measurement of the potential to form the stem loop structures. Positive values across genomes suggest that there is a general evolutionary pressure on DNA molecules to accept mutations that would promote the formation of stem loop structures, which is the design that initiates recombination. Second, he reported a heterogeneity of the distribution of the minimum free energy between introns and exons: the introns have a higher potential to form stem loop structures, as the higher FORS-D value in introns indicated. Because the recombination might have evolved before the coding of proteins, Forsdyke claimed that this model supports the introns-early hypothesis.

This is an interesting model that is based on the distribution of the initiation of recombination, the step loop. However, there are several aspects that need further investigation. The major question is whether the difference between introns and exons in FORS-D values will stand the test of more genes, especially the test of sequence data of genome projects. Furthermore, data from molecular population analysis have already shown that there is recombination within exons. The evolution of the organisms without introns may not be hampered by genetic recombination that takes place in introns.

B. Introns-Late Hypothesis

There are two forms of introns-late hypotheses. The first one is the insertional theory of genes, in which the nuclear intron is thought to be some form of transposable element or the descendent from group II introns in organelles. The second one assumes the relaxation of alternative splicing, as proposed by Dibb.[131] The insertional hypothesis is more influential because of the observed mobility of group I and group II introns.

The Insertional Theory of Intron Origin

The basic form of the theory is that nuclear introns originated after prokaryotes and eukaryotes diverged.[58,131–135] Stoltzfus et al.[136] summarized the theory. First, nuclear introns originated from uninterrupted genes by the insertion of introns; second, introns are not needed to assemble contemporary proteins; third, the splicing machinery of nuclear introns arose from self-splicing introns. Finally, nuclear introns were never present in the ancestors of those organisms that now lack them.

The possible sources of inserted introns were summarized by Palmer and Logsdon,[58] including (1) invasion of group II introns located in organellar genomes[132,135]; (2) transposable elements[137–139]; (3) the tandem duplication of exons, which may generate new introns by activating internal cryptic splice sites[132,140]; and (4) the *trans*-splicing of introns.[109–111]

To many investigators, the mobility of group I and group II introns makes the origin of introns via transposable elements an attractive hypothesis. In his early proposal, Doolittle[122] realized that the insertion of transposable elements is deleterious to organisms, hence it is unlikely that today's nuclear introns could be created by the insertion of so many transposable elements. Although the detrimental nature of transposable elements was not initially certain, later studies of the population genetics survey of the transposable elements on a large scale in yeast, bacteria, and especially in *Drosophila* strongly supported this conclusion.[141–143]

Cavalier-Smith, one of the major advocates of the introns-late theory, also indicated this difficulty.[135]

The invasion of group II introns from mitochondria proposed by Cavalier-Smith is a more plausible hypothesis. Palmer and co-workers observed many examples of gene movement from organelles to nuclear genomes.[105,106,144] However, if we assume that intron insertions occur continuously during eukaryotic evolution, we should expect to see some recent insertion of group II introns into nuclear genes. But there are still no such insertions observed in the more than 20,000 eukaryotic intron-containing genes sequenced so far.

Both the creation of introns from tandem duplication of exons and the reverse splicing of introns may explain the origin of a few nuclear introns, but obviously cannot be taken as general sources of nuclear introns.

By-products of Alternative Splicing Evolution

According to this hypothesis, which was first suggested by Tonegawa et al.[145] and Crick,[146] nuclear introns could originate from coding sequences. Dibb and Newman[147] proposed the concept of proto-splice sites, hypothetical consensus sequences in exons that flank introns, by analyzing the distribution of the sequence conservation in the functions of introns in actin, α-tubulin. They suggested that the distribution of introns in these two genes is more consistent with the possibility of insertions of introns into the proto-splice sites. After this influential proposal, Dibb[131] went further, describing a general scenario for the origin of nuclear introns. In addition to the assumption of proto-splice sites, he also assumed that in its early stage, splicing machinery was inefficient, all RNA was coding, and no nuclear introns existed. A protein contains more than one splicing site that is involved in alternative splicing. When two proto-splice sites work at a higher frequency simultaneously, it would be possible that the coding sequence between the sites would lose the coding constraint and evolve into nuclear introns. Once the nuclear introns had been created, they would spread among genes by inserting into other proto-splice sites of genes.

C.   Evidence for the Introns—Early Theory

Similar Intron Positions across Plant and Animal

The early work showed that there are some introns in identical positions of genes in plants and animals. Several introns in triosephosphate isomerase genes (TPI) were found in identical or equivalent positions in maize and chicken.[148] Similar examples were reported for actin,[149] RNA polymerase II,[150] plasma membrane H^+-ATPase,[151] ubiquitin conjugating enzyme,[152] and cytochrome c_1 genes.[104] These results from intron position analysis showed that the introns originated before the divergence of plants and animals, about a billion years ago. However, one may argue that these coincidences in intron positions do not demonstrate that introns existed in the progenote of eukaryotes and prokaryotes, because it might not be a short time from the divergence of prokaryotes and eukaryotes to the split of plant and animals, which might allow the origin of introns. However, these studies demonstrate at least that the origin of introns is not a recent event.

Similar Intron Positions between Ancient Duplicated Genes

The idea used in this method is similar to what is described above: searching for the equivalent intron positions between genes, so any correspondence of intron positions can be traced back to the time before the divergence of genes. But here the genes of choice are those genes originating from a duplication that took place before prokaryotes and eukaryotes split. The glyceraldehyde-3-phosphate dehydrogenase (GAPDH), malate dehydrogenase (MDH), and aspartate aminotransferase isomerase (AspAT) genes are examples. These genes have both cytosolic and organellar copies, which originated before the divergence of prokaryotic and eukaryotic genes. A number of introns were found in equivalent positions in both copies of these genes.[153–155] More recently, van den Hoff et al.[178] showed introns in identical positions in the gene for carbamoylphosphate synthetase, which is a product of ancient duplication.

Ancient Exon Shuffling

Long et al.[46] investigated the distribution of intron phases in an exon data base of ancient conservative regions (ACRs)[156] that are homologous between prokaryotic and eukaryotic genes. A significant correlation was observed among intron phases, suggesting that there was exon shuffling in these regions of ancient origin. It is obvious that any exon recombination, or exon shuffling, if happening after the divergence of eukaryotic and prokaryotic genomes, would destroy the colinear correspondences between the counterparts of these ACRs in eukaryotes and prokaryotes. Therefore, the observed exon shuffling in this ancient set of genes must have taken place before the divergence of eukaryotes and prokaryotes, which was facilitated by introns in their ancestors.

Excess of Introns in Linker Regions

The previous descriptions have provided several independent lines of evidence for the early origin of nuclear introns; however, they did not say anything about one of the important predictions of the exon theory of genes, that ancient exons should encode a minimum functional unit of proteins, or module,[124,125] and therefore there should be a correlation between the three dimensional structures and intron positions. This prediction is perhaps most controversial in recent literature regarding intron-exon evolution.[8,61,136]

In a recent statistical survey of such a relationship, de Souza and colleagues[176] observed a significant excess amount of introns in the linker regions, defined as module boundaries, of 30 eukaryotic proteins, all of them of ancient origin. A more remarkable observation made in this work is that the greater portion of conserved ancient introns are located in the linker regions, suggesting, for the first time, that the correlation between module and intron positions could be connected with the antiquity of introns. This is not predicted in other versions of the introns-early hypothesis.[117]

D.   Evidence for the Introns—Late Hypothesis

Phylogenetic Distribution

Dibb and Newman[147] showed that the distribution of introns in actin and tubulin was restricted to some lineages within plant-animal-fungus branches. They concluded that the patterns of intron distribution in these genes are more consistent with a scenario of intron gain rather than intron loss. Palmer and Logsdon58 further surveyed the phylogenetic distribution of introns in a larger scale. They observed a big variation in intron numbers in various lineages. In 1 kb of coding sequence, animal genes contain six introns, plant and fungi genes contain three introns, and the organisms like ciliates and plasmodium contain only one intron. In other early protists, these authors did not find any record of introns. A large number of parallel losses of introns are needed to explain this pattern in terms of the introns-early hypothesis. This distribution was interpreted as a consequence of intron gains. Recently a similar rationale was used to analyze the distribution of introns in a single gene, TPI, and a similar conclusion was reached.[157,158]

To date, this picture of intron distribution has gone through some changes; the phylogenetic tree as presented in Palmer and Logsdon[58] may also need revision because of the expanding data base. But in general, the distribution of introns may be uneven across organisms. The question is whether this uneven distribution can be explained solely as a result of intron gains. The answer is not certain. There is good evidence showing that introns can be lost in yeast S. cerevisae.[159] There is also good evidence for the movement of introns, as described in Section VB (Intron Movements). All of these factors can make the distribution of introns uneven. Therefore, the phylogenetic approach might not have enough power to make distinguishable predictions.

In relation to the results of Dibb and Newman,[147] in a more sophisticated statistical approach, Nyberg and Cronhjort[157] simulated the intron evolution and calculated the probability of intron arrangements in actin and α-tubulin. They reached a conclusion opposite that of Dibb and Newman: the intron distributions in these two genes are inconsistent with the introns-late hypothesis.

De Novo Creation of Introns from Ds Element in Maize

Giroux et al.[161] provided a possibility that the insertion of the transposable elements creates a novel intron. They found that in maize after the insertion of Ds (*Dissociation*) to *shrunken-2*, a gene that encodes a subunit of the starch-synthetic enzymes, an alternative RNA splicing pattern could be created. Although most of the transcripts are aberrant, about 2% of transcripts are normal when the Ds is precisely excised. Consensus splicing signals in the two sides of the Ds insertion, GT··AG, are generated as a consequence of an 8-bp duplication caused by Ds. This is the first interesting example to show that the transposable element can create a GT··AG consensus sequence of splicing signal and generate a small portion of normal transcript. Furthermore, the authors proposed that transposable element insertion might be a general mechanism for the formation of introns. However, it may be too early to conclude anything about intron origin in general from this experiment, because the insertion of Ds might be deleterious and could not be fixed in nature as a consequence of abnormal transcripts caused by Ds insertions.

E. Speculations on the Controversy

The critical point in the controversy on the origin of introns is the presence or absence of introns in the ancestor of all forms of life. In the last few years, group I as well as group II introns have been discovered in eubacteria, specifically in the groups that are the ancestors of mitochondria and chloroplasts. Interestingly, some of these introns are also present in the organelle in the same position, suggesting that their origin at least antedates the endosymbiotic event that gave rise to the organelles.

Researchers favoring introns-early theory believe that present day bacterial cells are in general intronless as a consequence of a pressure for rapid replication. Bacteria, however, must have retained their introns at least until the endosymbiotic event to account for the fact that some introns are shared between eubacteria and organelles. What kind of pressure could be responsible for intron loss in bacteria only after the endosymbiotic event? Recently we speculated that the acquisition of an organelle represented a powerful advantage for the emerging eukaryotic cell. We also proposed that fast replication and intron loss arose as a response leading to the appearance of a better adapted competitor. In the eukaryotic branch, however, introns evolved to a more complex type, depending on a spliceosomal machinery.[59]

Recently Patthy[101] argued that group I and group II introns are not suitable for exon shuffling because the recombination would disrupt the structure of the intron that is fundamental for its catalytic activity. However, there are examples of exon shuffling in bacteriophage T4, where recombination between self-splicing introns produced a chimeric gene containing a hybrid intron that displayed self-splicing activity.[162–164] In fact, as we discussed in Section II, the ORFs in group I and II introns are recent acquisitions, suggesting that some regions of the intron, probably flanking the core, are suitable for recombination events. Finally, events of *trans*-splicing in an RNA world would promote recombination between exons, without affecting the structure of the intron, as recently reported.[160]

VII.   EVOLUTION OF INTRON-EXON STRUCTURES

A.   Protein Diversity and the Exon Universe

It is clear that promotion of diversity was a fundamental step in the early stages of gene evolution and in several other stages in which an expansion of the number of variants is observed, as in the origin of multicellularity and the Cambrian explosion. One way to increase diversity is gene duplication. One copy, free of selective constraints, can accept substitutions that would be forbidden in the original gene. Another possibility is exon shuffling. All multicellular organisms are characterized by the presence of a rich extracellular matrix, the origin of which was a *sine qua non* for the development of multicellularity. It is clear that the origin of the components of extracellular matrix is intimately related to exon shuffling events. Most of the proteins, if not all, are mosaic proteins that share a limited number of domains that in general correspond to exons.[102]

It is possible that the same process was important in earlier phases of evolution. As we already discussed, according to the exon theory of genes, the scenario of gene evolution is totally dependent on exon shuffling. In fact, several lines of evidence suggest that exon shuffling *in fact* was important in the origin of protein diversity, as we discussed in the previous sections. The stronger evidence comes from an analysis of intron phase correlation, showing an excess of symmetric exons in genes that were present in the progenote. As discussed in a previous section, an excess of symmetric exons is an indicator of exon shuffling. Additional supportive evidence is the correlation between intron and module boundaries in ancient genes.[176]

In 1990, Dorit et al. estimated the number of exons that were used in the assembly of the genes that we see today.[4] The approach was based on the frequency with which exons are reused during evolution. They concluded that only 1000–7000 primordial exons were necessary to construct the current pool of genes. The fact that a small number of building blocks can be defined is strong evidence that exon shuffling was an important process in the origin of protein diversity. Let us assume, for example, 1000 primordial exons for a simple calculation. Let us also assume that in the origin of protein diversity exon shuffling produced only bimodular structures. This gives a repertoire of 10^6 different protein structures. If we add more modules, the universe of possibilities expands considerably. On the other hand, how was protein diversity created in a scenario without intron-mediated recombination? It is reasonable to think that in an inefficient replication machinery, variation would be produced in each replication step. This argument, however, is also valid for a scenario of intron-containing genes that would increase the 10^6 value reached above.

Natural selection works on protein functions, which is a consequence of their three-dimensional organization. How many shapes were necessary to cover all basic reactions in the progenote's environment? Certainly, the useful shapes were a small fraction of the original pool on which natural selection worked. Therefore, the numerical argument of the exon theory of genes is strong. It provides an efficient mechanism that produces diversity.

B.   Origin of New Genes

Evolutionary biologists have long been fascinated by the occurrence of novelty in nature. Novelties in morphology can to be traced back to changes at the molecular level. One of the most fundamental components contributing to evolutionary novelty is the origin of new genes. Haldane[166] speculated on the role of duplication in the origin of new genes. Ohno[167] systematically addressed the importance of gene duplications. Ohta[168–170] analyzed concrete models of evolution of new genes by gene duplications. The two major points can be extracted from the voluminous literature on gene duplications. First, the duplicated copy provides a "shelter" for the accumulation of new mutations that may lead to a new function. Second, a faster evolutionary rate is associated with the newly created duplicated gene. Today no one doubts the important role of gene duplication, but no one has tried to estimate its quantitative contribution to the origin of new genes.

The exon theory of genes[115] has painted a global picture of the origin and evolution of new genes. Although there are debates on some predictions made by this theory, one of its major components, exon shuffling, appears to have been widely accepted as a way of creating new genes in nature. This mechanism is so important that more than 20% of exons in eukaryotic genes are created, as we discussed in Section IV.

C.   Connection between Self-splicing and Spliceosomal Introns

Similarities between the splicing processes of group II and spliceosomal introns led to a general agreement that the two share a common ancestor. Several findings, however, in the last 5 years seem to indicate that spliceosomal introns also share some similarities with group I introns. In fact, when one looks at the splicing reaction in the three types of introns in general, one can find several similarities. The first step of the splicing is always an attack on the 5' splice site. In the same way, the second step is an attack on the 3' splice site that joins the two exons and releases the intron. Similarities can also be found when one looks at the chemical details of the splicing in the three groups of introns.[9] What is intriguing is the fact that the first splicing reaction occurring at the 5' splice site is more similar between group II and spliceosomal introns. The similarities include (1) the presence of a guanosine in the 5' end of the intron, (2) the nucleophylic attack, which is promoted by a 2'-OH derived from an adenosine in the branch site, and (3) the presence of an

Sp diastereometer as the chiral diastereometer in the catalytic center. The second step, however, seems to be more similar between group I and spliceosomal introns. In this case the similarities are (1) the presence of a guanosine in the 3' end of the intron and (2) the presence of an Sp diastereometer in the chiral center.

What are the evolutionary implications for these similarities? It is generally agreed that group II and spliceosomal introns share a common origin.[171] Can we extend this evolutionary relationship to the group I introns? And if so, what was the evolutionary pathway that led to the present-day intron distribution? We are confident that the huge amount of sequence information that will be accumulated in the next decade will shed some light on these questions.

D. Microevolution of Introns

How a population can cross a valley of deleterious states and evolve from one fitness peak to another remains a basic evolutionary problem.[172,173] One approach to this problem is to analyze the compensatory interaction of mutations, since the effects of the first deleterious mutation could be compensated by the second mutation and achieve a higher fitness. At the molecular level, the secondary structure of RNA provides an opportunity to detect compensatory mutations. Stephan and Kirby[174] and Kirby et al.[175] investigated the property of nucleotide changes in intron 2 of the *Adh* gene in *Drosophila*. The second intron forms a stem loop structure, which is believed to play some role in the splicing of the gene. These authors observed covariation of nucleotide substitutions in this intron in several *Drosophila* species, suggesting compensatory evolution in the intron.

VIII.   CONCLUDING REMARKS

Most research activities in the field have been carried out with several interests: description of the intron-exon structures from newly identified genes; exploration of splicing mechanisms; and the search for origin and evolution of intron-exon structures.

The most solid contribution from the field is probably the understanding of the distribution of intron-exon structures across present-day organisms. It is known that group I and group II introns exist in both prokaryotes and eukaryotes, whereas nuclear introns are present only in eukaryotes. The distribution of nuclear introns is uneven across various eukaryotic lineages. It is also well known that intron length is much more variable than that of exons.

The mechanism of RNA splicing is well understood in several model organisms, e.g., yeast and *Drosophila*. It is assumed that components and processes of splicing are conserved. How these components of the machinery evolved in various major lineages of eukaryotes has not been investigated thoroughly and is a worthy topic for further evolutionary study.

The mystery of the origin of introns has been a hot topic since their discovery in the late 1970s, with two conflicting hypotheses (introns-early and introns-late) giving contrary explanations. The short history of the field has been full of controversies about whether intron positions and three dimensional structures of proteins are correlated, and whether intron positions between remotely divergent lineages are conserved. Neither hypothesis can explain all of the available data, although it appears that the introns-early or exon theory of genes is consistent with more of the available evidence. There is not yet a final answer for the origin of introns.

In the last several years, the introns-late proponents have attempted to point out disparities between the introns-early theory and its supported evidence. Although such arguments are important, they would not alone, even if true, force acceptance of the introns-late hypothesis. General acceptance of the introns-late hypothesis requires direct evidence for the late origin, which, to date, has not been found. The hypothetical origin of transposable elements and transferred group II introns from organellar genomes is not well supported by data. After all, the introns-early theory has its own theoretical framework and is increasingly supported by the evidence, whereas the introns-late theory looks weaker in both regards. A theory cannot stop at a place where it only attacks the alternatives; it must build its own arguments.

The newly emerging approach, the analysis of gene populations from whole genomes coupled with expanding data bases of sequences and protein structures, as conducted in the intron-phase analysis and linker-region analysis, brings impetus to the field. After the sequence data from whole genomes of various organisms across major lineages become available around the turn of the century, decisive evidence for the origin of introns sould be forthcoming.

ACKNOWLEDGMENTS

We thank Drs. Walter Gilbert, Walter Taylor, and Liming Li for critical reading and valuable discussion.

REFERENCES

1. Berget, S.M.; Moore, C.; Sharp, P.A. Spliced segments at the 5' terminus of adenovirus 2 late mRNA. *Proc. Natl. Acad. Sci. USA* **1977**, *74*, 3171–3175.
2. Chow, L.T.; Gelinas, R.E.; Broker, T.R.; Roberts, R.J. An amazing sequence arrangement at the 5' ends of adenovirus 2 messenger RNA. *Cell* **1977**, *12*, 1–8.
3. Mount, S.M. A catalogue of splice junction sequences. *Nucleic Acid Res.* **1982**, *10*, 459–472.
4. Dorit, R.L.; Schoenbach, L.; Gilbert, W. How big is the universe of exons? *Science* **1990**, *250*, 1377–1382.
5. Fichant, G.A. Constraints acting on the exon positions of the splice site sequences and local amino acid composition of the protein. *Hum. Mol. Genet.* **1992**, *1*, 259–2647.
6. Fedorov, A.; Suboch, G.; Bujakov, M.; Fedorova, L. Analysis of nonuniformaty in intron phase distribution. *Nucleic Acids Res.* **1992**, *20*, 2553–2557.
7. Gilbert, W. Why gene in pieces? *Nature* **1978**, *271*, 501.

8. Long, M.; de Souza, S.J.; Gilbert, W. Evolution of the intron/exon structure of eukaryotic genes. *Curr. Opin. Genet. Dev.* **1995**, *5*, 774–778.

9. Moore, M.J.; Query, C.C.; Sharp, P.A. Splicing of precursors to mRNAs by the spliceosome. In: *The RNA world* (Gesteland, R.F.; Atkins, J.F.; eds.). Cold Spring Harbor Laboratory, Cold Spring Harbor, NY, 1993, 303–358.

10. Sharp, P.A. Split genes and RNA splicing. *Cell* **1994**, *77*, 805–815.

11. Nilsen, T.W. RNA-RNA interactions in the spliceosome: unraveling the ties that bind. *Cell* **1994**, *78*, 1–4.

12. Moore, M.J.; Sharp, P.A. Evidence for two active sites in the spliceosome provided by stereochemistry of pre-mRNA splicing. *Nature* **1993**, *365*, 294–295.

13. Fabrizio, P.; Abelson, J. Two domains of yeast U6 small nuclear RNA required for both steps of nuclear precursor messenger RNA splicing. *Science* **1990**, *250*, 404–409.

14. McPheeters, D.S.; Abelson, J. Mutational analysis of the yeast U2 snRNA suggests a structural similarity to the catalytic core of group I introns. *Cell* **1992**, *71*, 819–831.

15. Legrain, P.; Seraphin, B.; Rosbach, M. Commitment of yeast pre-mRNA to the spliceosome pathway does not require U2snRNP. *Mol. Cell. Biol.* **1988**, *8*, 3755–3760.

16. Zhaler, A.M.; Lone, W.S.; Stalk, J.A.; Roth, M.B. SR proteins: A conserved family of pre-mRNA splicing factors. *Genes Dev.* **1992**, *6*, 837–847.

17. Stephens, R.M.; Schneider, T.D. Features of spliceosome evolution and function inferred from an anlysis of the information at human splice sites. *J. Mol. Biol.* **1992**, *228*, 1124–1136.

18. Robberson, B.L.; Cote, G.J.; Berget, S.M. Exon definition may facilitate splice site selection in RNAs with multiple exons. *Mol. Cell. Biol.* **1990**, *10*, 84–94.

19. Hall, S.L.; Padgett, R.A. Conserved sequences in a class of rare eukaryotic nuclear introns with non-consensus splice sites. *J. Mol. Biol.* **1994**, *239*, 357–365.

20. Hall, S.L.; Padgett, R.A. Requirement for U12 snRNA for in vivo splicing of a minor class of eukaryotic nuclear pre-mRNA introns. *Science* **1996**, *271*, 1716–1718.

21. Tarn, W.-Y.; Steitz, J.A. A novel spliceosome containing U11, U12, and U5 snRNPs excises a minor class IAT-AC) intron *in vitro*. *Cell* **1996**, *84*, 801–811.

22. Mount, S.M. AT-AC introns: an attack on dogma. *Science* **1996**, *271*, 1690–1691.

23. Kreivi, J.-P.; Lamond, A.I. RNA splicing: unexpected spliceosome diversity. *Curr. Biol.* **1996**, *6*, 802–805.

24. Agabian, N. Trans-splicing in nuclear pre-mRNAs. *Cell* **1990**, *61*, 1157–1160.

25. Nilsen, T.W. Trans-splicing in nematodes. *Exp. Parasitol.* **1989**, *69*, 413–416.

26. Wissinger, B.; Schulster, W.; Brennicke, A. Trans-splicing in *Oenothera* mitochondria: nad1 mRNAs are edited in exon and trans-splicing group II intron sequences. *Cell* **1991**, *65*, 473–482.

27. Eul, J.; Graessmann, M.; Graessmann, A. Experimental evidence for RNA trans-splicing in mammalian cells. *EMBO J.* **1995**, *14*, 3226–3235.

28. Ullu, E.; Tschudi, C. Accurate modification of the tryanosome spliced leader cap structure in a homologous cell-free system. *J. Biol. Chem.* **1995**, *270*, 20365–20369.

29. Hannon, G.J.; Maroney, P.A.; Nilsen, T.W. U small nuclear ribonucleoprotein requirements for nematode *cis-* and *trans-*splicing *in vitro*. *J. Biol. Chem.* **1991**, *266*, 22792–22795.

30. McKeown, M. Alternative mRNA splicing. *Annu. Rev. Cell Biol.* **1992**, *8*, 133–155.

31. Hodges, D.; Bernstein, S.I. Genetic and biochemical analysis of alternative RNA splicing. *Adv. Genet.* **1994**, *31*, 207–228.

32. Inoue, K.; Shimura, Y. Mechanisms of alternative RNA splicing. In: Tracing Biological Evolution in Protein and Gene Structures (Go, M.; Schimmel, P.; eds.). Elsevier, Amsterdam, 1995, pp. 87–99.

33. Cech, T.R.; Bass, B.L. Biological catalysis by RNA. *Annu. Rev. Biochem.* **1986**, *55*, 599–629.

34. Cech, T.R. Self-splicing of group I introns. *Annu. Rev. Biochem.* **1990**, *59*, 543–568.

35. Burke, J.M. Molecular genetics of group I introns: RNA structures and protein factors required for splicing-a review. *Gene* **1988**, *73*, 273–294.

36. Lambowitz, A.M.; Perlman, P.S. Involvement of aminoacyl-tRNA synthetases and other proteins in group I and group II intron splicing. *Trends Biochem. Sci.* **1990**, *15*, 440–444.

37. Guo, Q.; Lambowitz, A.M. A tyrosil-tRNA synthetase binds specifically to the group I intron catalytic core. *Genes Dev.* **1992**, *6*, 1357–1372.

38. Bell-Pedersen, D.; Quirk, S.; Clyman, J.; Belfor,t M. Intron mobility in phage t4 is dependent upon a distinctive class of endonucleases and independent of DNA sequences encoding the intron core: mechanistic and evolutionary implications. *Nucleic Acids Res.* **1990**, *18*, 3763–3770.

39. Loizos, N.; Tillier, E.R.M.; Belfort, M. Evolution of mobile group I introns: recognition of intron sequences by an intron-encoded endonuclease. *Proc. Natl. Acad. Sci. USA* **1994**, *91*, 11983–11987.

40. Zimmerly, S.; Guo, H.; Eskes, R.; Yang, J.; Perlman, P.S.; Lambowitz, A.M. A group II intron RNA is a catalytic component of a DNA endonuclease involved in intron mobility. *Cell* **1995**, *83*, 529–538.

41. Yang, J.; Zimmerly, S.; Perlman, P.S.; Lambowitz, A.M. Efficient integration of an intron RNA into double-stranded DNA by reverse splicing. *Nature* **1996**, *381*, 332–335.

42. Hong, L.; Hallick, R.B. A group III intron is formed from domains of two individual group II introns. *Genes Dev.* **1994**, *8*, 1589–1599.

43. Christopher, D.A.; Hallick, R.B. *Euglena gracilis* chloroplast ribosomal protein operon: a new chloroplast gene for ribosomal protein L5 and description of a novel organelle intron category designated group III. *Nucleic Acids Res.* **1989**, *17*, 7591–7608.

44. Drager, R.G.; Hallick, R.B. A complex twintron is excised as four individual introns. Nucleic *Acids Res.* **1993**, *21*, 2389–2394.

45. Hawkins, J.D. A survey on intron and exon lengths. Nucleic Acids Res. 1988, 16, 9893–9898.

46. Long, M.; Rosenberg, C.; Gilbert, W. Intron phase correlations and the evolution of the intron/exon structure of genes. *Proc. Natl. Acad. Sci. USA* **1995**, *92*, 12495–12499.

47. Benian, G.M.; L'Hernaut, S.W.; Morris, M.E. Sequence of an unusually large protein implicaterd in regulation of myosin activity in *C. elegans. Nature* **1989**, *342*, 45–50.

48. Uphold, W.B.; Sandell, L.J. Exon/intron organization of the chicken type II procollagen: intron size distribution suggests a minimum intron size. *Proc. Natl. Acad. Sci. USA* **1986**, *83*, 2325–2329.

49. Mount, S.M.; Burks, C.; Hertz, G.; Stormo, G.D.; White, O.; Fields, C. Splicing signals in *Drosophila*: intron size, information content, and consensus sequences. *Nucleic Acids Res.* **1992**, *20*, 4255–4262.

50. Fu, X.-Y.; Cilgan, J.; Manley, J.L. Multiple *cis*-acting sequence elements are required for efficient splicing of simian virus 40 small-t antigen pre-mRNA. *Mol. Cell. Biol.* **1988**, *8*, 3582–3590.

51. Russell, C.B.; Fraga, D.; Henrichsen, R.D. Extremetly short 20-33 nucleotide introns are the standard length in *Paramecium tetraurelia. Nucleic Acids Res.* **1994**, *22*, 1221–1225.

52. Gilson, P.R.; McFadden, G.I. The miniaturized nuclear genome of a eukarytic endosymbiont contains genes that overlap, genes that are cotranscribed, and the smallest known spliceosomal introns. *Proc. Natl. Acad. Sci. USA* **1996**, *93*, 7737–7742.

53. Roberts, R.G. Dystrophin, its gene, and the dystrophinopathies. *Adv. Genet.* **1995**, *33*, 177–231.

54. Boyce, F.M.; Beggs, A.H.; Feener, C.; Kunkel, L.M. Dystrophin is transcribed in brain from a distant upstream promoter. *Proc. Natl. Acad. Sci. USA* **1991**, *88*, 1276–1280.

55. Field, C. Information content of *Caenorhabditis elegans* splice site sequence varies with intron length. *Nucleic Acids Res.* **1992**, *18*, 1509–1512.

56. Hughes, A.L.; Hughes, M.K. Small genomes for better flyers. *Nature* **1995**, *377*, 391.

57. Cohen, I.R.; Grassel, S.; Murdoch, A.D.; Lozzo, R.V. Structural characterization of the complete human perlecan gene and its promoter. *Proc. Natl. Acad. Sci. USA* **1993**, *90*, 10404–10408.

58. Palmer, J.D.; Logsdon, J.M. The recent origins of introns. *Curr. Opin. Genet. Dev.* **1991**, *1*, 470–477.

59. de Souza, S.J.; Long, M.; Gilbert, W. Introns and gene evolution. *Genes Cells* **1996**, *1* (in press).

60. Brown, N.P.; Whittaker, A.J.; Newell, W.R.; Rawlings, C.J.; Beck, S. Identification and analysis of multigene families by comparison of exon fingerprints. *J. Mol. Biol.* **1995**, *249*, 342–359.

61. Mattick, J.S. Introns: evolution and function. *Curr. Opin. Genet. Dev.* **1994**, *4*, 823–831.

62. Shaefer, B.; Wilde, B.; Massardo, D.R.; Manna, F.; Del Giudice, L.; Wolf, K. A mitochondrial group I intron in fission yeast encodes a maturase and is mobil in crosses. *Curr. Genet.* **1994**, *25*, 336–341.

63. Shub, D.A. The antiquity of group I introns. *Curr. Opin. Genet. Dev.* **1991**, *1*, 478–484.

64. Kuhsel, M.G.; Strickland, R.; Palmer, J.D. An ancient group I intron shared by eubacteria and chloroplasts. *Science* **1990**, *250*, 1570–1573.

65. Sharp, P.A. On the origin of RNA splicing and introns. *Cell* **1985**, *42*, 397–400.

66. Cech, T. The generality of self-splicing RNA: relationship to nuclear mRNA splicing. *Cell* **1986**, *44*, 207–212.

67. Saldanha, R.; Mohr, G.; Belfort, M.; Lambowitz, A.M. Group I and group II introns. *FASEB J.* **1993**, *7*, 15–24.

68. Carignani, G.; Groudinsky, O.; Frezza, D.; Schiavon, E.; Bergantino, E.; Slonimski, P.P. An mRNA maturase is encoded by the first intron of the mitochondrial gene for the subunit I of cytochrome oxidase in S. cerevisiae. *Cell* **1983**, *35*, 733–742.

69. Goguel, V.; Delahodde, A.; Jacq, C. Connections between RNA splicing and DNA intron mobility in yeast mitochongria: RNA maturase and DNA endonuclease switching experiments. *Mol. Cell. Biol.* **1992**, *12*, 696–705.

70. Matsuoka, M.; Matsubara, M.; Kakehi, M.; Imanaka, T. Homologous maturase-like protein are encoded within the group I introns in different mitochondrial genes specifying *Yarrowa lipolytica* cytochrome c oxidase subunit 3 and *Saccharomyces cerevisiae* apocytochrome b. *Curr. Genet.* **1994**, *26*, 377–381.

71. Du, J.; Portetelle, D.; Harvengt, L.; Dumont, M.; Wathelet, B. Expression of intron-encoded maturase-like polypeptides in potato chloroplasts. *Curr. Genet.* **1994**, *25*, 158–163.

72. Xiong, Y.; Eickbush, T.H. Origin and evolution of retroelements based upon their reverse transcriptase sequences. *EMBO J.* **1990**, *9*, 3353–3362.

73. Michel, F.; Lang, B.F. Mitochondrial class II introns encode proteins related to the reverse transcriptases of retrovirus. *Nature* **1985**, *316*, 641–643.

74. Kennell, J.C.; Moran, J.V.; Perlman, P.S.; Butow, R.A.; Lambowitz, A.M. Reverse transcriptase activity associated with maturase-encoding group II introns in yeast mitochondria. *Cell* **1993**, *73*, 133–146.

75. Belfort, M.; Perlman, P.S. Mechanisms of intron mobility. *J. Biol. Chem.* **1995**, *270*, 30237–30240.

76. Mueller, M.W.; Allmaier, M.; Eskes, R.; Schweyen, R. Transposition of group II intron al1 in yeast and invasion of mitochondrial genes at new locations. *Nature* **1993**, *366*, 174–176.

77. Sellem, C.H.; Lecellier, G.; Belcour, L. Transposition of a group II intron. *Nature* **1993**, *366*, 176–178.

78. Schmidt, W.R.; Schweyen, R.J.; Wilf, K.; Mueller, M.W. Transposable group II introns in fission and budding yeast. Site-specific genomic instabilities and formation of group II IVS plD-NAs. *J. Mol. Biol.* **1994**, *243*, 157–166.

79. Levra-Juillet, E.; Boulet, A.; Seraphin, B.; Simon, M.; Faye, G. Mitochondrial introns al1 and/or al2 are needed for the *in vivo* deletion of intervening sequences. *Mol. Gen. Genet.* **1989**, *217*, 168–171.

80. Benyajati, C.; Spoerel, N.; Haymerle, H.; Ashburner, M. The messenger RNA for alcohol dehydrogenase in *Drosophila melanogaster* differs in its 5' end in different developmental stages. *Cell* **1983**, *33*, 125–33.

81. Monaco, L.; Vicini, E.; Conti, M. Structure of two rat genes coding for closely related rolipram-sensitive cAMP phophodiesterase: multiple mRNA variants originate from alternative splicing and multiple start sites. *J. Biol. Chem.* **1994**, *269*, 347–357.

82. Irniger, S.; Egli, C.M.; Kuenzler, M.; Braus, G.H. The yeast actin intron contains a cryptic promoter that can be switched on by preventing transcriptional interference. *Nucleic Acids Res.* **1992**, *20*, 4733–4739.

83. Meyer, K.B.; Sharpe, M.J.; Surani, M.A.; Neuberger, M.S. The importance of the 3'-enhancer region in immunoglobulin kappa gene expression. *Nucleic Acids Res.* **1990**, *18*, 5609–5615.

84. Jonsson, J.J.; Converse, A.; McIvor, R.S. An enhancer in the first intron of the human purine nucleoside phosphorylase-encoding gene. *Gene* **1994**, *140*, 187–193.

85. Brooks, A.R.; Levy, W.B. Hepatocyte nuclear factor 1 and C/EBP are essential for the activity of the human apolipoprotein B gene second-intron enhancer. *Mol. Cell. Biol.* **1992**, *12*, 1134–1148.

86. Xie, Y.; Favot, P.; Dunn, T.L.; Cassady, A.I.; Hume, D.A. Expression of mRNA encoding the macrophage colony-stimulating factor receptor (c-fms) is controlled by a constitutive promoter and tissue-specific transcription elongation. *Mol. Cell. Biol.* **1993**, *13*, 3191–3201.

87. Lavett, D.K. Secondary structure and intron-promoter homology in globin-switching. *Am. J. Hum. Genet.* **1984**, *36*, 338–345.

88. Lavett, D.K. A model for transcriptional regulation based upon homology between introns and promoter regions and secondary structure in the promoter regions: the human globins. *J. Theor. Biol.* **1984**, *107*, 1–36.

89. Henikoff, S.; Keene, M.A.; Fechtel, K.; Fristrom, J.W. Gene within a gene: nested *Drosophila* genes encoded proteins on opposite DNA strands. *Cell* **1986**, *44*, 33–42.

90. Chalfie, M.; Eddy, S.; Hengartner, M.O.; Hodgkin, J.; Kohara, Y.; Plasterk, H.A.; Waterson, R.H.; White, J.G. Genome maps VI *Caenorhabditis elegans. Science* **1995**, *270*, 415–430.

91. Maxwell, E.S.; Fournier, M.J.A. The small nuclear RNAs. *Annu. Rev. Biochem.* **1995**, *35*, 899–934.

92. Sollner-Webb, B. Novel intron-encoded small nucleolar RNAs. *Cell* **1993**, *75*, 403–405.

93. Nicoloso, M.; Ferrer, C.M.; Michot, B.; Azum, M.C.; Bachellerie, J.P. U20, a novel small nucleolar RNA, is encoded in an intron of the nucleolin gene in mammals. *Mol. Cell. Biol.* **1994**, *14*, 5766–5776.

94. Tycowski, K.T.; Shu, M.-D.; Steitz, J.A. A mammalian gene with introns instead of exons generating stable RNA products. *Nature* **1996**, *379*, 464–466.

95. Li, W.-H.; Graur, D. *Fundamentals of Molecular Evolution.* Sinauer Associates, Sunderland, MA, 1991, pp. 67–98

96. Ohno, S. *Evolution by Gene Duplication.* Springer-Verlag, Berlin, 1970.

97. Sudhof, T.C.; Russell, D.W.; Goldstein, J.l.; Brown, M.S.; Sanchez-Pescador, R.; Bell, G.I. Cassette of eight exons shared by genes for LDL receptor and EGF precursor. *Science* **1985**, *228*, 893–895.

98. Patthy, L. Evolution of the proteases of blood coagulation and fibrinosis by assembly from modules. *Cell* **1985**, *41*, 657–663.

99. Long, M.; Langley, C.H. Natural selection and origin of jingwei: a chimeric processed functional gene. *Science* **1993**, *260*, 91–95.

100. Patthy, L. Modular exchange principles in proteins. Curr. Opin. Struct. Biol. 1991, 1, 351–361.

101. Patthy, L. Introns and exons. *Curr. Opin. Struct. Biol.* **1994**, *4*, 383–392.

102. Doolittle, R.F. The multiplicity of domains in proteins. *Annu. Rev. Biochem.* **1995**, *64*, 287–314.

103. Dorit, R.L.; Gilbert, W. The limited universe of exons. *Curr. Opin. Struct. Biol.* **1991**, *1*, 973–977.

104. Long, M.; de Souza, S.J.; Rosenberg, C.; Gilbert, W. Exon shuffling and the origin of the mitochondrial targetting function in plant cytochrome c1 precursor. *Proc. Natl. Acad. Sci. USA* **1996**, *93*, 7727–7731.

105. Nugent, J.M.; Palmer, J.D. RNA-mediated transfer of the gene coxII from the mitochondrion to the nucleus during flowering plant evolution. *Cell* **1991**, *66*, 374–381.

106. Gantt, J.S.; Baldauf, S.L.; Calie, P.J.; Weeden, N.F.; Palmer, J.D. Transfer of rpl22 to the nucleus greatly preceded its loss from the chloroplast and involved the gain of an intron. *EMBO J.* **1991**, *10*, 3073–3078.

107. Hurst, L.D.; McVean, G.T. Molecular evolution: a difficult phase for introns—early. *Curr. Biol.* **1996**, *6*, 533–536.

108. Cerff, R. The chimeric nature of nuclear genomes and the antiquity of introns as demonstrated in the GAPDH gene system. In: *Tracing Biological Evolution in Protein and Gene Structures* (Go, M.; Schimmel, P.; eds.). Elsevier, Amsterdam, 1995, pp. 205–227.

109. Takahashi, Y.; Urushiyama, S.; Tani, T.; Ohshima, Y. An mRNA-type intron is present in the *Rhodotorula hasegawae* U2 small nuclear RNA gene. *Mol. Cell. Biol.* **1991**, *5*, 1022–1231.

110. Tani, T.; Ohshima, Y. mRNA-type introns in U6 small nuclear RNA genes: implications for the catakysis in pre-mRNA splicing. *Genes Dev.* **1991**, *5*, 1022–1031.

111. Tani, T.; Ohshima, Y. The gene for the U6 small nuclear RNA in fission yeast has an intron. *Nature* **1989**, *337*, 87–90.

112. Muller, K.; Schmitt, R. Histone genes of Volvox carteri: DNA sequence and organization of two H3-H4 gene loci. *Nucleic Acids Res.* **1988**, *16*, 4121–4136.

113. Jellie, A.M.; Tate, W.P.; Trotman, C.N.A. Evolutionary history of introns in a multidomain globin gene. *J. Mol. Evol.* **1996**, *42*, 641–647.

114. Feller, W. *An Introduction to Probability Theory and Its Applications*, Vol. 1. John Wiley & Sons, New York, 1950, pp. 303–341.

115. Gilbert, W. The exon theory of genes. Cold Spring Harb. Symp. Quant. Biol. 1987, 52, 901–905.

116. Seidel, H.M.; Pomplianp, D.L.; Knowles, J.R. Exons as microgenes? *Science* **1992**, *257*, 1489–1490.

117. Senapathy, P.; Bertolaet, B.L.; Seidel, H.M.; Knowles, J.R. Introns and the origin of protein-coding genes. *Science* **1995**, *268*, 1366–13647.

118. Senapathy, P. Origin of eukaryotic introns: a hypothesis, based on codon distribution statistics in genes, and its implications. *Proc. Natl. Acad. Sci. USA* **1986**, *83*, 2133–2137.

119. Senapathy, P. Possible evolution of splicing-junction signals in eukaryotic genes from stop codons. *Proc. Natl. Acad. Sci. USA* **1988**, *85*, 1129–1133.

120. Forsdyke, D.R. Conservation of stem-loop potential in introns of snake venom phospholipase A2 genes: an application of FORS-D analysis. *Mol. Biol. Evol.* **1995**, *12*, 1157–1165.

121. Forsdyke, D.R. A stem-loop "kissing" model for the initiation of recombination and the origin of introns. *Mol. Biol. Evol.* **1995**, *12*, 949–958.

122. Doolittle, W.F. Gene-in-pieces: were they ever together? *Nature* **1978**, *272*, 581–582.

123. Darnell, J.E., Jr. Implications of RNA-RNA splicing in evolution of eukaryotic cells. *Science* **1978**, *202*, 1257–1261.

124. Blake, C.C.F. Do genes-in-pieces imply protein in pieces? *Nature* **1978**, *273*, 267.

125. Go, M. Correlations of DNA exonic regions with protein structural units in haemoglobin. *Nature* **1981**, *291*, 90–93.

126. Bertolaet, B.L.; Knowles, J.R. Complementation of fragments of triosephosphate isomerase defined by exon boundaries. *Biochemistry* **1995**, *34*, 5736–5743.

127. Shiba, K. Dissection of an enzyme into two fragments at intron-exon boundaries. In: *Tracing Biological Evolution in Protein and Gene Structures* (Go, M.; Schimmel, P.; eds.). Elsevier, Amsterdam, 1995, pp. 11–22.

128. Hawley, R.S.; Arbel, T. Yeast genetics and the fall of the classical view of meiosis. *Cell* **1993**, *72*, 301–303.

129. Tomizawa, J. Control of ColE1 plasmid replication: the process of binding of RNA I to the primer transcript. *Cell* **1984**, *38*, 861–870.

130. Kleckner, N.; Weiner, B.M. Potential advantages of unstable interactions for pairing of chromosomes in meiotic, somatic and premeiotic cells. *Cold Spring Harb. Symp. Quant. Biol.* **1993**, *58*, 553–565.

131. Dibb, N.J. Proto-splice site model of intron origin. *J. Theor. Biol.* **1991**, *151*, 405–416.

132. Rogers, J. How were introns inserted into nuclear genes? *Trends Genet.* **1989**, *5*, 213.

133. Patthy, L. Exons—original building blocks of proteins? *BioEssays* **1991**, *13*, 187–192.

134. Cavalier-Smith, T. Selfish DNA and the origin of introns. *Nature* **1985**, *315*, 283–284.

135. Cavalier-Smith, T. Intron phylogeny: a new hypothesis. *Trends Genet.* **1991**, *7*, 145–148.

136. Stoltzfus, A.; Spencer, D.F.; Zuker, M.; Logsdon, J.M., Jr.; Doolittle, W.F. Testing the exon theory of genes: the evidence from protein structure. *Science* **1994**, *265*, 202–207.

137. Wessler, S.R. The splicing of maize transposable elements from pre-mRNA—a mini-review. *Gene* **1989**, *82*, 127–133.

138. Menssen, A.; Hohmann, S.; Martin, W.; et al. The En/Spm transposable element of Zea mays contains splice sites at the termini generating a novel intron from a dSpm element in the A2 gene. *EMBO J.* **1990**, *9*, 3051–3057.

139. Fridell, R.A.; Pret, A.M.; Searles, L.L. A retrotransposon 412 insertion within an exon of the *Drosophila melanogaster* vermilion gene is spliced from the precursor RNA. *Genes Dev.* **1990**, *4*, 559–566.

140. Rogers, J.H. The role of introns in evolution. *FEBS Lett.* **1990**, *268*, 339–343.

141. Charlesworth, B.; Langley, C.H. The population genetics of *Drosophila* transposable elements. *Annu. Rev. Genet.* **1989**, *23*, 251–287.

142. Cameron, J.R.; Loh, E.Y.; Davis, R.W. Evidence for transposition of dispersed repetitive DNA families in yeast. *Cell* **1979**, *16*, 739–751.

143. Hartl, D.L.; Medhora, M.; Green, L.; Dykhuizen, D.E. The evolution of DNA sequences in *Escherichia coli. Philos. Trans. R. Soc. Lond. Biol.* **1986**, *312*, 191–204.

144. Palmer, J.D. Comparative organization of chloroplast genomes. *Annu. Rev. Genet.* **1985**, *19*, 325–354.

145. Tonegawa, S.; Maxam, A.M.; Tizard, R.; Bernard, O.; Gilbert, W. Sequence of a mouse germline gene for a variable region of an immunoglobulin light chain. *Proc. Natl. Acad. Sci. USA* **1978**, *74*, 1485–1489

146. Crick, F. Split genes and RNA splicing. *Science* **1979**, *204*, 264–271

147. Dibb, N.J.; Newman, A.J. Evidence that introns arose at proto-splice site. *EMBO J.* **1989**, *8*, 2015–2022.

148. Marchionni, M.; Gilbert, W. The triosephosphate isomerase gene from maize: Introns antedate the plant-animal divergence. *Cell* **1986**, *46*, 133–141.

149. Shah, D.M.; Hightower, R.C.; Meagher, R.B. Genes encoding actin in higher plants: intron positions are highly conserved but the coding sequences are not. *J. Mol. Appl. Genet.* **1983**, *2*, 111–126.

150. Nawrath, C.; Schell, J.; Koncz, C. Homologous domains of the largest subunit of eucaryotic RNA polymerase II are conserved in plants. *Mol. Gen. Genet.* **1990**, *223*, 5–75.

151. Pardo, J.M.; Serrano, R. Structure of a plasma membrane H+-ATPase gene from the plant *Arabidopsis thaliana. J. Biol. Chem.* **1989**, *264*, 8557–8562.

152. Wing, S.S.; Bonville, D. The 14-kDa ubiquitin conjugating enzyme: structure of the rat gene and regulation of mRNA levels upon fasting and by insulin. *Am. J. Physiol.* **1994**, *267*, E39–E48.

153. Kersanach, R.; Brinkmann, H.; Liaud, M.-F.; Zhang, D.-X.; Martin, W.; Cerff, R Five identical intron positions in ancient duplicated genes of eubacterial origin. *Nature* **1994**, *367*, 387–389.

154. Setoyama, C.; Joh, T.; Tsuzuk, T.; Shimada, K. Structural organization of the mouse cytosolic malate dehydrogenase gene: comparison with that of the mouse mitochondrial malate dehydrogenase gene. *J. Mol. Biol.* **1988**, *202*, 355–364.

155. Obaru, K.; Tsuzuki, T.; Setoyama, C.; Shimada, K. Structural organization of the mouse aspertate aminotransferase isozyme genes: introns antedate the divergence of cytosolic and mitochondrial isozyme genes. *J. Mol. Biol.* **1988**, *200*, 13–22.

156. Green, P.; Lipman, D.; Hillier, L.; Waterson, R.; States, D.; Claverie, J.-M. Ancient conserved regions in new gene sequences and the protein database. *Science* **1993**, *259*, 1711–1717.

157. Logsdon, J.M., Jr.; Tyshenko, M.G.; Dixon, C.; Jafari, J.D.; Walker, V.K.; Palmer, J.D. Seven newly discovered intron position in the triose-phosphate isomerase gene: evidence for the introns-late theory. *Proc. Natl. Acad. Sci. USA* **1995**, *92*, 8507–8511.

158. Kwiatowski, J.; Krawczyk, M.; Kornacki, M.; Bailey, K.; Ayala, F.J. Evidence against the exon theory of genes derived from the triose-phosphate isomerase gene. *Proc. Natl. Acad. Sci. USA* **1995**, *92*, 8503–8506.

159. Fink, G.R. Pseudogenes in yeast? *Cell* **1987**, *49*, 5–6.

160. Nyberg, A.M.; Cronhjort, M.B. Intron evolution: a statistical comparison of two models. *J. Theor. Biol.* **1992**, *157*, 175–190.

161. Giroux, M.J.; Clancy, M.; Baier, J.; Ingham, L.; McCarty, D.; Hannah, L.C. De novo synthesis of an intron by the maize transposable element Dissociation. *Proc. Natl. Acad. Sci. USA* **1994**, *91*, 12150–12154.

162. Hall, D.H.; Liu, Y.; Shub, D.A. Exon shuffling by recombination between self-splicing introns of bacteriophage T4. *Nature* **1989**, *340*, 574–576.

163. Shub, D.A. The antiquity of group I introns. *Curr. Opin. Genet. Dev.* **1991**, *1*, 478–484.

164. Bryk, M.; Belfort, M. Spontaneous shuffling of domains between introns of phage T4. *Nature* **1990**, *346*, 394–396.

165. Mikheeva, S.; Jarrell, K. Use of engineered ribozymes to catalyze chimeric gene assembly. *Proc. Natl. Acad. Sci. USA* **1996**, *93*, 7486–7490.

166. Haldane, J.B.S. *The Causes of Evolution*. Longmans, New York, 1932.

166a. Caffarelli, E.; Arese, M.; Santoro, B.; Fragapane, P.; Bozzoni, I. In vitro study of processing of the intron-encoded U16 small nucleolar RNA in *Xenopus laevis*. *Mol. Cell. Biol.* **1994**, *14*, 2966–2974.

167. Ohno, S. *Evolution by Gene Duplication*. Springer-Verlag, Berlin, 1970.

168. Ohta, T. On the evolution of multigene families. *Theor. Popul. Biol.* **1983**, *23*, 216–240.

169. Ohta, T. Simulating evolution by gene duplication. *Genetics* **1987**, *115*, 207–213.

170. Ohta, T. The role of gene duplication in evolution. *Genome* **1989**, *31*, 187–192.

171. de Souza, S.J.; Long, M.; Gilbert, W. Introns and Gene Evolution. *From Genes to Cells*, Vol. 1. 1996, in press.

172. Haldane, J.B.S. A Mathematical Theory of Natural Selection. VIII. Stable Metapopulation. *Proc. Cambridge Philos. Soc.* **1931**, *27*, 137–142.

173. Wright, S. The roles of mutation, inbreeding, crossbreeding, and selection in evolution. *Proc. Sixth Int. Cong. Genet.* **1932**, *1*, 356–366.

174. Stephan, W.; Kirby, D.A. RNA folding in *Drosophila* shows a distance effect for compensatory fitness interaction. *Genetics* **1993**, *135*, 97–103.

175. Kirby, D.A.; Muse, S.V.; Stephan, W. Maintenance of pre-mRNA secondary structure by epistatic selection. *Proc. Natl. Acad. Sci. USA* **1995**, *92*, 9047–9051.

176. de Souza, S.J.; Long, M.; Scheonbach, L.; Gilbert, W. Intron positions correlate with module boundaries in ancient proteins. *Proc. Natl. Acad. Sci. USA* **1996**, *93*, 14632–14636.

177. Desseyn, J.L.; Guyonnet Duperat, V.; Porchet, N.; Aubert, J.P.; Laine, A. Human mucin gene muc5B: The large, 10.7-kb-large central exon encodes various alternate subdomains resulting in a super-repeat—Structural evidence for a 11p15.5 gene family. *J. Biol. Chem.* **1997**, *272*, 3168–3178.

178. van den Hoff, M.J.B.; Jonker, A.; Beintema, J.J.; Lamers, W.H. Evolutionary relationships of the carbamoylphosphate synthetase. *J. Mol. Evol.* **1995**, *41*, 813–832.

GENOME ARCHITECTURE

Andrei O. Zalensky

Advances in Genome Biology
Volume 5A, pages 179-210.
Copyright © 1998 by JAI Press Inc.
All rights of reproduction in any form reserved.
ISBN: 0-7623-0079-5

I. INTRODUCTION

"The best new [idea] is the well forgotten old [one]." This Russian saying is true for research in genome architecture, not only because the century-old works by Rabl[1] and Boveri[2] top the citation list in the field, but also because researchers again and again pursue very similar, if not the same, questions. The works of predecessors may often be forgotten or unknown (especially if they are not covered by common computer data bases, which have a short memory of 5–10 years). In this chapter I will try to present a contemporary picture of chromosome topology within eukaryotic nuclei; however, in some instances I can not resist mentioning the "old goodies," in view of the fact that "the motions of the human mind are similar whatever optical means lie before it" and that "pure research is often overshadowed by technology."[3]

There is no doubt that spatial order exists within the cell nucleus, and genome architecture is a prominent constituent within this order. An extreme example is metaphase chromosomes, which reappear during each cell division in a reliable, recognizable, and reproducible pattern. A much more meaningful example is the chromosome structure in the nucleus during the interphase, during which DNA expresses and multiplies.[4] Interphase nuclei are arranged so that replication, transcription, repair, and RNA processing can occur at restricted sites.[5] These events are accompanied (and may be regulated) by dynamic changes in genome architecture at levels well above an individual gene and even chromatin structure.

Throughout the century and up to the present day studies concerning chromosome topology have been influenced by pioneer ideas and observations of Rabl[1] and Boveri.[2] While studying cell divisions in *Salamandra*, Rabl noticed that patterns of chromosome localization in telophase and early prophase were the same. He suggested that during intervening stages, the chromosomes, although not separately recognizable, do not lose their identity and remain relatively localized. Rabl believed that interphase nuclei should have a "pole" toward which the apices of chromosomes [centromeres] converge, whereas [telomeres] are located at the opposite "antipole," arranged in the so-called Rabl configuration (Figure 1A). Boveri extended these observations in more detailed studies of cell divisions in *Ascaris*. In modern terms, these works suggested (1) territorial organization, (2) nonrandom relative positioning, and (3) defined orientation of chromosomes in interphase nuclei.

Unfortunately, the organization of chromosomes in interphase cells is extremely difficult to study. Light microscopy, and even reconstructions made from conventionally stained electron micrographs of serial sections, in most cases did not allow the distinction of individual chromosomes.[6] Therefore, direct experimental validation of the hypothesis by Rabl and Bovery became possible only recently, mostly owing to development of the nonisotopic *in situ* hybridization techniques. Other important components of the modern approach are practically unlimited sources of DNA probe sequences, development of high-resolution microscopy,

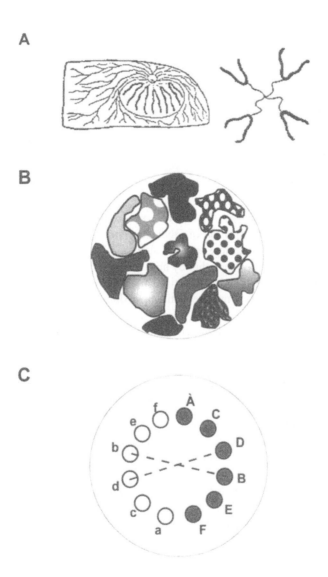

Figure 1. (**A**) Diagram of chromosome individuality according to Rabl.[1] Left, an interphase cell showing a nuclear "pole" toward which the apices of the V-shaped chromosomes converge. Right, view of the same from above. Only four chromosomes connected by the fibril network are shown. (**B**) Chromosome territories in interphase nuclei, simplified from Cremer et al.[9] (**C**) Scheme of chromosome positioning in prometaphase (compiled from refs. 47, 48, 50, 51). Capital and small letters designate homologous chromosomes; filled and opern circles correspond to two parental genomes.

and computerized three-dimensional reconstruction—methods that will be briefly discussed below.

The current study of genome architecture is a very dynamic and active field of research relevant to such topics as cell differentiation, carcinogenesis, development, and others, yet after more than a century this field is still in its adolescence. Many good and relatively fresh reviews have appeared in recent years that have undoubtedly influenced this chapter, and I would like to turn the reader's attention to these.[6–13]

II.   DEFINITION OF TERMS

Collectively, *genome architecture* refers to the spatial arrangement of chromosomes within the nuclear volume. (The same term is used in a different context to describe relative linear organization of DNA sequences of different types (e.g. unique, repetitive, etc.).) Therefore, on one side it is bordered by higher-order chromosome structure, and on the other by nuclear architecture (e.g., localization of transcription, processing and replication machineries). Several levels and aspects of genome architecture may be distinguished in greater detail:

1.   Chromosome territory. Volume or area in two-dimensional presentations, occupied by an individual chromosome.
2.   Chromosomal domain. Part of the chromosome determined by specific DNA sequence and protein–nucleic acid interactions. Chromosomal domains may have a specific chromatin structure different from that of bulk chromatin. Often chromosomal domains manifest particular genetic and structural functions, e.g., telomere and centromere domains.
3.   Chromosome path. Chromosome trajectory within nuclei, formed by the chromosome thread of the highest (suprachromatin) level of organization.
4.   Chromosome positioning. Spatial localization of chromosomes relative to each other or to defined nuclear structures, e.g., nuclear membrane.

Other authors[14] also use the term *functional domain* (active chromatin domain, transcript-processing domain, replication domain), which, in my opinion, is less suitable, especially for the structural perspective addressed in this chapter.

III.   METHODS OF INVESTIGATION

This section will present a brief outline of the experimental approaches that are used to study chromosome organization and topology. Specialized reviews are recommended for the detailed description of methods.

Earlier studies of genome architecture in interphase nuclei were mostly based on the extrapolation of patterns observed in smear or squash preparations of early prometaphase cells to the preceeding stage of the cycle. Classical cytochemical staining techniques (Feulgen, Giemsa, etc.) in several specific cases (e.g., some plant species, polytene nuclei) allowed discrimination between individual chromosomes and chromosome domains.

Three components are crucial for the study of genome architecture:

1. Ability to specifically delineate entire chromosomes, chromosomal domains and individual sequences *in situ*;
2. Adequate high resolution optical techniques for imaging;
3. Methods of data analysis, including reliable distance determination and three-dimensional reconstruction.

The development of nonisotopic and, in particular, fluorescent *in situ* hybridization (FISH) supported by the availability of enormous numbers of chromosomal DNA probes completely satisfies the first prerequisite. For the detailed discussion of methods of sample preparation, DNA labeling, and FISH, I refer the reader to recent reviews.[15,16] The ability to tag several regions of chromosome simultaneously using biotin, digoxigenin, and direct fluorochrome labeling of DNA probes makes the FISH technique extremely powerful. Detectable target sequences may be as small as 1 kb or as large as whole chromosomes or even an entire genome. Accordingly, several quite different types of DNA probes are explored in studies of genome architecture by FISH.

A. Chromosome-Specific Painting Probes

Chromosome-specific painting probes are a collection of DNA seqences belonging to an individual chromosome (e.g. isolated by sorting), which hybridize along the entire chromosome length (Figure 2A). Such probes have proved to be useful for a "large-scale" analysis, such as establishing chromosome territorial organization or relative positioning. They are less suitable for studies of "smaller-scale" details of chromosome organization (e.g., tracking chromosome path or packing), because the strong signals emerging from different regions of chromosome tend to overlap. Painting probes for most of the human chromosomes are commercially available.

B. Region-Specific Probes

Region-specific probes target long stretches (several Mbp) of a chromosome (Figure 2B). Such probes may be composed of overlapping or adjacent genomic DNA clones. Alternatively, DNA is obtained by mechanical microdissection of metaphase chromosomes and amplified by random primed PCR.[17] Region-specific probes, when used in combination in multicolor FISH experiments, may pro-

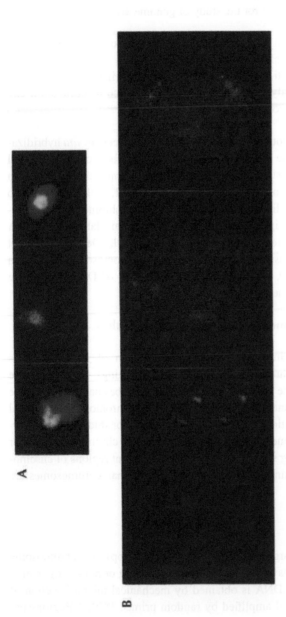

Figure 2. Types of the DNA probes most commonly used in FISH experiments for studies of genome architecture. **(A)** Chromosome painting probe. Hybridization to chromosome 2 in the nucleus of human sperm cells. Biotin-labeled DNA probe from Oncor. Hybridization detected with avidin-fluorescein isothiocyanate. Total DNA (in left and right cells) counterstained with propidium iodide. **(B)** Region-specific probes. Dual-color hybridization using microdissected probes corresponding to a particular regions of the p and q arms of human chromosome 6 (provided by M. Bittner, NIH). Left, hybridization to metaphase chromosomes of peripheral blood lymphocytes; right, hybridization to interphase nuclei of skin fibroblasts.

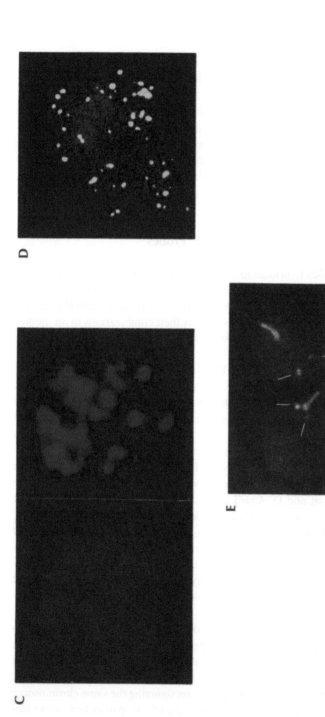

Figure 2. (C, D) Chromosome domain-specific probes. (C) HeLa nucleus, probe labeling all centromeres. Right, hybridization (red); left, the same nucleus with the total DNA counterstained with DAPI (blue). (D) Probe labeling all telomeres. Hybridization (yellow) localizes telomeres in highly decondensed nuclei of human sperm cell, total DNA is counterstained with propidium iodide (red). (E) Unique DNA probes. Human sperm cells were hybridized simultaneously with two cosmid probes labeled with biotin and digioxigenin (green and red hybridization signals, shown by arrows). DNA probes originating from human chromosome 4 were provided by B. Trask, University of Washington.

vide direct information on chromosome higher order structures and chromosome paths within nuclei.

C.   Domain-Specific Probes

Domain-specific probes are usually represented by α-satellite or other repetitive sequences corresponding to specific functional domains of the chromosome (Figure 2C,D). Most commonly used are centromere-specific and telomere-specific probes. These probes produce intense signals that are easy to register. For the human genome, both chromosome-specific centromeric and subtelomeric probes, in addition to probes targeting all telomeres or centromeres, are currently available commercially.

D.   Unique "Point" Probes

Unique "point" probes hybridize to a specific, unique DNA target sequence on a given chromosome (Figure 2E). The constantly growing number of DNA sequences being studied (e.g., cosmid or YAC clones emerging from the Human Genome Projects) makes these probes (especially their combinations in multicolor FISH experiments) very suitable for a detailed analysis of the genome arcitecture on different scales of distances.

The high sensitivity and specificity of FISH demands compatable techniques of microscopic detection. Epifluorescent microscopy, using multi-band-pass filters and conventional photoregistration or chilled charge-coupled device cameras, is the technique of choice for most applications.[18,19] There are more sophisticated and advanced approaches that may provide three-dimensional information on chromosome topography; these are laser scanning confocal microscopy[20,21] (LSCM) and optical sectioning microscopy. The latter approach, in combination with digital 3-D reconstruction, has been developed and used to study genome architecture by J. Sedat and D. Agard.[22,23] LSCM, although it has a number of intrinsic limitations,[24] (e.g., in the faithful determination of distances in the z-dimension, its ability to utilize multicolor data, etc.), has been successfully applied to the investigation of nuclear architecture in a number of laboratories. Methods of processing confocal data and three-dimesional reconstruction[25–29] (and the corresponding software) are currently being rapidly developed. Taken together, recent technical developments make it feasible to resolve the three-dimensional arrangement of a genome within the nuclear volume.

A quite different approach to the analysis of FISH data has been developed in the laboratories of B. Trask's and G. van der Engh's.[30,31] Their method is based on the use of stastistical analysis of many observations to study systematic patterns of chromosome organization. Chromatin folding on the scale of >50 kb is determined by distance measurements between two probes recognizing the same chromosome and with known genomic locations. This approach is particulary important because it bridges a huge gap between the rapidly accumulating data on large-scale genome architecture and the data concerning chromatin higher order structures.

Finally, the more traditional techniques, such as immunolocalization of proteins and electron microscopy sectioning, also provide important and unique information about genome architecture. Unfortunately, all present experimental approaches (with a few exceptions)[32,33] have a very limited ability to capture dynamic changes in genome architecture, which makes it more difficult to establish a direct correlations with such processes as transcription or replication.

IV. CHROMOSOME TERRITORIES

The chromosome was recognized as the Mendelian "hereditary particle" in the beginning of this century (theories of Weismann, Sutton, and Boveri).[34,35] Each eukaryotic chromosome is an independent structural/functional unit of genome organization. Obligatory elements required for chromosome formation and functioning—the centromere, the telomere, and origins of replication—were first recognized in yeast.[36] Both metaphase and interphase chromosomes are complex hierarchical structures in which the DNA double helix is compacted by proteins at several successive levels: nucleosomes, solenoids, chromatid fibers, loops, etc. A number of models for the interphase and the metaphase chromosome structures, sometimes mutually exclusive, have been proposed.[37–42] Whereas a metaphase chromosome can be identified cytogenetically by its distinctive banding pattern and size, an interphase chromosome, as an entity, has escaped such easy identification until recently. First, C. Cremer and T. Cremer,[43] using UV microbeam irradiation of local areas in the metaphase plate and immunofluorescence to localize damaged DNA, demonstrated distinct patches of damaged DNA in interphase. This experiment was interpreted as favoring topological individuality of chromosomes in interphase. The development of FISH and chromosome-specific painting probes allowed direct localization of individual human chromosomes in interphase nuclei of several cell types. These works,[44–47] which were carried out almost simultaneously in several laboratories, clearly demonstrated that chromosomes occupy distinct areas in the mammalian interphase nucleus (Figure 1B, see also example, Figure 2A). The shape of these compartments is irregular and highly variable among cells. The territorial organization of the interphase chromosomes has also been shown for plant cells[48]; therefore this feature of genome architecture is evolutionarily conserved. Additional support for territorial organization came from B. Trask's laboratory. Systematic distance measurements between probes belonging to the same chromosome indicate the existence of irregularly shaped chromosome territories with average sizes much less than the nuclear volume.[31,42]

In summary, one of the predictions of the Rabl-Bovery hypothesis concerning individuality of the interphase chromosomes was proved to be true. The interphase nucleus is not a sack filled with DNA; rather each chromosome represents a structural entity throughout the cell cycle. Possible functional roles of chromosome territories will be discussed later in this chapter.

V. CHROMOSOME POSITIONING, SOMATIC PAIRING, AND GENOME SEPARATION

The existence of the interphase chromosome territories evokes questions concerning their positioning within the nuclear volume. Spatial distribution of chromosomes (their locations relative to each other or to morphologically defined nuclear structures; e.g., nucleolus, nuclear membrane) may be either completely random or ordered. Once again, according to Rabl and Bovery, chromosome positioning in the interphase is dictated by their presumed ordered localization in the metaphase plate and is significantly nonrandom. There are a some genetic data obtained in *Drosophila* that demand exact chromosome positioning.[49] A particular locus of *Drosophila melanogaster* functions as a repressor of another locus (*zeste–white* interactions). Repression takes place only if the two alleles of the repressed locus are physically adjacent or paired. This demands a precise nuclear architecture in interphase nuclei and particularly close association of homologs. Another biological implication of chromosome positioning is connected with the initiation of meiotic chromosome synapsis.[50,51]

A. Positioning of Nonhomologous Chromosomes

Difficulties in determining interphase positioning using classical methods of cytogenetic analysis are evident—individual chromosomes are not visible, and the large number of chromosomes in most organisms adds to the confusion. The ordered position of chromosomes in the metaphase plate is itself questionable. For example, reconstruction of serial sections from electron micrographs of mitotic human male fetal lung fibroblasts did not show any significant order.[52] The authors did note some tendency toward a central location for chromosomes Y and 18 and a peripheral location for chromosome 6. Therefore, only partial sorting by size across the metaphase plate could be confirmed.

One of the more fruitful approaches to the analysis of chromosome positioning is connected with the analysis of early prophase cells, a stage where chromosomes are individually visible. In 1979, Ashley studied the localization of chromosomes in haploid generative nuclei and diploid prophase nuclei of the root tip in the plant *Ornithogalum virens*.[51] A nice feature of this organism is that in the haploid genome, it has only three chromosomes with distinguishable banding patterns. Using the simple classical cytogenetic approach, the author demonstrated that in both diploid and haploid nuclei, chromosomes are arranged in a specific order. Importantly, in prometaphase, homologous chromosomes lie opposite each other.

Later, the three-dimensional chromosome arrangement was studied by LSCM in mitotic prophase and anaphase of another plant (*Crepis cappilaris*; $2n = 6$).[53] A series of optical sections were obtained after Feulgen staining. In metaphase, the centromeres of all chromosomes in this species are circularly arranged in the equatorial plane. Theoretically, 11 circular permutations of the six centromeres are possible. But the observed distribution differed from random, in that one particular arrangement was prefered. Finally, 16 years later, after the work of Ashley (discussed above), Nagele and co-authors[54] come to exactly the same conclusions while studying localization of chromosomes in human skin and lung diploid fibroblasts and HeLa prometaphase cells. FISH with chromosome-specific centromeric probes was used to identify individual chromosomes, it was shown that chromosome homologs were consistently positioned on opposite sides of the prometaphase ring. Indications were that precise, cell-type-independent relative positioning of chromosomes exists within the rosette. Thus the data from both plant and animal systems, although they are not final, indicate an ordered relative localization of chromosomes in prometaphase (metaphase?) (Figure 1C). The mechanisms underlying the establishment and maintenance of such an order are completely obscure.

Data on chromosome positioning within interphase nuclei are contradictory. Most of the results have been obtained by using FISH with chromosome-specific centromeric DNA probes. Therefore, the spatial position of a chromosome was judged by the position of one relatively small domain, the centromere (CEN). It has been hypothesized that chromosomes may be distributed within interphase nuclei according to their size—larger chromosomes are located closer to the periphery. Indeed, multicolor FISH, followed by 2-D and 3-D image analysis, demonstrated that in human skin and lung fibroblasts, amniotic fluid cells, and lymphocytes, CEN-1 and CEN-X were much closer to the periphery than were CEN-18 and CEN-15.[55,56] The positions of the two Barr bodies (inactive X chromosomes) were examined in human XXX fibroblasts.[57] The authors used Feulgen staining, which identifies the Barr bodies in order to distinguish a highly nonrandom intranuclear distribution. The inactive X is preferentially localized at the nuclear periphery in a plane parallel to the plane of cell growth. In contradiction to all of the above results, the positions of CEN-7, CEN-11, and CEN-17 in interphase human lymphocytes are not spatially defined.[58] Here, 3-D arrangement of three chromosomes was analyzed simultaneously by measuring CEN–CEN distances after multicolor FISH and computer-assisted digital imaging fluorescent microscopy. Histograms of heterologous CEN separation were indistinguishable from that obtained for a random distribution. In summary, there is currently no firm support for the fixed relative positioning of nonhomologous chromosomes within interphase nuclei. In the absence of interphase positioning, the observed prometaphase order should be generated *de novo* during each cell division.

B.   Somatic Pairing of Homologous Chromosomes

There have been intense efforts to determine the relative spatial positioning of homologous chromosomes in diploid interphase nuclei in a number of laboratories. The close location of the homologs is suggestive of some sort of interaction, demanded by genetic data as described above. According to *in situ* hybridization results, individual homologous chromosomes within the interphase nucleus are separated in nonassociated territories in human skin and lung fibroblasts, lymphocytes, and amniotic fluid cells[55,56,58]; cells of the central nervous system[59]; and mouse T-lymphocytes.[60] In contrast, data based on the immunolocalization of centromeres in rat kangaroo ovary tumor cells and Indian muntjac cells indicate that homologous chromosomes occupy adjacent territories.[61] It is noteworthy that these two species have few chromosomes, which simplifies the analysis.

Very interesting data concerning the association of homologs have been obtained in Sedat's laboratory while studying *Drosophila* embryogenesis.[32,62] The authors used 3-D optical sectioning microscopy and whole mount embryos. At early developmental stages, all nuclei within the embryo divide synchronously. Each chromosome has a specific shape and a characteristic pattern of 4′,6-diamidino-2-phenylindole (DAPI) staining, so that it is possible to visualize the 3-D path of an individual chromosome. It was shown that during prophase, telomeres of homologous chromosomes were closely associated. At other stages, homologs are distinctly separate. In the following work,[62] the authors used a histone gene cluster located in chromosome 2 to target this chromosome and thus investigate the proximity of two homologs. Up to the 13th division, the two homologous histone loci are distinct and separate through all stages of the cell cycle. Surprisingly, during the interphase of the 14th division, the clusters became colocalized. At the same time, both clusters move from the position near the midline of the nucleus to the apical side. Exactly during this period the initiation of zygotic transcription takes place; this is accompanied by a dramatic nuclear reorganization and, in particular, close associations of homologous chromosomes.

Finally, many studies (reviewed by Avivi and Feldman),[63] which have been performed by classical cytologogical methods on a variety of plant species, indicate intimate asociations of homologous chromosomes in somatic interphase.

C.   Spatial Separation of Genomes

All somatic cells of diploid organisms have two haploid genomes inherited from the parents. In prometaphase, homologous chromosomes seem to be positioned opposite each other (Figure 1C). This means that parental genomes are spatially separated at this stage of the cell cycle. It has been suggested,[54] that a mechanism exists to maintain this order (e.g., spatial separation of parental haploid genomes) throughout the entire cell cycle. One possibility is a permanent attachment of chromosomes to one another. Localization of complete genomes has been established in some plants (reviewed by Heslop-Harrison).[10,11] A unique opportunity was provided by hybrids between wild and cultivated cereals. Genomes in such hybrids have been localized using FISH with total genomic DNAs as a probe.[11] Parental genome separation was shown to be maintained throughout the cell cycle in different tissues and preserved over time.[64] Thus genome separation is under genetic control.

In summary, from the data discussed above it is evident that both large-scale (complete genome) and small-scale (individual chromosome) positioning exists, at least at some stages of the cell cycle, in some cell types, and in some species. In the majority of experiments, the relative chromosome positioning (both homologous and nonhomologous) is judged based on the observations of a single chromosomal locus, which might be nonrepresentative. Simultaneous 3-D localization of many loci spanning the entire lengths of a chromosome pair (in an approach similar to that developed in the Trask laboratory to study the higher order structure of chromosomes)[42] is desirable. This is certainly a serious experimental challenge for the future.

VI. LOCALIZATION OF CHROMOSOME DOMAINS IN DIVIDING AND TERMINALLY DIFFERENTIATED CELLS

A. Mitotic Cell Cycle

The results describing dynamic modifications of the genome architecture during the mitotic cycle have been obtained primarily by following the localization of the centromere and telomere domains.

The most detailed information comes from the studies of the human and mouse lymphocyte cells, from the laboratories of D. Ward[60,65] and M. Schmid.[66] The first group used FISH with centromere- and telomere-specific probes, and the second group employed immunolocalization of kinetochore-specific antibodies. Because it is known that both centromere hybridization and localization of these antigens correspond throughout all stages of the cell cycle,[67,68] results are comparable and complement each other.

Isolated human T-lymphocyte nuclei were sorted into G_1, S, G_2 fractions based on their DNA content and then were subjected to FISH, followed by LSCM.[65] Centromeric probes, specifically labeling nine different chromosomes, probes targeting subtelomeric regions of the long arms of chromosomes 11 and 14, and the composite probe painting chromosome 8 were used. In G_1, all centromeres were distributed in the vicinity of the nuclear periphery, whereas telomeres were localised in the interior. Localization of chromosome 8 indicated that the chromosome arms were directed toward the nuclear interior. In G_2, centromeric signals were located with more bias toward the nuclear center, and chromosome 8 arms demonstrated outward orientation. Welmer and co-authors visualized centromeres in

G_0, G_1, S, G_2 and early M of human lymphocytes with serum from scleroderma patients recognizing kinetochore proteins.[66] In G_0/G_1, centromeres were partially associated in several clusters, each consisting of up to four centromeres. These clusters were localized in the nuclear periphery. In S the clusters separated into individual centromeres, which had a tendency to localize toward the nuclear center. The authors note that this stage is very heterogeneous; different arrangements of centromeres have been found: associations around the nucleolus, small clusters in different nuclear regions, and centromere chains. In G_2, seemingly random centromere distribution over the nucleus was observed, whereas in prometaphase the centromeres were located centrally.

In later work by Vourc'h et al.,[60] nuclei isolated from mouse T-cells were sorted into G_1, S, and G_2/M stages. FISH with major and minor satellite probes (both label all mouse chromosomes) and subcentromere satellite probes specific for chromosomes X, 8, and 14 were used. *In situ* hybridization was followed by LSCM and three-dimensional analysis to determine the distribution with respect to center of the nucleus. In parallel experiments, the distribution of telomeres was determined. In G_1, centromeres were found to have a tendency toward periphral location, whereas in S and G_2 signals were found disseminated throughout the whole nuclear volume, with an increasing bias toward a more interior position. Telomere signals in all fractions were found to be distributed throughout the entire nuclear volume. The authors observe some minor differences between mouse and human which may reflect differences in chromosome structure between these animals (e.g., in the mouse, all chromosomes are acrocentric).[60]

Peripheral localization of centromeres in G_1 is probably a consequence of their association during mitosis. At this stage, centromeres are collected at the spindle. After the nuclear envelope reforms, most of the centromeres in cultured cells remain closely associated with the nuclear membrane.[4] All three studies[65–67] provide consistent information on the repositioning of centromeres toward the nuclear center during the cell's progression from G_1 to G_2. Noticable movements of chromosomes (topological modifications of chromosome territory) do occur. These movements are not restricted to centromere and telomere domains, as has been demonstrated in experiments using a chromosome painting probe. Furthermore, a comparison of results obtained in human and mouse indicate that the character of topological changes during the cell cycle is conserved in mammalian evolution, at least for lymphocytes.

There are a few data that provide a detailed description of chromosome localization during the complete division cycle for the other mammalian cell types. In human fibroblasts and HeLa cells at the prometaphase stage, chromosome arms are oriented outward from the ring formed by centromeres.[54] In human fibroblasts, most of the centromeres were found to be fused into chromocenters in G_1, but unlike lymphocytes, no association of centromeres with the nuclear envelope was demonsrated.[68] The centromere clusters (chromocenters) disperse during S. Telomere–telomere associations located near one pole of the nucleus were described in both early and late replicating Syrian hamster cells.[13] Telomere domains were not located at the nuclear membrane. In cultured cells of several brain,[13] lung, and breast[69] tumors, unusual and complex patterns of centromere domain localization during the cell cycle have been observed. Recently, Haaf and Ward,[70] using FISH with the probe corresponding to the binding site of the centromeric protein B, demonstrated that centromeres are always clustred together at one pole in the interphase skin fibroblasts of the tree shrew *Tupaia*. Although the exact stage of interphase has not been determined in this study, the data illustrate a well-defined Rabl orientation of chromosomes, which is rare for mammals.

The cell-cycle-dependent positioning of telomeres has been investigated in fission yeasts.[71,72] *Schizosaccharomyces pombe* has a small genome consisting of three chromosomes, which simplifies the localization of individual chromosomal domains. Importantly, mutations affecting cell divisions are known; therefore, descriptive cytological studies may be combined with advanced genetic analysis. Funabiki and co-authors[72] studied the behaviour of centromere and telomere domains during the cell division cycle of wild-type and mutant cells using FISH, with the probe labeling all centromeres and chromosome I- and II-specific telomeric probes. Immunolocalization of spindle pole bodies (SPBs), which are analogs of centrosomes in higher eukaryotes, has been performed simultaneously. In G_2, centromeres are clustered at the periphery of the chromatin area, next to SPBs. Telomeres are also clustered and located at the nuclear periphery. During mitosis both centromere and telomere associations disperse to identifiable individual domains. The telomere dispersal continues into the S phase. Several mutations have been described that affect normal localization of these chromosomal domains. Pronounced telomere–telomere associations and their localization near nuclear envelope in another species of yeast, *Saccaromyces cerevisiae* were demonstrated by the Gasser laboratory using immunolocalization of RAP1 protein.[73]

In most examples reviewed above, polarized (Rabl) orientation of centromere and telomere domains was not observed. In contrast to mammalian or yeast cells, pronounced polarization has been shown in embryonic cells of *D. melanogaster*.[32] In the prophase nucleus of whole-mount embryos, centromeres are located in the top half of the nucleus near the embryo surface, and telomeres are in the bottom half. Chromosome arms are oriented aproximately perpendicular to the embryo surface.

To my knowlege, systematic investigations of chromosome topology during the cell cycle (especially interphase stages) in plants have not been carried out. At the same time, some nontrivial (specific to plant kingdom?) chromosome arrangements have been described. As early as the turn of the century, chromosome chains were noticed in the nuclei of some plants.[74] In 1969, while studying root tip cells in six species of plants ($2n = 6–16$), Wagenaar observed end-to-end chromosome attachments in mitotic interphase, resulting in formations of chromosome chains.[75] Later,[50,51] similar end-to-end chromosome connections were described in more detail, in both haploid generative and diploid root tip nuclei of the angiosperm plant *Ornithogalum virens*. As has been mentioned above, these

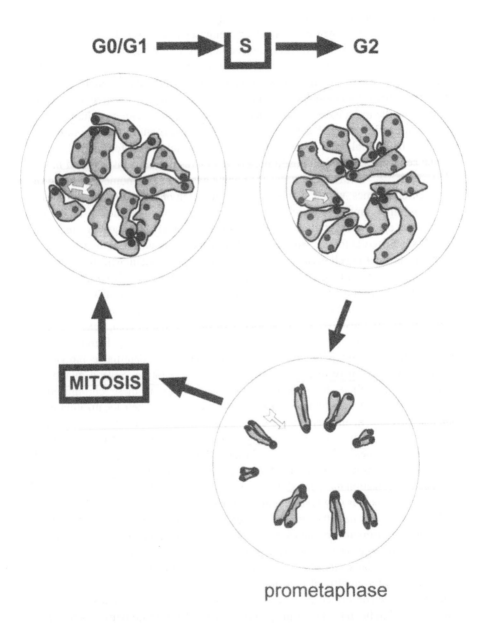

Figure 3. Scheme of the localization and movement of centromeres and telomeres during some stages of the cell cycle based on the data for lymphocyte cells obtained from Ward et al.[65,67] and Schmidt[66] et al. Centromeres are shown by black, telomeres by lighter filled circles. An arrow within a chromosome territory indicates the chromosome orientation from telomere to centromere.

works also demonstrate that homologs lie opposite one another and in specific relationship to other nonhomologous chromosomes. More recently, a three-dimensional arrangement of chromosomes was studied for yet another plant, *Crepis cappilaris* ($2n = 6$).[53] A series of optical sections obtained by LSCM after Feulgen staining were analysed. Easily distinguishable patterns of staining were used to recognize chromosomes. In metaphase, the centromeres of all chromosomes are circularly arranged at the equatorial plane. This arrangement is already visible in prophase and is prolonged throughout anaphase. Arrangements in prophase and anaphase are partly different.

In conclusion, the picture of genome architectural rearrangements ocurring from early G_0 throughout late G_2 phases of the cell cycle (Figure 3) is prelimimary. No doubt, complex and dynamic movements of entire chromosomes and particular chromosomal domains take place. The general view of events is further obscured by architectural changes connected with transcription and replication (see below). Difficulties in observing multiparameter three-dimensional patterns in nuclei nondisrupted by experimental treatments further exacerbate the problem. Architecture may be dependent on the cell type and species, although general demands defined by cell cycle events (formation of metaphase chromosomes, preparation for an ordered division, etc.) probably override the differences. Therefore, more similarities than differences will be demonstrated in future.

B.   Meiosis

Meiosis is a very specialized cell division from which haploid gametes are produced. There are numerous differences in the course of meiosis, both between sexes and between species; nevertheless, certain steps are remarkably constant throughout animals and plants. In most cases there are two "reduction" divisions, with the most crucial step at the first prophase. During this stage homologous chromosomes recognize each other, they pair, and recombination occurs. In spite of the great number of cytological and genetic studies (see refs. 35, 76, and 77 for the detailed discussion of meiotic events), few of them have examined the genome architecture as it is considered in this review. Because of the evident interest in recombination events, most detailed studies address the recognition and pairing of chromosome homologs.

As has been discussed above, the data concerning somatic pairing are very contradictory. The complete spatial separation of paretnal genomes was documented in some cases, but at best, the pairing of somatic homologs is cryptic. On the contrary, the homologous pairng during the first meiotic prometaphase is well documented cytologically. Thus establishment of this state of genome architecture (a prerequsite for recombination) should require a major rearrangement involving chromosome movement.

One of the most detailed and comprehensive descriptions of early meoitic prophase I was recently reported for maize.[78] The authors used three-dimensional optical microscopy and computerized epifluorescence light microscopy to localize large, clearly distinguishable chromosomes of this plant. At earlier stages, preceeding prometaphase I, telomere domains were widely dispersed. Prior to synapsis, telomeres moved to form homologous associations at the nuclear envelope. It was demonstrated that the pairing of at least one telomeric region preceeds pairing in internal sites of chromosomes. These data directly support the hypothesis[79] that chromosomes, during recombination, are first brought together by their telomeres. Bouquet formations, in which chromosomes are held together in meiotic prophase, has been described by methods of classical cytology in many animals and plants.[77,78] Large-scale relocations of telomeres have recently been observed during human spermatogenesis by using FISH in testis sections.[80] In spermatogonia, telomere hybridization loci were randomly dispersed in the nucleoplasm. In sharp contrast, in spermatocytes all telomeres were localized in clusters on the membrane.

In summary, telomere–telomere interactions and telomere association with membrane seem to be universal at certain stages of meiosis. At the same time, the overall topology of chromosomes remains unclear, especially for mammalian species, because there are few data describing the localization of centromere domains.

In another study, Brinkley analysed centromere localization during meiosis in mouse, using immunofluorescence with the anticentromere serum.[81] At meiotic prophase, centromeres are associated in big clusters around the periphery of pachytene nuclei. This behaviour indicates transient interactions of nonhomologous chromosomes. The authors note that this type of association was not seen in another rodent, the Chinese hamster. Our preliminary data indicate that in human spermatocytes, centromeres are localized internally relative to telomeres. The internal and clustered localization of centromeres has been documented in mouse spermatids[81] and mature sperm of human[82,83] and mouse,[84] but the stage of spermatogenesis (meiosis?) at which this localization is established is unknown.

Detailed information on genome architecture during meiosis has been obtained in two species of yeast. Scherthan and co-authors[85] studied meiotic chromosome condensation and pairing in *Saccharomyces cerevisiae*. The authors used *in situ* hybridization with ribosomal DNA (tagging the nucleolus organizing region on chromosome XII) and a collection of probes to paint chromosome V.[86] It was demonstrated that before synaptonemal complex formation, homologous chromosomes become condensed and aligned and then gradually fused. Chikashige and co-authors[33] studied chromosome dynamics during meiosis in living and fixed cells of fission yeast. The position of telomeres and centromeres was followed by computerized fluorescence microscopy. In fission yeast, under certain enviromental conditions, haploid zygotes fuse and undergo zygotic meiosis. Alternatively, an azygotic meiosis can be directly induced in diploid cells. Immediately before meiotic division, nuclear morphology changes to a specific "horse tail" shape. At this time, association of homologous chromosomes and recombination take place (meiotic prophase). In "horse tail" nuclei, telomeres were clustered next to the spindle pole body at the nuclear periphery, whereas centromeres were separated from this structure. Association of homologous chromosomes begins from the telomeres. At the first meiotic division, localization of these chromosome domains relative to the spindle pole body reversed. Interestingly, during the rearrangement of haploid nuclei toward each other before zygote fusion, chromosome movement is led by telomeres. During the later stages of meiotic division, chromosome movement is led by the centromere. These results demonstrate that drastic reorganization of genome architecture occurs, including large-scale chromosome movements during meiosis.

C. Terminally Differentiated and Specialized Cells

In this section I will review the data describing genome architecture in various types of specialized cells: cells of the central nervous system, spermatozoa, and polyploid cells posessing polytene chromosomes. The overall chromosome topology in these cases is generally well determined and very different from that of cycling cells.

Cells of the Central Nervous System

Human adult mammalian central nervous system (CNS) neurons are highly differentiated cells with little or no potential for replication. At the same time, there is high transcriptional complexity in them, which reflects the formation of a very large, and almost completely heterochromatic nucleus. Importantly for comparative studies of genome architecture, there are many functional and architectural subtypes among neuronal cells. Almost all data on chromosome topology in CNS cell nuclei discussed below have been obtained in a series of studies from the Manuelidis group, using *in situ* hybridization techniques followed by optical sectioning.[4,13,44,46,59,87]

In large (Purkinje) neurons of mouse[87] and human,[59] centromeres were shown to be clustered at the nucleolus, whereas in small, granular neurons they are more dispersed, and some are located near the nuclear membrane. In both cells, telomeres were apparently scattered in the nuclear interior. Thus these two types of CNS cells demonstrate rather different overall chromosome topologies. In astrocytes chromosomal arms protrud (radially) into interior nucleoplasm, whereas in large neurons telomeres and centromeres are often positioned together. The authors note that a specific pattern of centromere localization within a particular cell type is conserved during mammalian evolution, suggesting that such a pattern is nessesary. In several mouse and human tumor cell lines (glioma, neuroblastoma), highly variable patterns of centromere localization have been observed,[88] in contrast to that seen in normal cells of the same lineage.

Positions of centromeres and telomeres have been determined by FISH in dorsal root ganglion neurons.[89] In culture these cells form a single large central nucleolus and are permanently arrested in G_1 or G_2. Computer-assisted 3-D reconstruction indicated that a fraction of centromeres (41%) was associated with the nucleolus, whereas the rest were dispersed in the nucleoplasm. The majority of telomeres were also dispersed. A high percentage of centromeres associated with the nucleolus is surprising, because only five chromosomes (e.g., about 10% of centromeres) bear nucleolus organizing regions.

Polytene Nuclei

In some instances, individual chromatids remain closely paired during and after each round of replication to give rise to so-called polytene chromosomes, described in insects and some plants.[35] In the salivary gland of *Drosophila*, cells are arrested in the interphase, and each polytene chromosome (3–4 μm in diameter) is composed of identical copies aligned in register. Specific and individual banding patterns of polytene chromosomes enable their identification. Thus these polytene nuclei provide a perfect system for studying the high-order chromosome structures and the genome architecture. The detailed studies of *Drosophila* salivary gland chromosomes have come primarily from the Sedat and Agard groups.[22,90–92] The original technique of three-dimensional microscopy (see Section III above) has been developed and successfully used in these studies.[92]

It was shown that polytene chromosomes occupy topologicaly isolated territories, within which they fold in highly characteristic ways. The most pronounced motiffs of this folding are the specific loops, often forming roughly helical structures. The centromeres are collected into the chromocenter, which is localized within a restricted area next to the inner surface of the nuclear envelope. Telomeres are also partially clustered and localize around the opposite nuclear pole. Thus these chromosomal domains in polytene nuclei are in a well pronounced Rabl orientation.

Sperm Cells

The sperm cell is a highly differentiated cell type that results from a specialized genetic and morphological process termed *spermatogenesis*. During spermatogenesis in mammals, the somatic histones are gradually replaced with protamines or protamine-like proteins.[93,94] As a result, chromatin structure is reorganized,[95,96] DNA becomes supercondensed, and genetic activity is completely shut down.

Most mammalian sperm nuclei have a significant degree of asymmetry, which makes the study of the genome architecture much more feasible than in the spherical ovoid nuclei of somatic cells.[97] Recent data have demonstrated that a specific, unique, and organized architecture for chromosomes exists in sperm nuclei.

It has been shown by immunofluorescent localization of centromeric proteins (CENP-A) that centromeric domains preserve at least some elements of their nucleoprotein structure in the mature sperm of bull,[98] amphibians,[99] and human.[82,100] Both *in situ* hybridization and protein immunolocalization techniques indicate a nonrandom localization and prominent clustering of centromeres within sperm nuclei. In bovine sperm, centromeres are clustered across the entire width of the equatorial region of the sperm head.[101] The assembly of the 23 centromeres into a compact entity has been described in human sperm.[82,102] Similarly, centromere association was recently shown to occur in mouse sperm.[84] Centromere clusters in sperm have been named sperm chromocenters. Laser scanning confocal microscopy, followed by three dimensional reconstruction, demonstrated[83] that the human sperm chromocenter is located at the central portion of the nucleus (Figure 4). The internal structure of the chromocenter is emerging. Both CENP-A and satellite centromeric DNA have been localized in identical structures of dimers, tetramers, and linear arrays formed by centromeres of individual nonhomologous chromosomes in swollen human spermatozoa.[82] Arrays of centromeres have been also noticed in mouse[84] and centromere dimers have been found in amphibian sperm.[99]

Using FISH with the conserved human telomeric TTAGGG sequence as a probe, we demonstrate pronounced telomere–telomere interactions in the sperm cells of six mammalian species. In minimally swollen epididymal sperm of mouse and rat, all chromosomal ends were found to be clustered at a limited nuclear area while in sperm cells of human, bull, boar and stallion telomeres associate in tetramers and dimers. Recently the localization of telomeres was established in mouse testicular sperm,[84] where the telomeres were shown to be partially clustered. In sperm cells of human, bull, boar, and stallion, telomeres also form associations of tetramers and dimers.[80] Therefore, noticeable telomere–telomere interactions are characteristic of sperm in all mammals studied. At the same time, the extent of clustering and particulars of localization seem to be species-specific features, which may depend on the predominant type of chromosome (i.e., acrocentric, telocentric) and/or on other peculiarities of a karyotype (distribution of chromosome lengths).[97] Telomere–telomere dimers appear to be the most prominent and universal formation. In human sperm, telomere–telomere dimers correspond to interactions between two ends of one chromosome rather than random association between chromosomal ends. This has been demonstrated by the close proximity of the 3p and 3q subtelomere sequences.[83] Thus chromosomes in human sperm are looped. There are no data describing the nature of telomere dimers observed by us in the sperm of other mammals. Several alternative models have been proposed recently.[97]

LSCM data show that, in human sperm, telomeres are localized peripherally relative to the chromocenter (Figure 4A) and most probably are associated with nuclear membrane.[82,83] The membrane localization of telomeres is established early during human spermatogenesis, in spermatocytes.[80] Peripheral localization of telomeres has also been shown in mouse testicular sperm[84] and may be suggested from data obtained in highly decondensed hamster sperm.[103]

The existence of the internally localized chromocenter and the telomere associations at, or near, the nuclear membrane allows us to propose a model for chromosome packaging in mammalian sperm nuclei (Figure 5). This model is characterized by a pronounced spatial separation of telomere and centromere

A

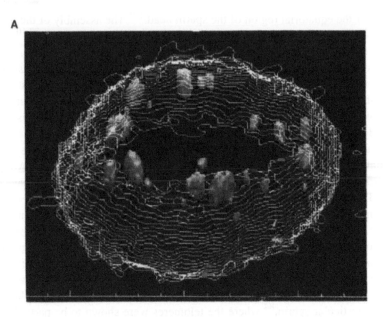

B

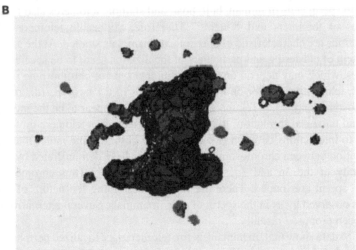

Figure 4. Elements of genome architecture in human sperm nuclei. (**A**) Localization of telomeres relative to total DNA. Three-dimensional rendering of images obtained by laser scanning confocal microscopy (LSCM), showing the positioning of telomere hybridization signals (orange) and total DNA (white countours). Upper and bottom parts of the nuclei, which do not have telomere signals, are not shown. (**B**) Relative localization of telomeres (green) and centromeres (red). Top view of the 3-D reconstruction of 22 LSCM z-sections. Cells were hybridized simulteneously with digoxigenin-labeled centromere and biotinylated telomere probes.

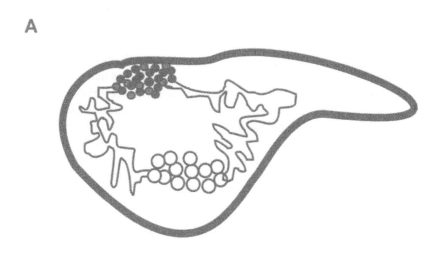

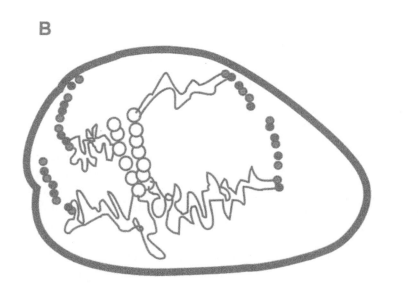

Figure 5. Models of genome architecture in mammalian sperm, showing relative localization and clustering of telomeres and centromeres (not in scale), modified from Zalensky et al.[82] and Ward and Zalensky.[97] (**A**) Mouse and rat sperm; (**B**) human sperm. Open circles, centromeres; filled circles, telomeres. The chromosome path (which is unknown) is indicated by solid lines for selected chromosomes only.

domains, so that a chromosome spans the entire nuclear volume. Indications of a linear (cylindrical) shape of chromosome territories in sperm are emerging from a few FISH data with painting probes.[83,84] Each chromosome may be folded in an overall hairpin-like structure, with the telomeres bound to each other in a dimer, as has been suggested for human sperm.[83] Alternatively, individual chromosomes may be in an extended conformation, with telomeres clustered into the telocenter or associated in smaller clusters.[97] In both cases, the telomere–telomere interactions may be either random (between the ends of different chromosomes) or specific. It is likely that the genome architecture in sperm will be species-specific. However, the basic principles of chromosome packing patterns, especially the major roles played by telomere–telomere interactions and the association of all centromeres in the chromocenter, are prominent aspects of sperm chromosome folding in many different species over a wide phylogenetic range.

VII. ROLE OF REPETITIVE DNA SEQUENCES

The repetitive DNA sequences comprise about 10% of the mammalian genome. It is known that heterochromatic regions (which are a cytological manifestation of repetitive DNAs) have a strong tendency to associate, and the association frequency increases with the size of the heterochromatic blocks involved.[14] Several chromosomal domains, characterized by the presence of repetitive DNA sequences, play a leading role in determining genome architecture. The most noticable examples are the centromere and telomere. The important structural role of these domains in different cell types and during all stages of the cell cycle has been illustrated in numerous examples noted above. Characteristics of reiterated DNA as important determinants of chromosomal positioning have previously been reviewed by Manuelidis.[13] In addition, direct evidence for the role of repetitive DNA sequences was provided by the same author.[104] In this study, the localization of an 11-Mb tandem repeat of β-globin sequence was determined in brain cells of transgenic mice. This DNA block in transgenics was integrated into the peritelomeric region of both chromosome 3 homologs and was shown to be transcriptionally silent. Importantly, transgenic domains were localized at the nucleolus and therefore behaved like (para)centromeric domains. Such localization differs from the "normal" localization characteristic of the peritelomeric domain, and therefore is the result of the repeated DNA insertion into chromosomes.

Another interesting example comes from studies on three-dimensional organization of the nucleolus by Kaplan et al.[105,106] The nucleolus is a specific area where rDNA is transcribed and preribosomes are assembled. In diploid human cells there are 10 nucleolus-organizing regions (NORs) consisting of 40–50 tandem copies of a 44-kb repeat. These blocks are located in the p-arms of the five acrocentric chromosomes 13–15, 21, and 22. NORs are flanked on both sides by long stretches of several megabases of satellite DNA. Dual-color FISH, followed by laser scanning confocal microscopy, provides information on the topography of the nucleolus.[105] Both the transcribed rDNA and the DNA of untranscribed intergenic spacer were colocalized at the periphery of the nucleolus, and the centromere of chromosome 21 was positioned nearby. Most interestingly, a large block of satellite III DNA, which lies between an rDNA cluster and the centromere in chromosome 15, was shown to play a leading role during formation of the nucleolus. The localization of satellite III and rDNA in human peripheral blood lymphocytes and HeLa cells was examined throughout mitosis.[106] Surprizingly, nucleolar DNA was less condensed in metaphase than in interphase. In telophase, rapid condensation occurs, during which rDNA clusters around satellite III heterochromatin.

VIII. GENOME ARCHITECTURE AND NUCLEAR FUNCTION

There are several reasons to believe that there is an intimate connection between the genome architecture and the functional state of the nucleus. Some connections seem to be evident. For example, the nonrandom, often peripheral, position of the nucleolus[107] may allow rapid transport of ribosomal particles to the cytoplasm. Other links, e.g., with replication, transcription, DNA repair, and RNA processing, are not as apparent, although the spatial arrangement of these events at restricted sites in interphase nuclei[5,108,109] seems to be meaningful. In the following section, I will illustrate the possible involvement of genome architecture in the processes of differentiation and transcription.

In cereal plants, the nonrandom, almost Rabl, orientation of chromosomes is well documented for some cell types.[10] The Heslop-Harrison group demonstrated that there is spatial separation of specific chromosomal domains and transcriptional activity across the nucleus (reviewed in refs. 10, 11). In the root tip cells, approximately one-third of the nuclear volume is filled with condensed chromatin. This region contains little genetic length, but does contain most of the physical length of chromosomes and includes all of the centromeres. All of the telomeres are located in the less filled region of the nucleus. Importantly, the majority of expressed genes are also found in this hemispere.

Specific patterns of centromere domain localizations have been demonstrated in cells of the central nervous system by the Manuelidis group (reviewed above). This group showed abnormal localization of telomeres in several brain tumors[110] and rearrangement of individual chromosomes in cells from human epileptic foci.[111] In the latter work, the authors used in situ hybridization with four chromosome-specific probes (1q12, 9q12, Yq12, and Xcen) to define the center of chromosome terrritoty. The "normal" position of each human chromosome domain was first evaluated in cells of normal cerebral cortex. In normal tissue, 1q and 9q were found only on the nuclear mebmrane or near the nucleolus, Xcen was found only on the nuclear membrane, and Yq12 was aggregatted with other centromeres on the nucleolus. In nuclei from a seizure foci in males, Xcen was repositioned to the interior of the nucleus. A similar type of repositioning was not observed for other domains studied. Movements of the X chromosome (one or both homologs) toward the nuclear interior was also documented in seizure foci of females. The authors suggest that dramatic repositioning of the X chromosome may participate in the creation of genetic memory for intractable seizure activity. Another example of chromosome movements associated with changes in transcription was demonstrated in peripheral blood lymphocytes.[112] Stimulation of lymphocytes by phytohemagglutinin triggers their development into large blastlike cells and causes transcriptional activation of previously repressed rDNA loci. During this process NOR-bearing acrocentric chromosomes move within the nucleus.

Recently genome rearrangement accompanying differentiation of rat myoblasts into myotubes in culture has been shown by comparison of the centromere localization in the nuclei of these cells.[113] Centromere positioning was determined with immunofluorescence. In myoblasts, centromeres were dispersed throughout the nucleoplasm. When the growth factors were withdrawn, a differentiation mimicing muscle developmenet *in vivo* took place, and genes for muscle-specific proteins were activated. Centromeres simulteneously were repositioned to the nuclear periphery.

An original model of dynamic and complex relationships between genome architecture and gene transcription was developed by T. Cremer, P. Lichter, C. Cremer, and their co-authors. This detailed model and its predictions have been described in several reviews from these authors[9,12], and I refer readers to the originals for in-depth information. This model is based on the existence of territorial organization of eukaryotic chromosomes in interphase (Figure 2B). The authors postulate that the combination of the space between adjacent chromosome territories and chromosome surface forms the functional nuclear compartment, the interchromosomal domain (ICD). The ICD contains the machineries for transcription, splicing, and macromolecular transport. The important feature of this model is the hypothetical three-dimensional network of channels starting at nuclear pores and expanding between surfaces of chromosomal territories. The channels also protrude into the chromosomal territory interior. The authors suggest that transcription regulatory factors may access only those target areas on chromosomes that are exposed to the ICD. Therefore, movements and spatial rearrangements of chromosomal domains, which have been experimentally observed during differentiation and the cell cycle (see above), could trigger (or block) transcriptional events. One of the predictions of this model is the spatial compartmentalization of transcribed chromosomal domains. So far, only a few genes have been found to be preferentially associated with the periphery of chromosome territory.[114] Some experimental data on the higher-order organization of interphase chromosome[42] contradict the compartmentalization into gene-rich and gene-poor subterritories, as suggested by this model.

IX. CONCLUSIONS

At this point we can say, with some degree of certainty, that eukaryotic chromosomes are nonrandomly organized within the nuclear volume and that genome architecture is a significant concept. Moreover, it is evident that this architecture is extremely dynamic and correlates with the cell type, stage of differentiation, stage of the cell cycle, and gene expression and is species specific. Some of the early speculations of Rabl and Bovery have come true (territorial organization of interphase chromosome has been directly proved) or have come true in particular cases (Rabl configuration), some are still under investigation (relative positioning of chromosomes). Current data are still inconclusive and therefore do not provide sufficient consensus. It is hoped that in the near future a more systematic and more straightforward view of the genome architecture will be formulated. Such is the pace of experimental pursuits over the past decade that we have already outdone those that took place over the previous 100 years.

At this point the establishment of a much more practical (i.e., clinical) significance of studies of genome architecture might be expected. The existing correlations between chromosome architecture and cell differentiation, including malignant transformation, are very exciting. Unfortunately, it is currently difficult to determine the extent to which the links are direct and meaningful. One of the most promising approaches to understanding the functional role(s) of genome architecture is to identify and characterize participating nuclear proteins; however, this falls beyond the scope of the current chapter.

ACKNOWLEDGMENTS

This work was supported by USDA grant 9601837 being given to the author and DOE grant DEFG03-88ER-60673 to Dr. E. M. Bradbury, in whose laboratory it was performed. I extend special thanks to Dr. Ray Teplitz and Dr. Alison O'Mahony for their critical reading of the manuscript and helpful comments.

NOTE ADDED IN PROOF

This review had been completed in the winter of 1995-96. Since that period a number of important publications have appeared which I have been unable to discuss here. I apologize both before the authors and the readers of this book. I have no doubt that new developments in the field will be scrutinized in coming years.

REFERENCES

1. Rabl, C. Uber zellteilung. *Morphol. Jahrbuch* **1885**, *10*, 214-330.
2. Boveri, K.T. Die befruchtung und teilung des eles von Ascaris magalocephala. In: *Zellen Studies, H 2.* G. Fisher, Jena, 1988, pp. 1–189.

3. Hughes, A. *A History of Cytology*. Abelard-Schuman, London, New York, 1959.
4. Manuelidis, L. A view of interphase chromosomes. *Science* **1990**, *250*, 1533–1540.
5. Spector, D.L. Macromolecular domains within the cell nucleus. *Annu. Rev. Cell Biol.* **1993**, 265–315.
6. Hilliker, A.J.; Apples, R. The arrangement of interphase chromosomes: structural and functional aspects. *Exp. Cell Res.* **1989**, *185*, 297–318.
7. Blobel, G. Gene gating: a hypothesis. *Proc. Natl. Acad. Sci. USA* **1985**, *82*, 8537–8529.
8. Comings, D.E. Arrangement of chromatin in the nucleus. *Hum. Genet.* **1980**, *53*, 131–143.
9. Cremer, T.; Kurz, A.; Zirbel, R.; Dietzel, S.; Rinke, R.; Schrock, E.; Speicher, M.R.; Mathieu, U.; Juach, A.; Emmerich, P.; Scherthan, H.; Ried, T.; Cremer, C.; Lichter, P. Role of chromosome territories in the functional compertmentalization of the cell nucleus. *Cold Sping Harb. Symp. Quant. Biol.* **1993**, *58*, 777–792.
10. Heslop-Harrison, J.S.; Leitch, A.R.; Schwarzacher, T. The physical organization of interphase nuclei. In: *The Chromosome* (Heslop-Harrison, J.S.; Schwarzacher, T.; eds.). Bios, Oxford, 1993, pp. 221–232.
11. Heslop-Harrison, J.S. Nuclear architecture in plants. *Curr. Opin. Genet. Dev.* **1992**, *2*, 913–917.
12. Cremer, T.; Dietzel, S.; Eils, R.; Licter, P.; Cremer, C. Chromosome territories, nuclear matrix filaments and interchromatin channels: a topological view on nuclear architecture and fuction. In: *Kew Chromosome Conference IV* (Brandham, P.E.; Bennet, M.D.; eds.). Royal Botanic Garden, Kew, England, 1995, pp. 63–81.
13. Manuelidis, L. Interphase chromosome positions and structure during silencing, transcription and replication. 1996, in press.
14. Haaf, T.; Schmidt, M. Chromosome topology in mammalian interphase nuclei. *Exp. Cell Res.* **1991**, *192*, 325–332.
15. Trask, B. Fluorescent *in situ* hybridization. *CSH Course* **1993**.
16. Scherthan, H.; Cremer, T. Nonisotopic *in situ* hybridization in paraffin embedded sections. *Methods Mol. Genet.* **1994**, *5*, 223–238.
17. Giuan, X.Y.; Meltzer, P.S.; Burgess, A.C.; Trent, T.M. Coverage of chromosome 6 by chromosome microdissection: Generation of 14 subregion-specific probes. *Human Genetics* **1995**, *95*, 637–640.
18. van Dekken, H.; Hulspas, R. Spatial analysis of intranuclear repetitive DNA regions by *in situ* hybridization and digital fluorescence microscopy. *Histochem. J.* **1993**, *25*, 173–182.
19. Lizard, G.; Chignol, M.C.; Souchier, C.; Schmitt D.; Chardonnet, Y. Laser scanning microscopy with a charge coupled device camera improve detection of human papillomavirus DNA revealed by fluorescence in situ hybridization. *Histochemistry* **1994**, *101*, 303–310.
20. Brakenhoff, G.J.; van Spronsen, E.A.; van der Voort, H.T.; Nanninga, N. Three-dimensional confocal microscopy. *Methods Cell Biol.* **1989**, *30*, 379–398.
21. Paddock, S.W. The boldly glow…applications of laser scanning confocal microscopy in developmental biology. *Bioessays* **1994**, *16*, 357–365.
22. Agard, D. A.; Sedat, J. W. Three-dimensional architecture of a polytene nucleus. *Nature* **1983**, *302*, 676–681.
23. Agard, D.A. Optical sectioning microscopy: cellular architecture in three dimensions. *Annu. Rev. Biophys. Bioeng.* **1984**, *13*, 191–219.
24. Laurent, M.; Johanin, G.; Gilbert, N.; Lucas, L; Cassio, D.; Petit, P.X.; Fleury, A. Power and limits of laser scanning confocal microscopy. *Biol. Cell* **1994**, *80*, 229–240.
25. Brakenhoff, G.J.; van der Voort, H.T.M.; Baarslag, M.W.; Mans, A.; Oud, J.L.; Zwart, R., van Driel, R. Visualization and analysis techniques for three dimensional information acquired by confocal microscopy. *Scanning Microsc.* **1988**, *2*, 1831–1838.
26. Houtsmuller, A.B.; Smeulders, A.W.M.; van der Voort, H.T.M.; Oud, J.L.; Nanninga, N. The homing cursor: a tool for three-dimensional chromosome analysis. *Cytometry* **1993**, *14*, 501–509.

27. Ikizyan, I.A.; Burde, S.; Leary, J.F. Interactive 3-D image analysis and visualization techniques for FISH-labelled chromosomes in interphase nuclei. *Bioimaging* **1994**, *2*, 41–56.

28. Salisbury, J.R. Three-dimensional reconstruction in microscopical morphology. *Histol. Histopathol.* **1994**, *9*, 773–780.

29. Shaw, P. Deconvolution in 3-D optical microscopy. *Histochem. J.* **1994**, *26*, 687–694

30. van der Engh, G.; Sachs, R.; Trask, B.J. Estimating genomic distance from DNA sequence location in cell nuclei by a random walk model. *Science* **1992**, *257*, 1410–1412.

31. Trask, B.J.; Allen, S.; Massa, H.; Fertitta, A.; Sachs, R.; Van der Engh, G.; Wu, M. Studies of metaphase and interphase chromosomes using fluorescent *in situ* hybridization. *Cold Spring Harb. Symp. Quant. Biol.* **1993**, *38*, 767–775.

32. Hiraoka, Y.; Agard, D.; Sedat, J.W. Temporal and spatial coordination of chromosome movement, spindle formation and nuclear envelope breakdown during prometaphase in *Drosophila melanogaster* embryos. *J. Cell Biol.* **1990**, *111*, 2815–2828.

33. Chikashige, Y.; Ding, D-Q.; Funabiki, H.; Haraguchi, T.; Mashiko, S.; Yanagida, M.; Hiraoka, Y. Telomere-led premeiotic chromosome movement in fission yeast. *Science* **1994**, *264*, 270–273.

34. Wilson, E.B. The Cell in Development and Heredity. Macmillan, New York, 1924.

35. Bostock, C.J.; Sumner, A.T. *The Eukaryotic Chromosome*. North-Holland Publishing Co., Amsterdam, New York, Oxford, 1978.

36. Tyler-Smith, C.; Willard, H.F. Mammalian chromosome structure. *Curr. Opin. Genet. Dev.* **1993**, *3*, 390–397.

37. Rattner, J.B.; Lin, C.C. Radial loops and helical coils coexist in metaphase chromosomes. *Cell*, **1985**, *42*, 291–296.

38. Saitoh, Y.; Laemmli, U.K. From the chromosomal loops and the scaffold to the classic bands of metaphase chromosomes. *Cold Spring Harb. Symp. Quant. Biol.* **1993**, *38*, 755–765.

39. Wolffe, A. *Chromatin*. Academic Press, San Diego, 1992.

40. Belmont, A.S.; Bruce, K. Visualization of G_1 chromosomes: a folded, twisted, supercoiled chromonema model of interphase chromatin structure. *J. Cell Biol.* **1994**, *127*, 287–302.

41. Cook, P. R. A chromomeric model for nuclear and chromosome structure. *J. Cell Sci.* **1995**, *108*, 2927–2935.

42. Yokota, H.; van der Engh, G.; Hearst, J.E.; Sachs, R. K.; Trask, B.J. Evidence for the organization of chromatin in megabase pair-sized loops arranged along a random walk path in the human G_0/G_1 interphase nucleus. *J. Cell Biol.* **1995**, *130*, 1239–1249.

43. Cremer, T.; Baumann, H.; Nakanishi, K.; Cremer, C. Correlation between interphase and metaphase chromosome arrangements as studied by laser-UV-microbeam experiments. *Chromosomes Today* **1984**, *8*, 203–215.

44. Manuelidis, L. Individual interphase chromosome domains revealed by in situ hybridization. *Hum. Genet.* **1985**, *71*, 288–293.

45. Schardin, M.T.; Cremer, T.; Hager, H.D.; Lang, M. Specific staining of human chromosomes in Chinese hamster × man hybrid cell lines demonstrate interphase chromosome territories. *Hum. Genet.* **1985**, *71*, 281–293.

46. Lichter, P.; Cremer, T.; Borden, J.; Manuelidis, L.; Ward, D.C. Delineation of individual human chromosomes in metaphase and interphase cells by *in situ* hybridization using recombinant DNA libraries. *Hum. Genet.* **1988**, *80*, 304–312.

47. Pinkel, D.J.; Landegent, J.; Collins, C.; Fuscoe, J.; Segraves, R.; Lucas, J.; Gray, J.W. Fluorescence *in situ* hybridization with human chromosome-specific libraries: Detection of trisomy 21 and translocation of chromosome 4. *Proc. Natl. Acad. Sci. USA* **1988**, 9138–9143.

48. Heslop-Harrison, J.S.; Bennet, M.D. Nuclear architecture in plants. *Trends Genet.* **1990**, *6*, 401–410.

49. Jack, J.W.; Judd, B.H. Allelic pairing and gene regulation: a model for the zeste-white interactions in *Drosophila melanogaster*. *Proc. Natl. Acad. Sci. USA* **1979**, *76*, 1368–1372.

50. Ashley, T.; Wagenaar, E.B. Telomeric associations of gametic and somatic chromosomes. *Can. J. Genet. Cytol.* **1974**, *16*, 61–76.

51. Ashley, T. Specific end-to-end attachment of chromosomes in *Ornithogalum virens. J. Cell Sci.* **1979**, *38*, 357–367.

52. Leitch, A.R.; Brown, J.K.M.; Mosgoler, W.; Schwarzacher, T.; Heslop-Harrison, J.S. The spatial localization of homologous chromosomes in human fibroblasts at mitosis. *Hum. Genet.* **1994**, *93*, 275–280.

53. Oud, J.L.; Mans, A.; Brakenhoff, G.J.; van der Voort, H.T.M.; van Spronsen, E.E.; Nanninga, N. Three-dimensional chromosome arrangement of *Crepis cappilaris* in mitotic prophase and anaphase as studied by confocal scanning laser microscopy. *J. Cell Sci.* **1989**, *92*, 329–339.

54. Nagele, R.; Freeman, T.; McMorrow, L.; Lee, H. Precise spatial positioning of chromosomes during prometaphase: evidence for chromosome order. *Science* **1995**, *270*, 1831–1835.

55. Emmerich, P.; Loos, P.; Jauch, A.; Hopman, A.; Wiegant, J.; Higgins, M.; White, B.; van der Ploeg, M.; Cremer, C.; Cremer, T. Double *in situ* hybridization in combination with digital image analysis: a new approach to study interphase chromosome topograpy. *Exp. Cell Res.* **1989**, *181*, 126–140.

56. Propp, S; Scholl, H.P.; Loos, P.; Jauch, A.; Stelzer, E.; Cremer, C.; Cremer, T. Distribution of chromosome 18 and X centric heterochromatin in the interphase nucleus of cultured human cells. *Exp. Cell Res.* **1990**, *189*, 1–12.

57. Belmont, A.S.; Bignone, F.; Ts'o, P. The relative intranuclear position of Barr bodies in XXX non-transformed fibroblasts. *Exp. Cell Res.* **1986**, *165*, 165–179.

58. Lesko, S.A.; Callahan, D.E.; LaVilla, M.E.; Wang, Z-P.; Ts'o, P. The experimental homologous and heterologous separation distance histograms for the centromeres of chromosomes 7, 11, and 17 in interphase human lymphocytes. *Exp. Cell Res.* **1995**, *219*, 499–506.

59. Manuelidis, L.; Borden, J. Reproducible compartmentalization of individual chromosome domains in human CNS cells revealed by in situ hybridization and three-dimensional reconstruction. *Chromosoma* **1988**, *96*, 397–410.

60. Vourc'h, C.; Tarusco, D.; Boyle, A.L.; Ward, D.C. Cell cycle-dependent distribution of telomeres, centromeres and chromosome-specific subsatellite domains in the interphase nucleuse of mouse lymphocytes. *Exp. Cell Res.* **1993**, *205*, 142–151.

61. Hadlaczky, G.; Went, M.; Ringertz, N.R. Direct evidence for the non-random localization of mammalian chromosomes in the interphase nucleus. *Exp. Cell Res.* **1986**, *167*, 1–15.

62. Hiraoka, Y.; Dernburg, A.F.; Parmelee, S.J.; Rykowski, M.C.; Agard, D.A.; Sedat, J.W. The onset of homologous chromosome pairing during *Drosophila melanogaster* embryogenesis. *J. Cell Biol.* **1993**, *120*, 591–600.

63. Avivi, L.; Feldman, M. Arrangement of chromosomes in the interphase nuclei of plants. *Hum. Genet.* **1980**, *55*, 281–295.

64. Leitch, A.R.; Schwarzacher, T.; Mosgoller, W.; Bennett, M.D.; Heslop-Harrison, J.S. Parental genomes are separated throughout the cell cycle in a plant hybrid. *Chromosoma* **1991**, *101*, 206–213.

65. Fergusson, M.; Ward, D.C. Cell-cycle dependent chromosome movement in pre-mitotic human T-lymphocyte nuclei. *Chromosoma* **1992**, *101*, 557–565.

66. Welmer, R.; Haaf, T.; Kruger, J.; Poot, M. Schmid, M. Characterization of centromere arrangements and test for random distribution in G_0, G_1, S, G_2, G_1 and early S' phase in human lymphocytes. *Hum. Genet.* **1992**, *88*, 673–682.

67. Haaf, T.; Ward, D.C. Structural analysis of (α-satellite DNA and centromere proteins using extended chromatin and chromosomes. *Hum. Mol. Genet.* **1994**, *3*, 697–709.

68. Bartholdi, M. F. Nuclear distribution of centromeres during the cell cycle of human diploid fibroblasts. *J. Cell Sci.* **1991**, *99*, 255–263.

69. Haaf, T.; Schmid, M. Centromeric association and non-random distribution of centromeres in human tumor cells. *Hum. Genet.* **1989**, *81*, 137–143.

70. Haaf, T.; Ward, D.C. Rabl orientation of CENP-B box sequences in *Tupaia belangeri* fibroblasts. *Cytogenet. Cell Genet.* **1995**, *70*, 258–262.

71. Uzawa, S.; Yanagida, M. Visualization of centromeric and nucleolar DNA in fission yeast by fluorescence *in situ* hybridization. *J. Cell Sci.* **1992**, *101*, 267–275.

72. Funabiki, H.; Hagan, I.; Uzava, S.; Yanagida, M. Cell cycle-dependent specific positioning and clustering of centromeres and telomeres in fission yeast. *J. Cell Biol.* **1993**, *121*, 961–976.

73. Gilson, E.; Laroche, T.; Gasser, S.M. Telomeres and the functional architecture of the nucleus. *Trends Cell Biol.* **1993**, *3*, 128–134.

74. Sharp, L.W. *An Introduction in Cytology*. McGraw-Hill, New York, 1926.

75. Wagenaar, E.B. End-to-end chromosome attachments in mitotic interphase and their possible significance to meiotic chromosome pairing. *Chromosoma* **1969**, *26*, 410–426.

76. John, B. *Meiosis*. Cambridge University Press, Cambridge, 1990.

77. Moens, P.B., Ed. *Meiosis*. Academic Press, Orlando, FL, 1987.

78. Dawe, R.K.; Sedat, J.W.; Agard D.A.; Cande, W.Z. Meiotic chromosome pairing in maize is associated with a novel chromatin organization. *Cell* **1994**, *76*, 901–912.

79. von Wettstein, D.; Rassmussen, S.W., Holm, P.B. The synaptonemal complex in genetic segregation. *Annu. Rev. Genet.* **1984**, *18*, 331–413.

80. Zalensky, A.O.; Tomilin, N.V.; Zalenskaya, I.A.; Teplitz, R.; Bradbury, E.M. Telomere–telomere associations and candidate telomere-binding proteins in mammalian sperm cells. *Exp. Cell Res.* **1997**, *232*, 29–41.

81. Brinkley, B.R.; Brenner, S.L.; Hall, A.M.; Tousson, A.; Balczon, R.D.; Valdivia, M.M. Arrangement of kinetochores in mouse cells during meiosis and spermiogenesis. *Chromosoma* **1986**, *94*, 309–317.

82. Zalensky, A.O.; Breneman, J.W.; Zalenskaya, I.A.; Brinkley, B.R.; Bradbury, E.M. Organization of centromeres in decondensed nuclei of mature human sperm. *Chromosoma* **1993**, *102*, 509–518.

83. Zalensky, A.O.; Allen, M.J.; Kobayashi, A.; Zalenskaya, I.A.; Balhorn, R.; Bradbury, E.M. Well-defined genome architecture in the human sperm nucleus. *Chromosoma* **1995**, *103*, 577–590.

84. Haaf, T.; Ward, D. Higher order structure in mammalian sperm revealed by *in situ* hybridization and extended chromatin fibers. *Exp. Cell Res.* **1995**, *219*, 604–611.

85. Scherthan, H.; Loidl, J.; Schuster, T.; Schweizer, D. Meiotic chromosome condensation and pairing in *Saccharomyces cerevisiae* studied by chromosome painting. *Chromosoma* **1992**, *101*, 590–595.

86. Scherthan, H.; Kohler, M.; Vogt, P.; von Malsch, K.; Schweizer, D. Chromosomal in situ hybridization with bilabeled DNA: signal amplification at the probe level. *Cytogenet. Cell Genet.* **1992**, *60*, 4–7.

87. Manuelidis, L. Different CNS types display distinct and non-random arrangements of satellite DNA sequences. *Proc. Natl. Acad. Sci. USA* **1984**, *181*, 3123–3127.

88. Manuelidis, L. Indications of centromere movement during interphase and differentiation. *Ann. N.Y. Acad. Sci.* **1985**, *450*, 205–220.

89. Billa, F; Boni, U. Localization of centromeric satellite and telomeric DNA sequences in dorsal root gandlion neurons, in vitro. *J. Cell Sci.* **1991**, *100*, 219–226.

90. Mathog, D.; Hochstrasser, M.; Gruenbaum, Y.; Saumweber, H.; Sedat, J. W. Characteristic folding pattern of polythene chromosomes in *Drosophila* salivary gland nuclei. *Nature* **1984**, *308*, 414–421.

91. Agard, D.A.; Hiraoka, Y.; Shaw, P.; Sedat, J.W. Fluorescence microscopy in three dimensions. *Methods Cell Biol.* **1989**, *30*, 353–377.

92. Urata, Y.; Parmelee, S.J.; Agard, D.A.; Sedat, J.W. A three-dimensional structural dissection of *Drosophila* polytene cromosomes. *J. Cell Biol.* **1995**, *131*, 279–295.

93. Kasinsky, H.E. Specificity and distribution of sperm basic proteins. In: *Histones and Other Basic Nuclear Proteins* (Hnilica, L.S.; Stein, G.S.; Stein, J.L.; Eds.). CRC Press, Boca Raton, FL, 1989.

94. Meistrich, M.L. Histone and basic nuclear protein transitions in mammalian spermatogenesis. In: *Histones and Other Basic Nuclear Proteins* (Hnilica, L.S.; Stein, G.S.; Stein, J.L.; Eds.). CRC Press, Boca Raton, FL, 1989, pp. 165–182.

95. Balhorn, R. A model of the structure of chromatin in mammalian sperm. *J. Cell Biol.* **1982**, *93*, 298–305.

96. Ward, W.S.; Coffey, D.S. DNA packaging and organization in mammalian spermatozoa: comparison with somatic cells. *Biol. Reprod.* **1991**, *44*, 569–574.

97. Ward, W.S.; Zalensky, A.O. The unique complex organization of the transcriptionally silent sperm chromatin. *Curr. Rev. Eukaryot. Gene Expression* **1996**, *6*, 139–147.

98. Palmer, D.K.; O'Day, K.; Margolis, R.L. The centromere specific histone CENP-A is selectively retained in discrete foci in mammalian sperm nuclei. *Chromosoma*, **1990**, *100*, 32–36.

99. Haaf, T.; Grunenberg, H.; Schmid, M. Paired arrangements of nonhomologous centromeres during vertebrate spermatogenesis. *Exp. Cell Res.* **1990**, *187*, 157–161.

100. Sumner, A.T. Immunofluorescent demonstration of kinetochores in human sperm. *Exp. Cell Res.* **1987**, *171*, 250–253.

101. Powell, D.; Cran, D.C.; Jennings, C.; Jones, R. Spatial organization of repetitive DNA sequences in bovine sperm nucleus. *J. Cell Sci.* **1990**, *97*, 185–191.

102. Jennings, C.; Powell, D. Genome organization in murine sperm nuclei. *Zygote* **1995**, *3*, 123–131,

103. de Lara, J.; Wynder, K.; Hyland, K.M.; Ward, W.S. Fluorescent in situ hybridization of the telomere repeat sequence in hamster sperm nuclear structures. *J. Cell. Biochem.* **1993**, *53*, 213–221.

104. Manuelidis, L. Heterochromatic features of an 11-megabase transgene in brain cells. *Proc. Natl. Acad. Sci. USA* **1991**, *88*, 1049–1053.

105. Kaplan, F.S.; Murray, J.; Sylvester, J.E.; Gonzalez, I.L.; O'Connor, P.; Doering, J.L.; Muenke, M.; Emanuel, B.S.; Zasloff, M.A. The topographic organization of repetitive DNA in the human nucleolus. *Genomics* **1993**, *15*, 123–132.

106. Kaplan, F.; O'Connor, J.P. Topographic changes in a heterochromatic chromosome block in humans (15P) during formation of the nucleolus. *Chromosome Res.* **1995**, *3*, 309–314.

107. Bourgeois, C.A.; Hubert, J. *Int. Rev. Cytol.* **1988**, *111*, 1–52.

108. Fey, E.G.; Bangs, P.; Sparks, C.; Odgren, P. The nuclear matrix: defining structural and functional roles. *Crit. Rev. Eukaryot. Gene Expression* **1991**, *1*, 127–143.

109. Jackson, D.A. Nuclear organization: uniting replication foci, chromatin domains and chromosome structure. *BioEssays* **1995**, *17*, 587–591.

110. Manuelidis, L. Genomic stability and instability in different neuroepithelial tumors. A role for chromosome structure? *J. Neuro-Oncology* **1994**, *18*, 225–239.

111. Borden, J.; Manuelidis, L. Movement of the X chromosome in epilepsy. *Science* **1988**, *242*, 1613–1732.

112. Wachtler, F.; Hopman, A.H.N.; Wiegant, J.; Schwarzacher, H.G. On the position of nucleolus organizer regions (NORs) in interphase nuclei. *Exp. Cell Res.* **1986**, *167*, 227–240.

113. Chaly, N.; Munro, S.B. Centromere repositin to the nuclear periphery during L6E9 myogenesis in vitro. *Exp. Cell Res.* **1996**, *223*, 274–278.

114. Kurz, A.; Lampel, S.; Nikolencko, J.; Bradl, J; Benner, A.; Zirbel, R.; Cremer, T.; Lichter, P. Active and inactive genes localize in the periphery of chromosome territories. *J. Cell Biol.* **1996**, *135*, 1195–1205.

THE MITOTIC CHROMOSOME

Adrian T. Sumner

Advances in Genome Biology
Volume 5A, pages 211-261.
Copyright © 1998 by JAI Press Inc.
All rights of reproduction in any form reserved.
ISBN: 0-7623-0079-5

I. INTRODUCTION:
THE NECESSITY OF HAVING CHROMOSOMES

Chromosomes were recognized as important components of cells in the latter half of the nineteenth century, and by 1903, only a few years after the rediscovery of Mendel's classic work, it was realised that the behavior of chromosomes was exactly that required to explain the behavior of genes, the Mendelian factors.[1] Chromosomes thus determine the behavior of genes, and their position in genetics is therefore central. It is not, however, my intention in this review to look at chromosomes simply as carriers of genes, but to describe the organization of the chromosome as a whole. The behavior of genes is then simply an inevitable consequence of how chromosomes are organized and how they behave, and therefore genetics cannot be understood without a proper knowledge of chromosomes.

We tend to take the existence of chromosomes so much for granted that it is worthwhile taking a little time to consider why they should be necessary. Could there be some other ways of distributing the genes to daughter cells, and if so, how effectively would they work? And could it be that chromosomes have to distribute something other than genes between the daughter cells? After all, the human genome has about enough DNA for 3 million genes, but probably has only about 100,000.[2] In fact, we have some good examples of alternative ways of segregating the genetic material, although explanations for what seems to be the great excess of DNA—the C-value paradox—remain somewhat speculative.

It seems clear that it is necessary to group genes into chromosomes to make them manageable. After all, if 100,000 genes were each attached separately to a spindle microtubule (each of which is about 25 nm in diameter), the microtubules would occupy the greater part of the cell, leaving little room for the other components of the mitotic apparatus, let alone all of the essential cellular organelles not involved in chromosome segregation. The macronuclei of ciliated protozoa do appear to contain what are effectively isolated genes, but have dispensed with microtubules; although there are multiple copies of each gene, eventually, as a result of the random distribution of the genes at the amitotic nuclear division, macronuclei are formed that lack certain essential genes, and the cell senesces unless new macronuclei can be formed from the micronuclei, which divide in a more conventional way.[3] Clearly, handling the genes individually is not a viable proposition, either metaphorically or literally.

On the other hand, why don't eukaryotes link all of their genes together on one chromosome, as bacteria do? There is, indeed, no simple answer to this. It seems probable that the 2 m of DNA in the human genome, if not suitably compacted into condensed chromatin, would be quite unmanageable and subject to random breakage when segregation occurred. However, would not a single chromosome (or rather a pair of chromosomes in a diploid organism) be sufficient? There is at least one animal, a species of ant, with only one pair of chromosomes,[4] so this is clearly not an impossible situation. Nevertheless, although there is a very wide range of chromosome numbers among living organisms, ranging up to several hundreds in some butterflies and ferns,[5] most organisms have intermediate numbers. For example, most eutherian mammals have chromosome numbers between 40 and 56.[6] There could well be some selective pressure to have a reasonable number of chromosomes to keep linkage groups small and to ensure independent assortment of most genes; however, to some extent, larger chromosomes have more chiasmata, so that their genes are not necessarily closely linked, and it is not known whether organisms with few chromosomes are, in fact, less variable. One can also imagine that a maximum limit to the number of chromosomes might be set by the large number of spindle microtubules required. On the other hand, if the number of chromosomes were too small, they might be too large individually to be moved efficiently by the spindle, although there is really no evidence for this. For example, certain urodele amphibia have very large amounts of nuclear DNA distributed among a small number of chromosomes, the average size of which is comparable to that of the whole human genome. Very large chromosomes do not, in themselves, appear to cause any problems in chromosome segregation. Nevertheless, the fact remains that most eukaryotes have their genomes divided among a substantial number of chromosomes, which come in all shapes and sizes. We really have no clues as to what factors determine chromosome morphology and number, although one might imagine, at least, that once a particular karyotype has evolved, drastic changes would not be tolerated, if only because of difficulties that might occur at meiosis. Although this is true to some extent, it is surprising how much change can be tolerated. In what has been described as the "Muntjac Scandal,"[7] it has been observed that the Chinese and Indian muntjacs are very similar morphologically, can hybridize, and even produce fertile offspring occasionally. Yet the former species has 46 chromosomes, whereas the latter has only seven in the male, and six in the female. Clearly there is a need for research on factors that determine the number, size, and shape of chromosomes; nevertheless, whether chromosomes are large or small, few or numerous, the basic principles of their organization are generally similar, and it is these common principles that will be considered in the remainder of this chapter.

II. THE PACKING OF DNA INTO CHROMOSOMES

The immense length of DNA (approximately 2 m) in each human nucleus has already been remarked upon. This has to be reduced to a length of approximately 200 μm or less for the condensed chromosomes. DuPraw[8] introduced the useful concept of the packing ratio, that is, the ratio between the length of fiber (whether DNA, chromatin, or whatever), and the length of the object into which it is packed. Thus the packing ratio of DNA in the chromosome is 2 m ÷ 200 μm, which is 10,000. This high degree of compaction is accomplished in several stages, each

Table 1. Packing Ratios of DNA and Chromatin in Chromosomes

Structure	Length/cell	Breadth	Packing Ratio
DNA Molecule	2 m	2 nm	1
Nucleosome Fiber	0.28 m	10 nm	7
Solenoids	0.04 m	30 nm	50
Loops	1 mm	0.26 μm	2000
Final Condensation	200 μm	2 μm	10,000

Table 2. Evidence for Uninemy of Chromosomes

Semi-conservative Replication of DNA and Chromosomes (ref. 11)
Measurements of Sizes of Chromosomal DNA Molecules
 - Ultracentrifugation (ref. 13-15)
 - Visco-elastic Retardation (ref. 16-17)
Breakage of a Single DNA Molecule by X-rays Induces Chromatid Aberrations (ref. 18)
Kinetics of Digestion of Lampbrush Chromosomes (ref. 19)
Width of Axial Fiber in Lampbrush Chromosomes (ref. 20,21)

with its own characteristic packing ratio (Table 1); some of these stages are well understood, while others are still the subject of argument.

A. Uninemy of the Chromosome

It is now generally accepted that there is only a single DNA molecule running throughout the length of a chromatid; that is, the undivided chromosome is unineme. Nevertheless, it is probably worth summarizing the evidence for uninemy, since it is not immediately obvious from looking at chromosomes. Even in the narrowest parts of chromosomes, viewed as whole mounts by transmission electron microscopy, there appear to be several fibers, not a single one.[9,10] In fact, there is no direct visual evidence for uninemy, but a variety of observations that can only be reasonably interpreted as showing that there is only one DNA fiber per chromosome (Table 2).

DNA replicates semiconservatively, and it has been shown repeatedly using autoradiography[11] and more recently by BrdU incorporation (Figure 1), that chromosomes themselves replicate semiconservatively. Uninemy is the simplest explanation for this. Direct measurements of a DNA molecule from a chromosome are not practicable, but indirect procedures, such as ultracentrifugation[13–15] and viscoelastic retardation,[16,17] indicate that DNA molecules from species with small chromosomes have the sizes that would be expected if there were only a single molecule per chromosome.

Calculations of the kinetics of chromosome breakage by various agents are also consistent with uninemy. Low energy x-rays, sufficient to break only a single

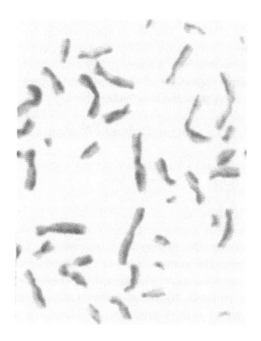

Figure 1. Evidence that chromosomes replicate semi-conservatively. When cells are grown in the presence of a suitable marker, in this case bromodeoxyuridine (BrdU), the marker is incorporated into the newly synthesised strand of DNA. At the first mitosis after BrdU incorporation, each daughter DNA molecule contains the same amount of BrdU. After a second round of replication without BrdU, one daughter DNA molecule, replicated from the DNA strand that did not contain BrdU, remains unsubstituted; the other daughter DNA molecule, however, still contains one strand that has incorporated BrdU. This differentiation is reflected at the chromosomal level, one chromatid being darkly stained (no BrdU) and the other pale (substituted with BrdU) after staining by the FPG method (Ref. 12).

DNA molecule, are nevertheless capable of causing chromatid aberrations.[18] Similarly, kinetics of digestion of lampbrush chromosomes (which are meiotic rather than mitotic) by DNase are consistent with one DNA molecule per chromatid.[19] The axial fiber of lampbrush chromosomes is, in fact, only wide enough to contain two DNA molecules, one for each chromatid.[20,21]

B.   Lower Levels of Packing

The DNA fiber is packed into nucleosomes, and the 10-nm nucleosomal fiber is, in turn, packed into a 30-nm fiber, which may consist of a regular solenoid, or a somewhat less regular kind of aggregation of nucleosomes (ref. 22, Chapter 8). Each of these levels of packing produces a packing ratio of approximately 7, giving a total of about 50 (Table 1). The 30-nm fiber produced by these two levels of condensation appears to be the basic unit of chromosome organization, and is also found in interphase nuclei. The remaining 200-fold compaction to form the metaphase chromosome appears to involve essentially further folding of the 30-nm fiber, and not its reorganization into thicker fibers.

C. Higher Orders of Packing

Studies of whole mounts of metaphase chromosomes[8-10] showed only an apparently random tangle of chromatin fibers running in all directions. Such images led DuPraw[8] to propose his "folded-fiber" model, in which the chromatin fiber is folded in a variety of ways throughout the body of the chromosome. However, a body of evidence, in particular the reproducibility of detailed banding patterns of various types along the chromosomes, indicates that chromatin fibers must be arranged in a reproducible pattern, at least on a microscopic scale.

Although observations on whole chromosomes, whether by light or electron microscopy, gave no indication that chromosomes were composed of anything but chromatin fibers, various authors proposed models of chromatin structure in which a core, distinct from the chromatin fibers, held the chromosome together.[23-25] Indeed, there were observations in the early 1970s, in which a core was left after the removal of the majority of the chromatin by a variety of means.[10,26] However, these results were not taken seriously, and it was not until Laemmli and his colleagues published a series of papers,[27,28] showing that dehistonized chromosomes consist of a proteinaceous "scaffold" from which radiate loops of DNA, that the idea of a chromosome core became widely accepted. Although many features of the core or scaffold organization remain to be elucidated, it is now possible to give a reasonably comprehensive account of the subject.

D. The Nature of the Scaffold

Electron micrographs of whole mounts of dehistonized chromosomes show a relatively dense structure running along the middle of the chromatids (Figure 2). The details of this structure are quite variable; sometimes it is a compact structure, and sometimes it is much more diffuse, consisting of numerous distinct but interconnected fibers. This is perhaps not surprising, since dehistonized chromosomes are an artefact (albeit a useful one), and they are usually swollen several times compared with intact chromosomes.[29] One feature of the scaffolds in dehistonized chromosomes is a pair of dense bodies, one per chromatid, in the region of the centromere, which remain condensed regardless of how much the rest of the chromosome and scaffold is dispersed (Figure 3). Whether these bodies represent the

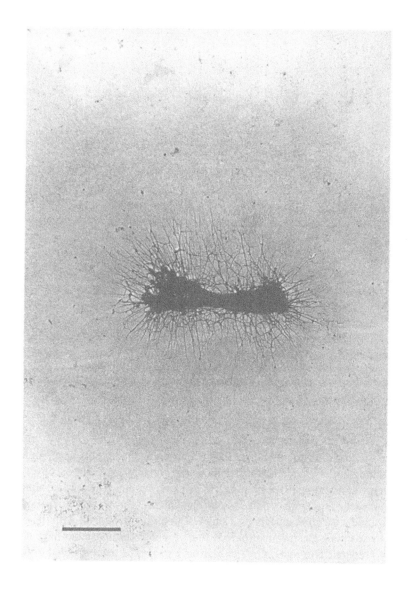

Figure 2. A chromosome scaffold or core, prepared by extracting an isolated CHO chromosome with 2M NaCl, spreading on a grid, and examining by transmission electron microscopy. The dense scaffold is surrounded by numerous loops of DNA which extend for 5-6 µm in all directions. Scale bar = 2µm. Reprinted with permission from Hadlaczky et al. (1981) *Chromosoma* 81: 537-555. © Springer-Verlag.

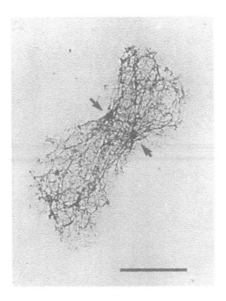

Figure 3. Chromosome scaffold, showing a pair of dense bodies (arrows), possibly corresponding to the kinetochores, at the centromeric region. Isolated CHO chromosome, treated with 0.7 M NaCl. Scale bar = 2μm. Reprinted with permission from Hadlaczky et al. (1981) *Chromosoma* 81: 537-555. © Springer-Verlag.

kinetochores, or a specialized region of the scaffold associated with the kineto-chores, has not yet been established.

Since the first concrete evidence for the existence of scaffolds was published, various attempts have been made to demonstrate them by light microscopy. Under appropriate conditions, silver staining can be used to produce a dark line along the middle of each chromatid, in both animal[30] and plant[31] chromosomes. In prophase chromosomes, the core appears as a single structure,[32] consistent with the observation by scanning electron microscopy that prophase chromosomes have not yet split into separate chromatids.[33] The scaffold can also be demonstrated immuno-cytochemically. One of the principal components of the scaffold (see below) is the enzyme topoisomerase II; antibodies to this enzyme also label a line along the chromatids[34,35] (Figure 4). Although in some reports the scaffold appears as a continuous line, in others it is broken up into a series of dots, and it is doubted by some whether the scaffold is really a continuous structure.

Two main proteinaceous components of the scaffold have been identified: Sc1, with a molecular weight of 170,000, and Sc2, with a molecular weight of 135,000[36]. Certain other scaffold proteins have been identified, but so far little is

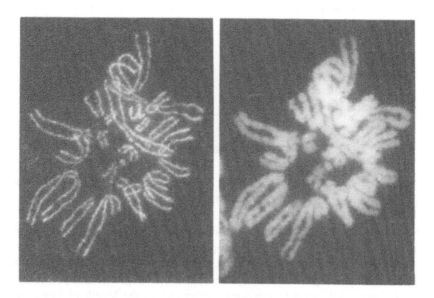

Figure 4. CHO chromosomes immunolabelled to show the distribution of topoisomerase II (Topo II). The Topo II forms a line along the middle of the chromatids (left), corresponding to the core. (Right) chromosomal DNA shown by ethidium fluorescence.

known about them. The scaffold structure appears to be stabilized by copper (Cu^{2+}) ions; dehistonized chromosomes become dissociated when treated with metal chelators, and scaffolds reconstituted in the presence of Cu^{2+} consist largely of Sc1 and Sc2.[36] Sc1 has been identified as the important nuclear enzyme topoisomerase II (Topo II), which is involved in many processes that require topological alterations of the DNA molecule; these include replication, transcription, DNA repair, and especially the decatenation (separation) of newly replicated DNA molecules.[37–39] Topo II also seems to be necessary for chromosome condensation.[40–42] It may be that the function of Topo II in the chromosome scaffold is primarily structural, but there is evidence (see below) that specific sequences in the chromosomal DNA may attach themselves to Topo II.

Sc2 belongs to a class of proteins known as the *SMC1* family, which appear to have ATPase activity, and it has been shown that Sc2 can form complexes with Topo II.[43] A role in chromosome condensation has been suggested on the basis of its presence in the mitotic chromosome scaffold, but also its absence from the nuclear matrix (which does, however, contain Topo II).

Table 3. Properties of an Average Chromosome Loop
(from Pienta & Coffey, Ref. 44)

Pairs of DNA Bases/Loop	63,000
Length of DNA /Loop	21.4 μm
Number of Nucleosomes/Loop	315
Packing Ratio of Loops	40
Number of Loops/Diploid Human Cell	100,000

E. Chromatin Loops

Although, as we have seen above (Section II.A), all of the evidence indicates that there is a single DNA molecule running throughout each chromatid, it appears to behave as if it consists of many much smaller units. At frequent intervals, the DNA is attached to the scaffold, and forms loops with characteristic properties (Table 3). As seen by electron microscopy, loops of DNA extend for 6–30 μm from the scaffold,[28,45] or up to 100 kb. Although some of the variability in the measured size of loops is no doubt the result of differing technical procedures, there is some evidence for systematic differences in the length of loops (see below). Again, although the standard methods of spreading scaffold preparations for electron microscopy give the impression of a two-dimensional array surrounding the scaffold, it is more plausible, since the chromatid is approximately cylindrical, that the loops radiate in all directions from the scaffold, to form a rosette. Electron micrographs of partly swollen chromosomes support this view (Figure 5).[46,47] It has been estimated that there would be 18 loops with an average length of 63,000 bp per rosette.[44]

When a 30-μm-long loop is complexed with histones to form a 30-nm diameter fiber, it will be compacted about 50-fold, and therefore radiate out from the scaffold for about 0.6 μm. This would produce a diameter for the chromatid in the region of 1 μm, which is, in fact, about what is actually observed. However, it has been claimed that the structure formed by loops attached to a scaffold is no more than 0.2–0.3 μm in diameter.[48] It has been estimated that the degree of compaction produced by the loops would be in the region of 40-fold,[44] to produce a total packing ratio for the DNA of about 2000.

The DNA loops appear to be attached to the scaffold by special sequences, the scaffold attachment regions (SARs) (reviewed in ref. 49). SARs have been identified as DNA sequences that remain attached to scaffolds after exhaustive digestion: they are AT-rich and generally appear to contain a consensus sequence for topoisomerase cleavage. This fits well with the observation that Topo II is a major scaffold protein; however, not all SARs appear to contain a Topo II-cleavage consensus sequence, and apparently SARs can bind to scaffolds deficient in Topo II. AT-richness alone is not a sufficient criterion for binding to scaffolds. SARs are never found in coding sequences, and are spaced about 3 to 140 kb apart. In a num-

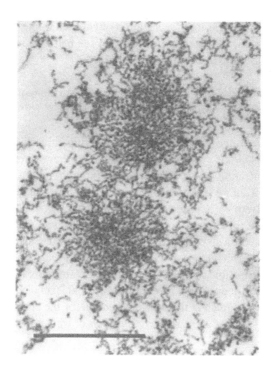

Figure 5. Cross-sections of a pair of chromatids from a CHO chromosome, swollen before fixation according to Adolph (Ref. 46), to show chromatin fibers radiating out from a central dense region. Transmission electron micrograph. Scale bar = 1 μm.

ber of cases, SARs were found to flank genes, and to coincide with the boundaries of nuclease-sensitive domains associated with active genes.[49] They have also been associated with origins of replication,[49] and it has been noted that the sizes of loops are similar to the sizes of replicons.[50] The idea has therefore developed that SARs may be functional units of chromatin and chromosome organization.

As mentioned above, there is evidence for substantial variation in the size of loops. Much of this involves variation around a mean, perhaps reflecting, for example, variation in sizes of replicons or transcribed domains. There is, however, some evidence for more systematic differences in loop sizes. It has been suggested that the existence of a secondary constriction at the sites of the nucleolar organizer regions (NORs) could be explained by the more frequent attachment of ribosomal DNA to the scaffold, and the resultant shorter loops; similarly, the centromeric constriction might occur because of the tendency of alphoid satellite (in humans) to associate with the scaffold rather than with the loops.[45] The tendency of a segment of yeast chromosome, inserted in a mouse chromosome, to form a constriction was also interpreted in terms of shorter loops in yeast than in mouse DNA.[51] However, it is far from clear that organization of chromatin into loops is the final stage of chromosome condensation (ref. 48, and see below), and other explanations of the phenomena just described are possible.

F.   The Final Stages of Chromosome Condensation: Coiling or Chromomeres?

Early prophase chromosomes appear as long, thin threads, and these grow shorter and fatter as the cell proceeds to metaphase. Although this is the aspect of chromosome organization that is most accessible to light microscopy, and which has therefore been available for study the longest, we know remarkably little about the processes involved. We have just seen (above) that it is possible to explain variations in the diameter of chromosomes (i.e., primary and secondary constrictions) in terms of differences in size of the chromatin loops; however, a simple model of the chromosomes in which loops of a fixed size would be attached to a scaffold of fixed length is incompatible with the observations. In any case, it is not at all clear that the scaffolds-and-loops structure could be the final level of chromosome condensation; an estimate given in the previous section for the packing ratio of such a structure gives a figure of 2000, still fivefold short of what is needed, and Rattner

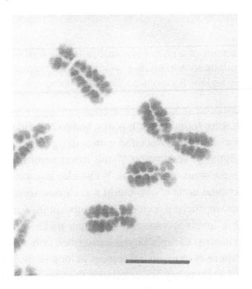

Figure 6.   Coiling of chromosomes. Human chromosomes, prepared using osmium impregnation, and examined by scanning electron microscopy, with back-scattered electron imaging in negative contrast. Scale bar = 5 μm. Reprinted with permission from Sumner (1991) *Chromosoma* 100: 410-18. © Springer-Verlag.

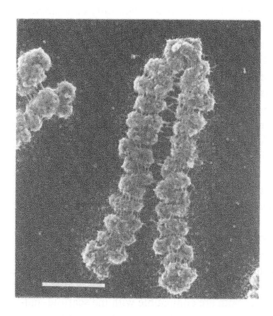

Figure 7. Chromomeres in a mitotic metaphase chromosome. Scanning electron micrograph of a mouse chromosome, prepared using osmium impregnation. Scale bar = 2 μm.

and Lin[48] believe that this results in a structure of only 0.2–0.3 μm diameter. In fact, there are several models, not necessarily mutually exclusive, for explaining the final stage of chromosome condensation, and some evidence for each of them.

Many workers have reported coiling in chromosomes, but although it is not difficult to demonstrate (Figure 6), and can not only be induced by special treatments,[52] but can also be seen in living cells,[53] most chromosomes do not, in fact, display coils. One obvious explanation of this would be that chromosomes are normally too tightly coiled for the individual gyres to be visually separable. Nevertheless, intermediate stages in the process of coiling should be discernible, but are not generally seen during chromosome condensation. Coiling is not incompatible with a scaffold-and-loops model; both Rattner and Lin[54] and Boy de la Tour and Laemmli[55] have presented evidence for a coiled scaffold. In the former model, the chromosome would be composed of a 200–300-nm fiber, made up of radial loops, which would form the metaphase chromosome by coiling; the coiling process would result in at least a ninefold compaction.[54] The latter authors described coiling of the scaffold, immunolabeled for Topo II, but only in about 1% of the chromosomes examined.[55] It should be noted that in such structures, constrictions would be the result of differences in coiling rather than in loop size, and it would also imply that the loops themselves, when fully condensed, would be considerably smaller than indicated in the previous section.

Other workers have described the condensation of chromosomes into a series of chromomeres. Chromomeres can most easily be demonstrated at the pachytene stage of meiotic prophase; however, they can also be seen from time to time in mitotic chromosomes, in both prophase and metaphase (Figure 7). Nevertheless, just as coils are generally visible in only a small proportion of chromosomes, so the majority of mitotic chromosomes do not show any clear sign of chromomeric structure. Condensation of chromosomes into a series of chromomeres, which then fuse to form a uniform cylinder, does not seem to be incompatible with a scaffold-and-loops model. In fact, in some reports, the scaffold appears as a series of discontinuous spots of material,[56] and it is easy to imagine these aggregating as the chromosome contracts, to form a continuous structure. Cook[57] has in fact proposed a model in which loops radiate spherically from so-called transcription factories to form chromomeres; these aggregate to form strings of chromomeres, which in turn fuse to form the familiar cylindrical chromatid.

It is difficult to see how the helical and chromomeric models of chromosome condensation can be reconciled. Although there is plenty of observational evidence for both, and both are compatible with the scaffold-and-loops model, more often than not, neither type of structure is visible. In turn, it is not, in fact, clear whether a further level of condensation is actually needed beyond that provided by loops attached to a scaffold. For the time being, it has to be admitted that there is no consensus on the highest level of chromosome structure.

G. The Chromosome Periphery

Chromosomes do not simply consist of chromatin fibers radiating from a protein core, and for a good many years a characteristic surface layer has been reported on chromosomes, although it has not attracted much attention. This has been called the chromosome periphery, or perichromosomal material, and consists of closely packed fibrils and granules.[58] Early reports indicated that this layer consisted of ribonucleoprotein (RNP),[59] and more recently, a considerable number of proteins have been identified at the surface of chromosomes by immunocytochemistry.[60,61] It turns out that different proteins are bound to the surface of chromosomes for different periods during mitosis. Some (including, for example, snRNPs), are bound only during metaphase and anaphase, and appear to be "passenger proteins"[62]; others are present from early prophase until telophase, and this class includes a number of nucleolar proteins. Finally, in anaphase and telophase, the lamin B receptor covers the chromosomes,[63] as an essential stage in the re-formation of the nuclear envelope at the end of mitosis.

Various functions have been proposed for the surface coating of chromosomes (reviewed in ref. 61). These include a role in chromosome organization, particularly in condensation. Another possibility is that it protects the chromosomes in the absence of the nuclear envelope during mitosis. So far, there is no overwhelming evidence that either of these is the essential reason for the existence of the surface layer. Clearly, however, the attachment of proteins to the chromosome surface is an efficient way of segregating the proteins to daughter cells, and an involvement in the reformation of the nuclear envelope also seems to be well established.

III. STRUCTURALLY DIFFERENTIATED REGIONS OF CHROMOSOMES

Although the greater part of the length of all chromosomes appears fairly uniform, certain parts show "nonstandard" packing of their chromatin, resulting in constrictions of one sort or another. The three types of constriction are the primary constriction, or centromere; the secondary constrictions, which are normally the sites of nucleolus organizers (NORs); and fragile sites. Often these are only clearly visible at certain phases of mitosis or are induced only under special conditions. For example, the centromeric constriction is usually inconspicuous in the thin elongated prophase chromosomes, and only becomes obvious after the rest of the chro-

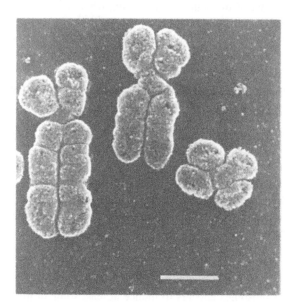

Figure 8. Metaphase human chromosomes, with the arms divided into separate chromatids, but with the centromeres remaining undivided. Osmium impregnated chromosomes viewed using the scanning electron microscope. Scale bar = 2 μm. Reprinted with permission from Sumner (1991) *Chromosoma* 100: 410-18. ©Springer-Verlag.

mosome has shortened and fattened. Fragile sites are not normally visible, but result from culture in media containing special additives.

A. The Centromere

The centromere appears as a constriction in the chromosome, and is the part of the chromosome that is attached to the spindle at metaphase and anaphase. In fact, these simple statements conceal a complex organization that is far from being completely understood, although considerable progress has been made in recent years.

Observations by scanning electron microscopy have shown that chromosomes split in half to form two chromatids during late prophase or early metaphase, apparently starting at their ends, and that the centromere is the last region to split, at the beginning of anaphase[33] (Figure 8). This led to the suggestion that an essential centromeric function is to hold sister chromatids together until anaphase, to ensure proper segregation to daughter cells. In fact, this function may be restricted to specific parts of the centromeric regions. It has been known for many years that, in large plant and insect chromosomes, there is a gap between the sister chromatids at the very center of the centromeric region, whereas the chromatids remain joined (until anaphase) on either side of this region,[64,65] at sites that, although they might be regarded as the proximal regions of the chromosome arms, are probably better thought of as special structures involved in centromeric function. Recently this type of structure has also been identified in mammalian chromosomes (Figure 9). In *Drosophila*, a specific satellite DNA sequence has been described that appears to hold sister chromatids together.[66]

The other centromeric function, attachment to the spindle, is mediated through a special structure, the kinetochore. (It should be noted that in the past, there has been some confusion over the nomenclature of this region of the chromosome, but it is now generally accepted that the term *centromere* applies to the actual constricted region of the chromosome, and *kinetochore* indicates the specialized structures to which the spindle microtubules are attached.) Electron microscopy showed that in many organisms, the kinetochore is a three-layered (trilaminar) structure (Figure 10).

Specific DNA fractions have been located in the centromeric regions of chromosomes immediately below the kinetochores. These include such DNAs as the human α-satellite and the mouse minor satellite. Insertion of human α-satellite into monkey chromosomes appears to induce at least some of the functions of a centromere.[67] In *Drosophila*, centromere function seems to involve a 220-kb region that includes a significant amount of complex DNA.[68]

In recent years, a number of proteins have been described that are associated with the centromere or kinetochore, the CENPs, or centromeric proteins (Table 4), although the precise location and function of these are still not known in many cases. CENP-A is a histone H3 variant that appears to be required for the forma-

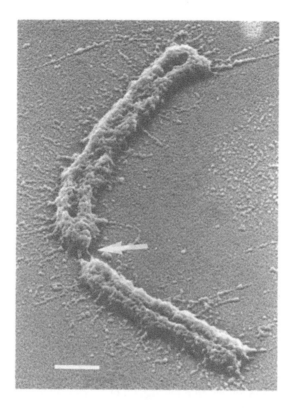

Figure 9. Metaphase CHO chromosome, with the centromere (arrow) split into two separate fibers, the chromatids remaining joined on either side of the centromere. Scanning electron micrograph. Scale bar = 1 μm.

tion of the specific chromatin structure of centromeric chromatin.[69] CENP-B binds to the DNA of the centromeric constriction—α-satellite DNA in humans. It recognises a specific 17-bp motif, the CENP-B box, in the α-satellite of humans.[70] Mouse chromosomes also contain a specific, centromeric, minor satellite that contains the CENP-B box.[71] However, it appears that not all chromosomes have centromeric DNA that contains a CENP-B,[72] although proteins that can bind to the CENP-B box have also been reported from *Drosophila*.[73] CENP-C is a component of the inner kinetochore plate,[74] and in stable dicentric chromosomes is only found at the active centromere;[75] CENP-E, a kinesin-like motor protein, is also found only at active centromeres.[75] CENP-F, like CENP-E, is associated with the outer kinetochore plate.[77] If mitotic cells at the appropriate stage are injected with antibodies to certain CENP proteins, mitosis is disrupted, and cells become arrested in metaphase.[78–82]

Figure 10. Transmission electron micrograph of a CHO chromosome, showing the trilaminar kinetochore (arrow) attached to the spindle microtubules. Scale bar = 0.5 μm.

In addition to the CENPs, which in any case may form only a small proportion of the centromeric and kinetochore proteins, a number of other proteins are known to be associated with the centromeric region of chromosomes. These include such proteins as the INCENPs (inner centromeric proteins)[83] and CLiPs (chromatid linking proteins),[84] which appear to lie in a region between the sister kinetochores, and at other regions where the sister chromatids remain in contact. It is not clear whether such proteins are important for centromere function, with perhaps a role in holding sister chromatids together, or ensuring their separation at anaphase, or whether they are "chromosomal passenger proteins"[62] that use the chromosomes as a means of transport to the appropriate parts of the cell to carry out nonchromosomal functions during the later stages of cell division. Another intriguing protein appears to form a ring that totally encircles the centromeric constriction.[85] Such a protein, which appears to have no structural counterpart, looks as if it could well be holding the sister chromatids together.

Table 4. CENPs (CENtromeric Proteins)

Protein	Molecular Weight	Properties
CENP-A	17 kDa	Histone H3 variant (ref. 69)
CENP-B	80 kDa	Binds to DNA of primary constriction (α-satellite) (ref. 70,71)
CENP-C	140 kDa	Inner kinetochore plate; binds DNA. Only in active centromeres (ref. 74, 75)
CENP-D	50 kDa	GTP-binding protein (ref. 76)
CENP-E	312 kDa	Kinesin-like motor protein; in outer kinetochore plate. Active centromeres only (ref. 75)
CENP-F	400kDa	Outer surface of outer kinetochore plate (ref. 77)

The suggestion that proteins such as INCENPs and CLiPs may have a role in sister centromere "pairing" at mitotic metaphase raises questions of centromeric structure and the mechanism of chromatid separation. To take the second point first, there is now abundant evidence that the enzyme topoisomerase II (Topo II), which is required to decatenate (disentangle) newly replicated DNA molecules, is also required for chromatid separation. Although much of the decatenation of newly replicated DNA may well occur shortly after replication, it appears that at least a segment of centromeric DNA does not become decatenated immediately, but remains entangled until the metaphase/anaphase transition. In yeasts with mutations that inhibit topoisomerase II function, chromosome segregation is inhibited,[86,87] and the same effect can be produced in *Drosophila*[88] and mammalian[89–92] cells with drugs that inhibit Topo II. Immunofluorescence studies show that Topo II is concentrated in the centromeric regions of mammalian chromosomes at metaphase, but is lost from these regions when the cells pass into anaphase, consistent with a role for Topo II in chromatid separation[35]. On the other hand, reports have been published that specific proteins are required for chromatid separation,[93–95] but this is hardly surprising; it is likely that some proteins must be involved in controlling the activity of Topo II, so that it operates only at the appropriate time. In fact, a large number of functions have been implicated in the anaphase separation of chromosomes, including also proteolysis and protein dephosphorylation, and the processes involved must be highly complex.[96]

If the DNA of the sister chromatids is still physically linked at metaphase, this has implications for centromeric structure. Evidently there could not be two sister chromatids in this region, held together by proteins such as the INCENPs and CLiPs, but only a single, undivided centromere, and it would not be appropriate to refer to "pairing" of sister chromatids. Structural observations support such an interpretation. Whereas the chromosome arms are clearly split into two separate chromatids, the centromere, even at the highest resolution of electron microscopy, appears as a single, undivided segment (Figure 8) (but note the observation above, in Figure 9, that the region in the middle of the centromere may split, and it may be that it is segments on either side of this that remain unsplit). The presence of a constriction at the centromere might itself be a consequence of the centromere not having divided. It has been proposed that, once the chromosome arms have split into sister chromatids, they are free to contract and become thicker by forming spirals; on the other hand, the centromere cannot condense by forming spirals, since it would then be unable to separate into two at anaphase; it must therefore remain as an uncondensed constriction.[33,48,97] (Note, however, that, as mentioned above, the presence of a centromeric constriction has also been explained as a result of a specific chromatin structure at the centromere, involving shorter loops).

The centromere is clearly a chromosomal region that has distinctive DNAs, proteins, and structures. Much still has to be learned about how it works, and further discussion is beyond the scope of this chapter; for further details the reader is referred to recent reviews (e.g., ref. 98 and 99; and the chapter by Sunkel in this volume).

B.   Nucleolar Organizer Regions

Certain chromosomes show another, secondary, constriction as well as the centromere or primary constriction. Some secondary constrictions are sites of heterochromatin, but in general they indicate the positions of nucleolus organizer regions (NORs). These are regions of chromosomes that contain the highly repeated genes for 18S and 28S ribosomal RNA, and participate in the formation of nucleoli. Over the years there have been numerous reviews of the organization of nucleolar genes and of nucleoli (e.g., 100–105), and it would be superfluous to repeat such information here. Instead, I shall restrict myself to a few comments on the nature of the constrictions formed by NORs, and on their staining reactions.

Gene activity is normally associated with decondensed chromatin. The genes for ribosomal RNA are highly active, and a nucleolus is often still present during early prophase, when it can be seen that the NOR-bearing chromosomes are still attached to the nucleoli. The tendency of NORs to appear as secondary constrictions may therefore be a consequence of the high degree of activity of the genes that they contain, and the consequent decondensation of their chromatin. However, Bickmore and Oghene[45] have suggested that the NORs form a constriction because the chromatin loops are smaller in these regions, although they do not give any figures. Without such information it is, of course, impossible to judge whether loop size alone would be adequate to explain the observed secondary constrictions.

The method of choice for staining NORs involves silver,[106] which under appropriate conditions reacts with one or more nucleolar proteins,[107] to give a black deposit (Figure 11). The precise chemical groups that are responsible for the reduction of silver are still a matter for debate, but it is clear that the stainable material is a remnant of the interphase nucleolus that stays attached to the chromosomes, and is not part of the chromosome itself.[108] As such, nucleolar silver

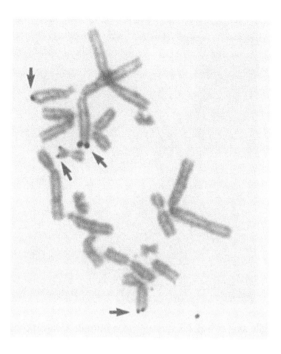

Figure 11. Silver (Ag-NOR) staining of nucleolus organizer regions (arrows) on CHO chromosomes.

staining reflects the activity of ribosomal DNA during the preceding interphase, rather than simply the presence of ribosomal genes. In humans, with 10 NOR-bearing chromosomes (five pairs), no more than six or seven NORs normally stain with silver. In hybrids, silver staining of the NORs from one of the parents is often suppressed, and this can be correlated with the suppression of transcription of rDNA from one of the parental genomes.[109] Silver staining has proved valuable for locating the sites of NORs on chromosomes of numerous species (with the proviso that not all sites may be stained in any one individual), and it also has applications in the prognosis of cancer, since the amount of silver staining can be correlated with the rate of division of the cells.[110,111]

NOR regions of chromosomes can often also be stained selectively with fluorochromes such as chromomycin that show specificity for G+C-rich DNA. This is a consequence of the 18S and 28S sequences of rDNA being G+C-rich;[112] nevertheless, not all NORs show clear differential fluorescence with G+C-specific fluorochromes, and of course such staining cannot distinguish between NORs and, for example, blocks of G+C-rich heterochromatin.

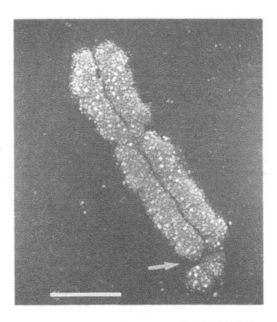

Figure 12. Fragile site at Xq27.3 (arrow) on a human X chromosome. Note that the segment of the chromosome distal to the fragile site has become displaced, as a result of the tendency of fragile sites to break. Osmium impregnated chromosome viewed using the scanning electron microscope. Scale bar = 2 μm.

C. Fragile Sites

Fragile sites are characteristic features of chromosomes, but only appear when cells are grown under special conditions. They are specific locations on chromosomes that have a tendency to break, and usually appear as a constriction or a complete interruption in one or both of the chromatids of a metaphase chromosome (Figure 12). In humans, fragile sites are divided into two classes depending on their frequency of occurrence. Common fragile sites are probably present in all individuals, and most are induced by culture in the presence of aphidicolin (an inhibitor of DNA polymerase α), while others are induced by bromodeoxyuridine or 5-azacytidine. Rare fragile sites, on the other hand, only occur in a small proportion of individuals, usually less than one percent of the population, and are normally detected after culture in the absence of folic acid, although a few can be induced by bromodeoxyuridine or distamycin A. (For listings of fragile sites on human chromosomes, see refs. 113 and 114). Fragile sites have also been detected on the chromosomes of various other mammals,[115–119] and are probably a universal feature of chromosome organization in this group. It is not known, however, if fragile sites are inducible in all groups of eukaryotes.

It is not the intention to give an account of the clinical correlates of human fragile sites here. The association of the fragile site at Xq27.3 with the commonest form of X-linked mental retardation is well known, and has been thoroughly reviewed;[120,121] other fragile sites have been identified as breakpoints in cancer cells,[122,123] and recently it has been shown that one fragile site is the location of an oncogene.[124] However, information on the mechanism of formation of fragile sites is rather scanty, and generally involves a good deal of speculation.

It was noted several years ago that in general fragile sites are induced by agents that perturb some aspect of DNA metabolism. Distamycin and similar compounds are known to bind to A+T-rich DNA and to prevent its condensation,[125] and bromodeoxyuridine can be incorporated into DNA and can also inhibit condensation.[126] Aphidicolin, as already mentioned, is an inhibitor of DNA polymerase α. Laird et al.[127] proposed, on the basis of many observations, not only in mammals, but also in *Drosophila*, that fragile sites were regions of delayed replication, and that because they replicated very late, they did not have time to condense properly. It was proposed that compounds that induced fragile sites might shorten the G2 phase, during which condensation of chromosomes should be occurring. The fragility of these sites could be caused by incomplete replication.

Most efforts toward understanding fragile sites have been focused on the fragile X site at Xq27.3, and some remarkable findings have been obtained. A gene (FMR-1) has been located at this fragile site, and a trinucleotide (CCG) repeat is associated with this gene.[128] The number of copies of the repeat is variable, being lowest (5-53) in normal individuals, and highest (>200) in those with X-linked mental deficiency.[129] When the number of trinucleotide repeats is high, a CpG island associated with the FMR-1 gene becomes methylated, and the gene is therefore inactivated, presumably leading to the observed mental deficiency. More interesting from the structural point of view is that the degree of expression of the fragile site (which can vary from a few percent up to about 50%) is related to the number of copies of the trinucleotide repeat.[130] The expression of the fragile site therefore seems to be related to the degree of DNA amplification; perhaps some protein required for DNA condensation is unable to bind to long stretches of $(CCG)_n$. It is not known if this apparent mechanism of formation of the fragile site at Xq27.3 is applicable to any other fragile site. Several other diseases have now been identified that are associated with trinucleotide repeats, but none of them appear to be associated with fragile sites. Conversely, there does not seem to be any evidence for trinucleotide repeats at the majority of fragile sites, although, because such fragile sites are not associated with any clinical condition, there has not been any great incentive to look for trinucleotide repeats in them.

IV. HETEROCHROMATIN

Heterochromatin was defined by Heitz,[131] as parts of chromosomes that did not decondense at telophase, but remained in a compact form throughout interphase.

More recently, it has been recognized that there are two main classes of hetero-chromatin: constitutive and facultative.[132] Facultative heterochromatin is perma-nently condensed chromatin that is found only in one of a pair of homologous chromosomes, and which necessarily has the same DNA composition as its uncondensed homologue. The best known example of facultative heterochromatin is the inactive X chromosome of female mammals. Constitutive heterochromatin, on the other hand, occurs in both of a pair of homologues and generally contains DNA that differs from that in the rest of the chromosome. While constitutive het-erochromatin can generally be stained distinctively with appropriate banding tech-niques, facultative heterochromatin does not normally show a characteristic staining reaction during mitosis.

A. Constitutive Heterochromatin

Heterochromatic bands are found in virtually all chromosomes of all animals and plants and appear to be universal. Almost all can be demonstrated by C-band-ing (for exceptions, see ref. 106), but can also be demonstrated in part by a wide variety of other methods. It should be emphasized that methods for staining het-erochromatic bands color only constitutive heterochromatin differentially. Facul-tative heterochromatin, such as the inactive X chromosome in female mammals, is not stained distinctively by routine banding methods, but can be distinguished by replication banding methods because of its characteristic late replication.

Heterochromatic bands are localized, and are normally found at and around cen-tromeres, often at telomeres and adjacent to nucleolar organizers, and sometimes in interstitial sites. They generally comprise a relatively small proportion of the genome, but in exceptional cases over 50% of the cell's DNA may consist of het-erochromatin; a few completely heterochromatic chromosomes are known. The DNA of heterochromatic bands is additional to the fixed amount of DNA found in euchromatin, and the quantity of heterochromatin can vary not only between indi-viduals of the same species, but also between homologous chromosomes in the same individual.

Constitutive heterochromatin can now be defined in terms of its DNA composi-tion and its reaction to specific banding techniques. It generally consists largely of highly repeated, "satellite" DNA sequences (for a review, see ref. 133). There is little, if anything, in common between the repeated sequences in different organ-isms, which show a wide range of base composition and repeat length. Many sat-ellite DNAs show a high degree of methylation of their cytosines.[133] This methylation appears to be important for the condensation of the chromosome seg-ments that contain such DNA because demethylation, whether naturally occurring or induced by culture in the presence of 5-azacytidine,[134] results in decondensa-tion of heterochromatin. However, other organisms, notably *Drosophila*, lack 5-methylcytosine in their DNA, not only in their heterochromatin, but also in their euchromatin. Certain species, including the Chinese hamster[135] and the field vole

Microtus agrestis,[136] appear to lack highly repetitive DNA in much of their heterochromatin, but instead have only moderately repetitive sequences. In no case is it possible to state that any block of heterochromatin contains exclusively one type of DNA, and indeed, in some cases, there is good evidence for the presence of more than one class of satellite DNA.

Much less is known about the proteins of constitutive heterochromatin. Probably the best characterized protein that is specific for heterochromatin is HP1 from *Drosophila,*[137] which contains a highly conserved sequence, the chromobox. This protein appears to be essential for assembly of heterochromatin, a process which is associated with the phosphorylation of this protein.[138] A similar protein has been identified in mammalian chromosomes, and is also localised to heterochromatin.[139,140] Various *Drosophila* proteins have been identified that are associated with position effect variegation, but it is not clear whether they are normal constituents of heterochromatin itself; the protein described by Gerasimova and colleagues[141] appears to act as an "insulator" to delimit the boundary between heterochromatin and euchromatin. As mentioned above, constitutive heterochromatin in mammals often, perhaps always, has highly methylated DNA. Proteins that interact specifically with methylated DNA have been recently identified, and one of these, MeCP2, is concentrated in the C-banded regions of mammalian chromosomes.[142] Another protein, HMG-I, is concentrated in heterochromatin,[143] but like MeCP2, is not located there exclusively. Like euchromatin, constitutive heterochromatin contains histones, but there are important differences between eu- and heterochromatin in the types of modifications that occur. In particular, centromeric heterochromatin in mammals[144,145] and in *Drosophila*[146] is deficient in acetylated histone H4, a form that is associated with regions of chromatin where active transcription occurs. Evidently our knowledge of the proteins of constitutive heterochromatin is still very fragmentary, and it is still not possible to give a satisfactory account of the class of proteins that cause a segment of chromatin to be heterochromatin (though this is also true of euchromatin).

Heterochromatin has traditionally been considered genetically inert. In *Drosophila,* the heterochromatin contains very few genes compared with the euchromatin,[147] and extensive losses or additions of heterochromatin are without apparent effect. During polytenisation of *Drosophila* chromosomes, the bulk of the heterochromatin does not replicate[148] implying that it is not required, at least for somatic interphase functions. A similar situation is found in organisms as diverse as nematodes,[149] copepods[150] and hagfish,[151] which eliminate C-banded heterochromatin in somatic cells, although it is retained in the germ line. Similarly in man, extensive variation in the amount of heterochromatin can occur without any phenotypic effect.[152,153] The presence of highly repetitive, short-sequence ("satellite") DNA in the heterochromatin of most organisms that have been studied has lent further support to the idea of the genetic inertness of heterochromatin because such sequences could not be translated into proteins. (See Ref. 133, for a review of satellite DNA). Nevertheless, an absence of genes in a stretch of DNA does not exclude the possibility that the DNA might have some other function that might not depend directly on the precise sequence of the DNA. In fact, heterochromatin appears to have several effects and also a few functions.

It has long been known that heterochromatin can have profound effects on the number and distribution of chiasmata at meiosis,[154–156] although these effects vary from one species to another, and are sometimes absent.[157] Meiotic chiasmata are virtually unknown in heterochromatin, and indeed these regions often fail to pair at meiosis. Another effect of constitutive heterochromatin is position effect variegation (PEV), that is, the tendency of genes to become inactivated when placed in close proximity to heterochromatin.[158–161] In PEV the condensed structure of heterochromatin spreads into the adjacent euchromatin.[162] It has recently been shown that proteins normally associated with heterochromatin can be found associated with genes that are inactivated in position-effect variegation.[163] For a detailed review of these and other "effects" of heterochromatin, such as ectopic pairing, see ref. 164. What seems clear is that such effects do not constitute an adequate reason for the existence of heterochromatin. When we start to look for an actual function of constitutive heterochromatin (something that the heterochromatin does, rather than merely influences), the evidence becomes much weaker. The observations that heterochromatin is deliberately eliminated in somatic cells of certain species (see above), while it is retained in the germ line, or is not amplified, as in the polytene chromosomes of *Drosophila*,[148] suggest that the essential functions of heterochromatin must be in oogenesis or spermatogenesis. However, apart from the effects on meiotic pairing and chiasmata, described above, there is little evidence as to what these germ-line functions could be. Nevertheless, it has become clear in recent years that the constitutive heterochromatin of *Drosophila* contains several "functions" that are not conventional genes.[165] These include the Y-chromosome fertility factors; the collochores, necessary for the accurate pairing and disjunction of the sex chromosomes in the achiasmate males; the *Responder* locus,[166] which is part of the *Segregation Distorter* system and which consists of up to 2,500 copies of a 120-bp repeat; and *ABO*, which can rescue the maternal effect mutation *abnormal oocyte (abo)*. Homology of heterochromatin is necessary to ensure accurate meiotic segregation in *Drosophila melanogaster*, in which the small 4th chromosome is always achiasmate and the X chromosome sometimes fails to recombine; increasing the amount of heterochromatin increases the effectiveness of this system.[167] Few organisms have been studied genetically as intensively as *Drosophila*, and it is not known if such "non-conventional" factors are widely distributed; however, even if they were, it would probably not account for the greater part of the heterochromatin found in most organisms.

If constitutive heterochromatin has a function, it is presumably related to its position (commonly centromeric) and its composition (usually highly repetitive DNA sequences). The actual sequence and base composition of the heterochromatic DNA are so variable[106,133] that if there is a general function for heterochromatin, it is clearly not dependent on DNA sequence. However, it has been shown that several satellite DNAs show characteristic curvature,[168–171] and one can therefore postulate that such curvature, rather the sequence itself, would be important for any function of satellite DNA in centromeric heterochromatin.

As discussed above, the essential functions of centromeres appear to be, on the one hand, to provide the point of attachment of the chromosome to the spindle microtubules, and on the other, to ensure that the sister chromatids are held together until the end of metaphase and then separate in a co-ordinated fashion at the beginning of anaphase. It appears that the sites of these functions may be physically separate: in certain plants the kinetochores may appear to be separated by a hole between them as early as prophase, and the sister chromatids are held together at late metaphase only in the regions on either side of the kinetochores.[64,65,172] While DNA fractions such as human α-satellite and mouse minor satellite are associated with the kinetochore, it may be that the "classical" satellites function to hold the sister chromatids together. Indeed, it has been proposed that the highly repeated sequences of satellite DNAs form a favourable substrate for the action of Topo II to separate the sister chromatids at the start of anaphase.[33]

In most cases, there is far more C-band material at the centromeres than seems to be needed for known centromere functions. Although a few species such as

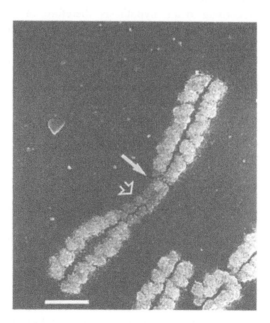

Figure 13. Human chromosome no. 1 from a patient with the ICF syndrome, showing extended heterochromatin (open arrow). Note that the heterochromatin is split into separate chromatids, although the centromere (solid arrow) remains undivided. Scale bar = 2 μm.

many Carnivora have minimal centromeric heterochromatin,[173] most species have quite substantial blocks of centromeric and para-centromeric heterochromatin, and some also have large blocks of non-centromeric heterochromatin. So far, no satisfactory function has been ascribed to non-centromeric heterochromatin, although Vig[174] has presented evidence that the size of blocks of heterochromatin may influence the order of separation of sister centromeres in different chromosomes at anaphase that could in extreme cases lead to aneuploidy (an effect rather than a function). While the sister chromatids are generally more closely apposed in heterochromatin than in euchromatin, in human chromosomes the pericentromeric heterochromatin does appear to split before the centromeres themselves do (Figure 13). It may be, after all, that non-centromeric constitutive heterochromatin is without function and has merely accumulated because there is insufficient selective pressure against its amplification. Macgregor and Sessions[175] have proposed that blocks of centromeric heterochromatin become amplified, probably by repeated unequal exchanges in these regions, which, because of their repetitive nature, are highly susceptible to such amplification.[176] In time, these blocks could become broken up and dispersed, first to pericentromeric regions, then to interstitial and terminal sites; in the process there would be gradual divergence of the DNA sequence so that eventually the material might no longer be recognizable as heterochromatin. According to this hypothesis, therefore, the accumulation of non-centromeric heterochromatin would simply be a consequence of the characteristics of the cell's replication machinery, combined with a lack of selection against such heterochromatin. Although this is not a satisfactory explanation, we have nothing better at present.

B. Facultative Heterochromatin

Facultative heterochromatin, as remarked above, involves only one of a pair of homologues.[132] The two best known types are the inactive X chromosome of female mammals, and the paternal chromosome set in mealy bugs.

It is assumed that the inactivation of one X in female mammals is a dosage compensation mechanism; if both X chromosomes remained active, they would presumably produce twice as much of their gene products as the single X chromosome in XY males. Such a situation might well be dangerous for the organism, with either an excess of some gene product in females, or an insufficiency in males. It should nevertheless be noted that in birds, with a ZZ/ZW sex determining mechanism, there is no evidence for inactivation of the second Z chromosome. (Note that in *Drosophila*, dosage compensation is achieved by the single X chromosome of males producing twice as much of its gene products as each of the two X chromosomes in females; see ref. 177). The hypothesis that one of the two X chromosomes in female mammals is inactivated, and that inactivation affects either the paternal or maternal X at random, is known as the Lyon hypothesis.[178] In fact, the inactivation of the X chromosome occurs early during embryonic development, and is only random in the tissues of the embryo itself; in the extra-embryonic tissues of the placenta, it is always the paternal X chromosome that is inactivated.[179-181] In the somatic tissues of marsupials, it is also the paternal X that is inactivated.[182] In interphase nuclei, the inactivated X chromosome appears as a densely staining spot against the nuclear membrane, the Barr body (183). The study of hair color in certain mammals shows that the inactivation of the X chromosome is random, and that female mammals are mosaics. For example, tortoiseshell cats, which are almost invariably females, have patches of orange and black fur. This is because the relevant gene for hair color is on the X chromosome. In some cells, the X chromosome bearing the allele for black hair is active, while in others the X that bears the allele for orange hair is active. Because inactivation occurs quite early in development, relatively large patches of skin develop with one type of hair color or the other. Normally, male cats cannot show the tortoiseshell coloration, since they have only one X chromosome; the very rare male tortoiseshell cats have the abnormal XXY chromosome constitution.

Individuals with an abnormal sex chromosome constitution—that is, those with more than 2 X chromosomes—always have only one active X chromosome, and the number of Barr bodies is always one less than the total number of X chromosomes (Table 5). In human female tetraploids (which, however, die before or immediately after birth), two X chromosomes are active and two inactive, while in female triploids, which are also lethal, either one or two X chromosomes are active, so that there is always an attempt to maintain the proper proportion of active X chromosomes. It seems that there must be some mechanism whereby all X chromosomes except one become inactivated. In fact, it has been shown that X chromosome inactivation diffuses from a locus somewhere in the middle of the X chromosome that has been called the X inactivation center, XIC. The position of

Table 5. Number of Inactive X Chromosomes and Barr Bodies
in Cells with Different Sex Chromosome Constitutions

Male	Female	Number of X Chromosomes	Number of Inactive X Chromosomes/Barr Bodies
XY/XYY	XO	1	0
XXY	XX	2	1
XXXY	XXX	3	2
XXXXY	XXXX	4	3
Triploid			
XXY		2	0/1
XYY		1	0
	XXX	3	1/2
Tetraploid			
XXYY		2	0
	XXXX	4	2

the X inactivation centre was discovered by studying rearranged and translocated
X chromosomes. Certain fragments of the X have the property of inducing inactivation, while others do not, and by this means the X inactivation centre has been
located to band Xq.13. When a person has a normal X chromosome and an X
translocated to an autosome, one finds that the normal X is usually inactivated.
This is because the fragment of the X with the inactivation centre also inactivates
the fragment of the autosome to which it has been translocated, producing, in
effect, monosomy for that part of the autosome. Since monosomy of an autosome
appears to be lethal, cells in which the fragment of the autosome is inactivated
would die.

It appears that, at the appropriate stage of development, one X inactivation centre
in each cell receives a signal that blocks its function, while all the other X inactivation centres (normally one, in an XX female) begin the process of inactivation.
The nature of the blocking signal remains unknown, but in recent years important
discoveries have been made about the nature of the inactivation signal. In the same
region of the X chromosome that contains the X inactivation centre, a gene has
been found that is known as *XIST* (X-inactive specific transcript), which is transcribed only from inactive X chromosomes, and not from the active X. The *XIST*
gene product has been shown to be essential for X chromosome inactivation.[184]
The sequence of this gene is not, surprisingly, highly conserved between mouse
and man,[185] and the gene does not code for a protein. Instead it produces an RNA
that seems to bind throughout the inactive X chromosome.[186] CpG islands in inactivated X chromosomes of eutherian mammals are methylated;[187] this does not
seem to be a primary mechanism of X chromosome inactivation, but rather a means
of stabilizing the inactivation. In marsupials, the CpG islands of the inactive X are
not methylated, and in fact the inactivation is less stable in these animals.[187]

In reality, not all of the inactive X chromosome is inactivated. If this were so,
then individuals with sex chromosome aneuploidy might be expected to be normal, when in fact they are not (although the consequences of sex chromosome
aneuploidy are far less serious than the effects of autosomal aneuploidy). This
implies that there must be some genes on the inactive X chromosome that are still
active. In fact, as well as genes in the pseudoautosomal region, which pairs with
the Y at male meiosis and forms an obligatory chiasma with it, there are some three
regions of the human X chromosome that remain active when the rest of the chromosome is inactivated.[188] These active regions of the inactive X contain a high
level of acetylated histone H4, which is characteristic of chromosomal segments
containing active genes;[189] (see also below under Euchromatin). Active genes on
the inactive X do not have their CpG islands methylated. It should be noted that in
the laboratory mouse, there are no active genes on the inactive X except in the
pseudoautosomal region, and that XO mice, unlike XO women, are almost normal
and fertile.

The other reasonably well known system of facultative heterochromatin is in the
male mealy bug, in which the whole of the paternal set of chromosomes becomes

heterochromatic. The main characteristics of this system were summarized by Brown & Nur;[190] in males, the whole paternal set of chromosomes becomes condensed at the blastula stage, forming a compact mass of chromatin in interphase, although all the chromosomes, both euchromatic and heterochromatic, are condensed to the same extent at metaphase. Although there are no sex chromosomes in these insects, the heterochromatinisation does not appear to be a sex determining mechanism, and unlike the mammalian situation, it occurs in the male. At male meiosis, two euchromatic nuclei and two heterochromatic nuclei are formed; the former develop into spermatozoa, while the latter degenerate. In somatic tissues, the heterochromatic set of chromosomes sometimes fails to replicate as often as the euchromatic set, so that the proportion of heterochromatin decreases;[191] in other tissues, the heterochromatic chromosomes may revert to a euchromatic state and become transcriptionally active.[192] RNA synthesis is only found in the euchromatin, but not the heterochromatin, of mealy bug nuclei; however, if cells are treated with a polyanion (which was believed to disrupt DNA-histone linkages, but could well have more widespread effects), the heterochromatin decondenses and can be transcribed.[193] Recently, a nuclease-resistant fraction of DNA has been described, associated with the facultative heterochromatin in males (194). Clearly this is a system with considerable potential for elucidating the mechanisms of formation of facultative heterochromatin, but unfortunately it has not been investigated thoroughly using modern techniques.

Many years ago, it was argued over whether heterochromatin was a substance or a state. We are now in a position to answer this question, and it is clear that in general, constitutive heterochromatin is a substance in that it contains specific types of DNA and has specific proteins associated with it. On the other hand, facultative heterochromatin is equally clearly a state because its DNA is essentially identical to that of its euchromatic homologue. It is derived from euchromatin, and in rare cases can revert to an active, euchromatic state.[192,195] In the case of position effect variegation, in which inactivation spreads from constitutive heterochromatin into the adjoining euchromatin, we are again dealing with a state rather than a substance. On the other hand, there are some situations in which perfectly good constitutive heterochromatin can become decondensed. In the early stages of *Drosophila* development, when the embryo is still syncytial and cell division is very rapid, no heterochromatin can be seen.[196] This, presumably, is an adaptation for rapid growth and division; interphase is extremely short at this stage, and there is probably no time to condense the heterochromatin. Another example is from *Microtus agrestis*, in which the gonosomal heterochromatin is decondensed in the nuclei of certain types of cells.[136] In spite of these exceptions, however, it is convenient and valuable to maintain the distinction between constitutive heterochromatin, a substance that is permanently condensed, and facultative heterochromatin, which is a state that can be derived from euchromatin.

V. EUCHROMATIN

All the chromatin that is not heterochromatin is euchromatin, which decondenses normally at the end of mitosis. When chromosomes are treated with suitable banding procedures, the euchromatic regions of chromosomes are found to show characteristic and reproducible patterns of alternating dark and light bands. The principal staining methods for euchromatic bands are G-, Q- and R-banding, supplemented by a variety of other methods of which the most important are replication banding and banding with a variety of fluorochromes that show specificity for various bases in the chromosomal DNA.[106]

Whatever banding method is used, euchromatic bands always consist of a series of positively and negatively staining bands. However, since, for example, a positive G-band is equivalent to a negative R-band, it is not possible to refer to a negative or positive euchromatic band. In fact, there is so far no general terminology to specify the type of euchromatic band being referred to, independently of the staining method used.

A. Euchromatic Bands in Mammals

When methods such as G-, Q-, and R-banding are applied to mammalian chromosomes, they reveal characteristic patterns of alternating dark and light bands throughout the euchromatic parts of the chromosomes. The patterns produced by G-, Q- and R-banding methods are essentially the same, although bands that are G-band and Q-band positive are R-band negative and vice versa; moreover, the

Table 6.   Characteristics of Euchromatic Bands in Mammalian Chromosomes

Positive G-bands	Negative G-bands	
Positive Q-bands	Negative Q-bands	
Negative R-bands	Positive R-bands	
Pachytene Chromomeres	Interchromomeric Regions	ref. 199-202
Late-Replicating DNA	Early-Replicating DNA	ref. 203-205
Early Condensation	Late Condensation	ref. 206,207
A+T-rich DNA	G+C-rich DNA	See text
Low Concentration of Genes	High Concentration of Genes	See text
DNase Insensitive	DNase Hyper-Sensitive	ref. 208,209
Low Level of Histone Acetylation	High Level of Histone Acetylation	ref. 189
High Level of Histone H1 Subtypes	Low Level of Histone H1 Subtypes	ref. 230
HMG-I Protein Present	HMG-I Protein Absent	ref. 143
Rich in LINEs (Long Intermediate Repetitive DNA Sequences)	Rich in SINEs (Short Intermediate Repetitive DNA Sequences)	ref. 210,211
Low Level of Chromosome Breakage	High Level of Chromosome Breakage	ref. 212-215
Little Recombination	Meiotic Pairing and Recombination	ref. 216

number of bands that can be demonstrated is greatest in the most elongated prophase chromosomes, and decreases as the chromosomes contract. Apart from the immense value of these patterns for identifying chromosomes, they have drawn our attention to the fact that the euchromatin is divided up into compartments with different properties, and is therefore not a single substance, and it is this latter aspect of euchromatic bands that will be discussed here (for a review of practical applications of banding methods, see ref. 106).

Numerous correlations have been shown between G-, Q- and R-bands on the one hand, and various functional or compositional attributes of chromosomes on the other (Table 6).[197, 198] Not all of these require detailed discussion, but a few comments will be in order. It must be recognized that the correlations listed in Table 6 are not perfect, although this does not detract from their general importance. One striking example of this is the inactive X chromosome in female mammals, discussed above (facultative heterochromatin). Although its G-banding pattern is identical with that of the active, early-replicating X, it shows a generally late pattern of replication (i.e. the negative G-bands of the inactive X are not early replicating, although different parts of the inactive X do replicate at different times during the latter part of the S phase); it also shows a high level of histone acetylation in only a very limited number of sites, many fewer than the negative G-bands.[189] On the autosomes, although pachytene chromomeres generally correspond to segments of A+T-rich DNA, many of the terminal regions of chromosomes consist of G+C-rich chromomeres.[202] The pairing segments of the human X and Y chromosomes show an homologous replication pattern, although there is no obvious resemblance in their G-banding patterns.[217]

Evidence for the division of mammalian chromosomes into segments of A+T-rich and G+C-rich DNA comes from a variety of sources. The most direct is the evidence that a variety of fluorochromes that produce banding patterns on chromosomes show enhanced fluorescence when bound to A+T-rich DNA (e.g. quinacrine, Hoechst 33258, DAPI) or to G+C-rich DNA (e.g. chromomycin, mithramycin, 7-amino-actinomycin D) (See ref. 106, Chapter 8, for a full discussion). Similarly, antibodies to different DNA bases produce banding patterns according to their base specificity.[218] An entirely different approach, using density gradient centrifugation, showed that early replicating DNA (later shown to be equivalent to negative G-bands) was relatively G+C-rich, while late replicating DNA was relatively A+T-rich.[219,220]

Evidence for the distribution of genes has come from a variety of techniques, which have become more sophisticated and precise over the years. In the very first years of human chromosome banding, it was pointed out that those trisomies that are compatible with live birth involve chromosomes (13,18, and 21) that have a relatively small proportion of negative G-bands;[203] it was inferred that such chromosomes carried relatively few genes, and that the extensive G-bands in these chromosomes were deficient in genes. Sequences complementary to messenger RNA also appeared to be localized in negative G-bands,[221] although the autorad-

Table 7. Number of Genes in G- and
R-bands (January 1995) (from Craig, ref. 228)

G-bands	354 (20%)
R-bands	1417 (80%)
T-bands*	808 (57%)
Total	1771

Note: *T-bands are a subfraction of R-bands

iographic methods that had to be used at that time did not allow high resolution mapping. More recently, sites of hypersensitivity to DNase I (which is regarded as a marker for potentially active genes) have been localized to negative G-bands.[208,209] Earlier work suggested that housekeeping genes (i.e. those that are required for basic cellular metabolism and that are therefore active in all cells) were in early replicating DNA, corresponding to negative G-bands, while tissue-specific genes appeared to be concentrated in late replicating DNA (equivalent to positive G-bands).[222] However, the work of Bernardi and his colleagues[223,224] indicated that most genes were concentrated not merely in G+C-rich DNA (corresponding to negative G-bands, or R-bands), but in the most G+C-fraction, the T-bands, a class of R-bands located largely, but not exclusively, at the ends of chromosomes.[225,226] With the rapid advance of high resolution gene mapping, a large number of human genes have now been mapped to specific bands;[227] and the majority of genes turn out to be in the R-bands, with the greatest concentration of genes in the T-bands (Table 7).

Much less is known about the distribution of specific proteins on chromosomes. Of particular interest is the observation that acetylated histone, which is associated with active genes,[229] is distributed primarily in the negative G-bands, which, as we have just seen, are rich in genes. On the other hand, histone H1 subtypes, which may be concerned with chromosome condensation, and therefore with the inactivity of genes, appear to be concentrated mainly in the positive G-bands.[230] However, the distribution of HMG-I is more difficult to explain. This protein binds preferentially to A+T-rich DNA, and therefore its occurrence in positive G-bands is not unexpected. However, this protein seems to be especially abundant in proliferating cells, suggesting an association with gene activity; indeed, HMG-I has been shown to bind to certain A+T-rich sequences associated with genes.[231] Until the function of this protein is known better, it will be difficult to understand its chromosomal distribution.

To summarize the situation in mammalian chromosomes, the negative G-bands are relatively G+C-rich, with relatively high concentrations of genes and a relatively loose structure, which presumably not only allows access of the transcription machinery to the genes (in interphase), but would also be necessary for meiotic pairing and recombination. However, lack of condensation may well lead to an increased level of chromosome breakage. It has been suggested that early replication, characteristic of these gene-rich regions, may in fact be a necessary condition for gene activity.[232] Positive G-bands, on the other hand, show opposite characteristics and appear to be poor in genes and relatively condensed so that some authors have even regarded them as a type of heterochromatin. It should be emphasized that the difference in condensation between positive and negative G-bands is essentially an interphase condition; at metaphase the degree of condensation appears to be uniform throughout the chromosomes (see ref. 106 for a discussion of this point).

B. Euchromatic Bands in Nonmammals

Mammals are the most intensively studied organisms from the point of view of euchromatic banding; however, when other groups of organisms are studied, it is clear that mammals are, in fact, rather exceptional. So far, it has not proved possible to obtain G- or R-bands, or banding with fluorochromes, in the majority of lower vertebrates, invertebrates and plants that have been studied. In the past this was explained by technical difficulties: the effort that has gone into developing reliable chromosome preparation methods and banding techniques for mammals, especially man, has not been available for many other organisms. Moreover, the reliance on squash techniques for preparing chromosomes from, for example, many insects and plants, must militate against successful banding because freedom from cytoplasmic proteins is clearly necessary to produce satisfactory banding in mammalian chromosomes. Nevertheless, it is now difficult to believe that after nearly 30 years' experience of banding the lack of euchromatic bands in the chromosomes of many organisms is simply due to technical problems. Although a few reports of G-banding of plant chromosomes have been published in recent years (e.g., refs. 233-236), it is not clear that the patterns produced by these methods are reproducible, or that they correspond to the G-bands found in mammalian chromosomes.

Table 8 summarizes the distribution of different types of longitudinal differentiation in different groups of animals and plants. Adequate information is not available for most groups of invertebrates and plants, although it is possible that investigation of these groups could provide valuable insights into the nature of chromosome banding. It is immediately obvious that both pachytene chromomeres and replication bands are virtually universal: although there are still many groups in which neither has yet been reported, there are no cases in which the absence of chromomeres or of replication bands has been shown conclusively. At the other extreme, banding with base-specific fluorochromes is largely confined to higher vertebrates, with a few scattered reports from other groups. This distribution corresponds quite well, but not perfectly, with the distribution of G+C-rich isochores—DNA fractions within the main band which show different average base compositions.[223,224] In between are the G-bands, which seem to occur in a greater range of organisms than base-specific bands and isochores, but

Table 8. Distribution of different types of banding and of G + C-rich isochores in different groups of eukaryotes

| | Pachytene Chromomeres | Replication Bands | DNase Hypersensitivity | G-bands | Base-specific Fluorochromes | | G+C-rich Isochores |
					Q-bands (A + T-rich DNA)	Chromomycin R-bands (G+C-rich DNA)	
Mammals	Yes	Yes	Yes (208,209)	Good	Good	Good	Yes (237)
Birds	Yes (238)	Yes (239)	-	Good	Moderate (240)	Moderate (240)	Yes (237)
Reptiles	Yes (241)	Yes (242)	-	Good	Poor (240)	Poor or absent (240)	Poor (243)
Amphibia	Yes (244)	Yes (245,246)	Yes (247)	No	No (240)	No (240)	No (243)
Xenopus	-	Yes (248)	-	Good (249)	-	-	No (243)
Fish	Yes (250)	Yes (251)	-	A few spp. (252)	No (240,253)	No (240)	No (254)
Eels (*Anguilla*)	-	-	-	Good (255)	Poor (253)	No	Yes (253)
Thermophilic spp.	-	-	-	-	-	-	Yes (256)
Insects	Yes (257)	-	Yes (258)	No	No	No	-
Drosophila	-	-	-	No	No	No	No (243)
Plants							
Monocotyledons	Yes (257)	Yes (259)	-	A few spp (234-236)	No	No	Yes (260)
Lilium	Yes	-	-	-	Yes (261)	-	-
Dicotyledons	Yes (257)	Yes (262)	-	-	No	No	No (260)
Vicia	-	-	-	Yes (233)	-	-	-

Note: Information not available for "-" entries.

246

which nevertheless do not seem to be universal, unlike chromomeres and replication bands. There is no evidence that G-banding is caused by differences in DNA base composition (either directly or indirectly), but on the other hand there are suggestions that it could be connected with the process of chromosome condensation (see Chapter 5 in ref. 106). Unfortunately, in spite of many years research, we are still largely ignorant not only of the functional significance of G-banding, but also of its mechanism of staining, which might be expected to throw light on the function of these chromosomal segments.

C. The Evolution of the Longitudinal Differentiation of Chromosomes

It is not known if the earliest eukaryotes showed any longitudinal differentiation of their chromosomes (other than, presumably, centromeres and telomeres), but if phyletic distribution is any guide, the differentiation into chromomeric and non-chromomeric regions, and into early and late replicating regions, must have occurred quite early on. In general, the chromosomal distribution of genes is not known except in mammals, but evidence is beginning to emerge that they may be distributed nonhomogeneously both in plants[263] and in lower animals.[247,258] While the relationship between the distribution of genes in these organisms, and their patterns of chromosomal replication and of pachytene chromomeres is not yet known, it seems quite plausible that plants, invertebrates and lower vertebrates, like mammals, would have their genomes divided into two compartments, one, gene-rich, early replicating and inter-chromomeric, and the other, poor in genes, late replicating, and forming chromomeres. Why the higher vertebrates, and probably certain other organisms, should have superimposed on this basic pattern a further level of differentiation based on DNA base composition is not clear. Bernardi and Bernardi[256] have suggested that G+C-rich isochores evolved in warm-blooded vertebrates because they conferred greater thermal stability on the DNA, and indeed proteins coded by G+C-rich genes also appear to show greater thermal stability, but it remains to be shown whether this is the complete explanation.

The other important question to be asked, but not yet answered, is why should eukaryotic chromosomes contain large segments which are apparently largely devoid of genes? The believers in the "junk DNA" hypothesis would have it that the chromosomes simply contain a large amount of useless material of neutral selective value. Nevertheless, it may be necessary for a chromosome to be more than simply a string of genes attached to a centromere. For example, Cavalier-Smith[264] suggested that the quantity of DNA alone, regardless of its composition or sequence, could be a vital regulator of nuclear size, with many implications for cellular physiology. Evidence is accumulating that in some types of cells both telomeres[265,266] and DNase-sensitive regions of chromatin,[267,268] which correspond to regions rich in active genes, tend to be preferentially located close to the nuclear envelope. In other words, the ends of the chromosomes, which appear to be richest in genes, are close to the interface with the cytoplasm, where their products will act; on the other hand, the interior of the nucleus appears to show much less gene activity. It may be necessary for the area of the nuclear envelope to be proportional to the number of active genes. At the same time, the area of the nuclear envelope will be determined by the volume of the nucleus, which depends, at least partly, on the quantity of DNA contained within it. Such DNA, having no genic function, would probably show no stringent base sequence requirements. Obviously this is an oversimplification because not only can nuclear volume vary within tissues of the same individual without changes in amount of nuclear DNA, but also, in some circumstances gene activity, as shown by DNase-sensitivity, can occur in the interior of nuclei.[268] Nevertheless, the correlation between peripheral nuclear location and gene activity is intriguing, not only for nuclear physiology, but also for our understanding of chromosome organization.

VI.   CONCLUDING REMARKS

In this review, I have tried to present the chromosome as a cellular organelle of central importance. Inevitably, several topics have been largely ignored, chiefly because they are covered thoroughly elsewhere in this volume, so that repetition would be pointless (nucleosome structure, Chapter 6; telomeres, Chapter 7; chromatin structure, Chapter 8); others are considered only briefly here, and again, are discussed at length in other chapters (Centromeres, Chapter 5; Topoisomerases, Chapter 13; Satellite DNA, Chapter 15). Important subjects such as the structure of genes, their replication and transcription, and the process of chromosome condensation have been deliberately omitted, the former because too much is known to include here, the latter rather because so little has been discovered yet. Nevertheless, it is hoped that a reasonably comprehensive view of chromosome organization has been given.

There is, indeed, no doubt that a vast amount has been learned in the last 20 years or so about the way chromosomes are constructed. For example, it is now well established that the basic organization of chromosomes consists of chromatin loops attached to a proteinaceous scaffold, even though there is much more to be learned about these components, particularly the loops. In spite of this, there is, as yet, no consensus whether the loops and scaffold represent the highest level of chromosome organization, or whether this structure is condensed further, either by coiling or by condensation into chromomeres. There is substantial evidence for all three possibilities, yet they seem, to a large extent, to be mutually exclusive.

A great effort has been put into understanding the organization of centromeres and kinetochores so that we now know a good deal about the proteins and DNA fractions involved in attaching the chromosomes to the spindle. Inevitably, at this stage, there are seeming inconsistencies, but no doubt in a few years' time the roles of the different centromeric proteins, whether structural or dynamic, and of the associated DNAs, will have become much clearer. The mechanism of chromosome segregation is also becoming understood; while it is clear that topoisomerase II has an essential role in disentangling the centromeric regions of sister chromatids at the metaphase/anaphase transition, it is nevertheless obvious that many other factors are involved in the process of chromosome segregation. One important point that has not yet been addressed seriously is the substrate for topoisomerase II in separating chromosomes; does this enzyme act on specific DNA fractions (e.g., a particular class of satellite DNA) and are these separated spatially from the kinetochore-organizing parts of the centromere? It seems possible that this could well be so.

Some 25 years ago, the first practical chromosome banding techniques were just becoming available. Not only have these revolutionized the identification of chromosomes and had vital applications in many fields, especially in clinical cytogenetics, but they have also drawn attention to the fact that the apparently uniform chromosome arms are, in reality, compartmentalized. We now have a large body of information about the composition of these compartments, which differ in their DNA and protein composition and in the concentration of genes that they contain, as well as in many other characteristics; it is beginning to emerge that the differentiation of chromosomes into regions having different properties is probably universal, even including species in which it does not seem possible to produce banding patterns using routine methods. However, the functional significance of this differentiation remains obscure, with only a few glimmers of light here and there; one possibility is that constraints imposed by the requirements of the interphase nucleus could be a significant factor. Evidently there is still much to be discovered about the organization of the mitotic chromosome, but we have come a long way in the past 20 years or so, and there seems to be no doubt that in the next few years we shall at last have a much clearer understanding of how chromosomes are constructed and how they function.

REFERENCES

1. Sutton, W.S. The chromosomes in heredity. *Biol. Bull.* **1903**, *4*, 231–251.
2. Antequera, F.; Bird, A. Number of CpG islands and genes in human and mouse. *Proc. Natl. Acad. Sci. USA* **1993**, *90*, 11995–11999.
3. Blackburn, E.H.; Karrer, E.M. Genomic reorganization in ciliated protozoans. *Annu. Rev. Genet.* **1986**, *20*, 501–521.
4. Crosland, M.W.J.; Crozier, R.H. *Myrmecia pilosula*, an ant with only one pair of chromosomes. *Science* **1986**, *231*, 1278.
5. White, M.J.D. *The Chromosomes*, 5th ed. Methuen, London, 1961.
6. Matthey, R. The chromosome formulae of eutherian mammals. In: *Cytotaxonomy and Vertebrate Evolution* (Chiarelli, A.B.; Capanna, E; eds.). Academic Press, London, 1973, pp. 531–616.
7. Capanna, E. Concluding remarks. In: *Cytotaxonomy and Vertebrate Evolution* (Chiarelli, A.B.; Capanna, E.; eds.). Academic Press, London, 1973, pp. 681–695.
8. DuPraw, E.J. *DNA and Chromosomes*. Holt, Rinehart and Winston, New York, 1970.

9. DuPraw, E.J.; Bahr, G.F. The arrangement of DNA in human chromosomes, as investigated by quantitative electron microscopy. *Acta Cytol.* **1969**, *13*, 188–205.

10. Golomb, H.M.; Bahr, G.F. Human chromatin from interphase to metaphase. *Exp. Cell. Res.* **1974**, *84*, 79–87.

11. Taylor, J.H.; Woods, P.S.; Hughes, W.L. The organization and duplication of chromosomes as revealed by autoradiographic studies using tritium-labeled thymidine. *Proc. Natl. Acad. Sci. USA* **1957**, *43*, 122–128.

12. Perry, P.; Wolff, S. New Giemsa method for the differential staining of sister chromatids. *Nature* **1974**, 251, 156–158.

13. Kavenoff, R. One chromatid, one piece of DNA. *J. Cell Biol.* **1970**, *47*, 103a.

14. Blamire, J.; Cryer. D.R.; Finkelstein, D.B.; Marmur, J. Sedimentation properties of yeast nuclear and mitochondrial DNA. *J. Mol. Biol.* **1972**, *67*, 11–24.

15. Petes, T.D.; Fangman, W.L. Sedimentation properties of yeast chromosomal DNA. *Proc. Natl. Acad. Sci. USA* **1972**, *69*, 1188–1191.

16. Kavenoff. R.; Zimm, B.H. Chromosome-sized DNA molecules from Drosophila. *Chromosoma* **1973**, *41*, 1–27.

17. Kavenoff, R.; Klotz, L.C.; Zimm, B.H. On the nature of chromosome-sized DNA molecules. *Cold Spring Harb. Symp. Quant. Biol.* **1973**, *38*, 1–8.

18. Neary, G.J.; Savage, J.R.K.; Evans, H.J. Chromatid aberrations in *Tradescantia* pollen tubes induced by monochromatic X-rays of quantum energy 3 and 1.5 keV. *Int. J. Radiat. Biol.* **1964**, *8*, 1–19.

19. Gall, J.G. Kinetics of deoxyribonuclease action on chromosomes. *Nature* **1963**, *198*, 36–38.

20. Miller, O.L. Fine structure of lampbrush chromosomes. *Natl. Cancer Inst. Monogr.* **1965**, *18*, 79–99.

21. Ullerich, F-H. DNS-Gehalt und Chromosomenstruktur bei Amphibien. *Chromosoma* **1970**, *30*, 1–37.

22. Van Holde, K.E. *Chromatin*. Springer, New York, 1989.

23. Dounce, A.L.; Chanda, S.K.; Townes, P.L. The structure of higher eukaryotic chromosomes. *J. Theor. Biol.* **1973**, *42*, 275–285.

24. Sobell, H.M. The stereochemistry of actinomycin binding to DNA and its implications for molecular biology. *Prog. Nucleic Acid Res. Mol. Biol.* **1973**, *13*, 153–190.

25. Sobell, H.M. Symmetry in protein-nucleic acid interaction and its genetic implications. *Adv. Genet.* **1973**, *17*, 411–490.

26. Stubblefield, E.; Wray, W. Architecture of the Chinese hamster metaphase chromosome. *Chromosoma* **1971**, *32*, 262–294.

27. Adolph, K.W.; Cheng, S.M.; Laemmli, U.K. Role of nonhistone proteins in metaphase chromosome structure. *Cell* **1977**, *12*, 805–816.

28. Paulson, J.R.; Laemmli, U.K. The structure of histone-depleted metaphase chromosomes. *Cell* **1977**, *12*, 817–828.

29. Hadlaczky, Gy.; Sumner, A.T.; Ross, A. Protein-depleted chromosomes. I. Structure of isolated protein-depleted chromosomes. *Chromosoma* **1981**, *81*, 537–555.

30. Howell, W.M., Hsu, T.C. Chromosome core structure revealed by silver staining. *Chromosoma* **1979**, *73*, 61–66.

31. Stack, S.M. Staining plant cells with silver. II. Chromosome cores. *Genome* **1991**, *34*, 900–908.

32. Giménez-Abián, J.F.; Clarke, D.J.; Mullinger, A.M.; Downes, C.S.; Johnson, R.T. A post-prophase topoisomerase II-dependent chromatid core separation step in the formation of metaphase chromosomes. *J. Cell Biol.* **1995**, *131*, 7–17.

33. Sumner, A.T. Scanning electron microscopy of mammalian chromosomes from prophase to telophase. *Chromosoma* **1991**, *100*, 410–418.

34. Gasser, S.M.; Laroche, T.; Falquet, J.; Boy de la Tour, E.; Laemmli, U.K. Metaphase chromosome structure. Involvement of topoisomerase II. *J. Mol. Biol.* **1986**, *188*, 613–629.

35. Sumner, A.T. The distribution of topoisomerase II on mammalian chromosomes. *Chromosome Res.* **1996**, *4*, 5–14.

36. Lewis, C.D.; Laemmli, U.K. Higher order metaphase chromosome structure: evidence for metalloprotein interactions. *Cell* **1982**, *29*, 171–181.

37. Smith, P.J. DNA topoisomerase dysfunction: a new goal for antitumour chemotherapy. *Bioessays* **1990**, *12*, 167–172.

38. Gasser, S.M.; Walter, R.; Qi, D.; Cardenas, M.E. Topoisomerase II: its functions and phosphorylation. *Antonie Van Leeuwenhoek* **1992**, *62*, 15–24.

39. Watt, P.M.; Hickson, I.D. Structure and function of type II DNA topoisomerases. *Biochem. J.* **1994**, *303*, 681–95.

40. Wood, E.R.; Earnshaw, W.C. Mitotic chromosome condensation *in vitro* using somatic cell extracts and nuclei with variable levels of endogenous topoisomerase II. *J. Cell Biol.* **1990**, *111*, 2839–2850.

41. Adachi, Y.; Luke, M.; Laemmli, U.K. Chromosome assembly *in vitro*: Topoisomerase II is required for condensation. *Cell* **1991**, 64, 137–148.

42. Sumner, A.T. Inhibitors of topoisomerases do not block the passage of human lymphocyte chromosomes through mitosis. *J. Cell Sci.* **1992**, *103*, 105–115.

43. Saitoh, N.; Goldberg, I.G.; Wood, E.R.; Earnshaw, W.C. ScII: An abundant chromosome scaffold protein is a member of a family of putative ATPases with an unusual predicted tertiary structure. *J. Cell Biol.* **1994**, *127*, 303–18.

44. Pienta, K.J., Coffey, D.S. A structural analysis of the role of the nuclear matrix and DNA loops in the organization of the nucleus and chromosome. *J. Cell Sci.* **1984**, *Suppl. 1*, 123–135.

45. Bickmore, W.A.; Oghene, K. Visualizing the spatial relationships between defined DNA sequences and the axial region of extracted metaphase chromosomes. *Cell* **1996**, *84*, 95–104.

46. Adolph, K.W. A serial sectioning study of the structure of human mitotic chromosomes. *Eur. J. Cell Biol.* **1981**, *24*, 146–153.

47. Marsden, M.P.F.; Laemmli, U.K. Metaphase chromosome structure: Evidence for a radial loop model. *Cell* **1979**, *17*, 849–858.

48. Rattner, J.B.; Lin, C.C. The higher order structure of the centromere. *Genome* **1987**, 29, 588–593.

49. Gasser, S.M.; Amati, B.B.; Cardenas, M.E.; Hofmann, J.F.-X. Studies on scaffold attachment sites and their relation to genome function. *Int. Rev. Cytol.* **1989**, *119*, 57–96.

50. Razin, S.V., Hancock, R.; Iarovaia, O.; Westergaard, O., Gromova, I.; Georgiev, G.P. Structural-functional organization of chromosomal domains. *Cold Spring Harb. Symp. Quant. Biol.* **1993**, *58*, 25–35.

51. McManus, J.; Perry, P.; Sumner, A.T.; et al. Unusual chromosome structure of fission yeast DNA in mouse cells. *J. Cell Sci.* **1994**, *107*, 469–486.

52. Ohnuki, Y. Demonstration of the spiral structure of human chromosomes. *Nature* **1965**, *208*, 916–917.

53. Manton, I. The spiral structure of chromosomes. *Biol. Rev.* **1950**, *25*, 486–508.

54. Rattner, J.B.; Lin, C.C. Radial loops and helical coils coexist in metaphase chromosomes. *Cell* **1985**, *42*, 291–296.

55. Boy de la Tour, E.; Laemmli, U.K. The metaphase scaffold is helically folded: sister chromatids have predominantly opposite helical handedness. *Cell* **1988**, *55*, 937–944.

56. Earnshaw, W.C.; Heck, M.M.S. Localization of topoisomerase II in mitotic chromosomes. *J. Cell Biol.* **1985**, *100*, 1716–25.

57. Cook, P.R. A chromomeric model for nuclear and chromosome structure. *J. Cell Sci.* **1995**, *108*, 2927–2935.

58. Gautier, T.; Masson, C.; Quintana, C.; Arnoult, J.; Hernandez-Verdun, D. The ultrastructure of the chromosome periphery in human cell lines. An *in situ* study using cryomethods in electron microscopy. *Chromosoma* **1992**, *101*, 502–510.

59. Moyne, G.; Garrido, J. Ultrastructural evidence of mitotic perichromosomal ribonucleoproteins in hamster cells. *Exp. Cell Res.* **1976**, *98*, 237–247.

60. Gautier, T.; Dauphin-Villemant, C.; André, C.; Masson, C.; Arnoult, J.; Hernandez-Verdun, D. Identification and characterization of a new set of nucleolar proteins which line the chromosomes during mitosis. *Exp. Cell Res.* **1992**, *200*, 5–15.

61. Hernandez-Verdun, D., Gautier, T. The chromosome periphery during mitosis. *Bioessays* **1994**, *16*, 179–185.

62. Earnshaw, W.C.; Bernat, R.L. Chromosomal passengers: toward an integrated view of mitosis. *Chromosoma* **1991**, *100*, 139–146.

63. Chaudhary, N.; Courvalin, J.-C. Stepwise reassembly of the nuclear envelope at the end of mitosis. *J. Cell Biol.* **1993**, *123*, 295–306.

64. Lima-de-Faria, A. The division cycle of the kinetochore. *Hereditas* **1955**, *41*, 238–240.

65. Lima-de-Faria, A. The role of the kinetochore in chromosome organisation. *Hereditas* **1956**, *42*, 85–160.

66. Carmena, M.; Abad, J.P.; Villasante, A.; Gonzalez, C. The *Drosophila melanogaster* dodecasatellite sequence is closely linked to the centromere and can form connections between sister chromatids during mitosis. *J. Cell Sci.* **1993**, *105*, 41–50.

67. Haaf, T.; Warburton, P.E.; Willard, H.F. Integration of human α-satellite DNA into simian chromosomes: centromere protein binding and disruption of normal chromosome segregation. *Cell* **1992**, *70*, 681–696.

68. Murphy, T.D.; Karpen, G.H. Localization of centromere function in a *Drosophila* minichromosome. *Cell* **1995**, *82*, 599–609.

69. Sullivan, K.F.; Hechenberger, M.; Masri, K. Human CENP-A contains a histone H3 related histone fold domain that is required for targeting to the centromere. *J. Cell Biol.* **1994**, *127*, 581–92.

70. Masumoto, H.; Masukata, H.; Muro, Y.; Nozaki, N.; Okazaki, T. A human centromere antigen (CENP-B) interacts with a short specific sequence in alphoid DNA, a human centromeric satellite. *J. Cell Biol.* **1989**, *109*, 1963–1973.

71. Wong, A.K.C.; Rattner, J.B. Sequence organization and cytological localization of the minor satellite of mouse. *Nucl. Acids Res.* **1988**, *16*, 11645–1161.

72. Wong, A.K.C.; Biddle, F.G.; Rattner, J.B. The chromosomal distribution of the major and minor satellite is not conserved in the genus *Mus*. *Chromosoma* **1990**, *99*, 190–195.

73. Avides, M.C., Sunkel, C.E. Isolation of chromosome-associated proteins from *Drosophila melanogaster* that bind a human centromeric DNA sequence. *J. Cell Biol.* **1994**, *127*, 1159–1171.

74. Saitoh, H.; Tomkiel, J.; Cooke, C.A.; et al. CENP-C, an autoantigen in scleroderma, is a component of the human inner kinetochore plate. *Cell* **1992**, *70*, 115–125.

75. Sullivan, B.A.; Schwartz, S. Identification of centromeric antigens in dicentric Robertsonian translocations: CENP-C and CENP-E are necessary components of functional centromeres. *Hum. Mol. Genet.* **1995**, *4*, 2189–2197.

76. Bloom, K. The centromere frontier: kinetochore components, microtubule-based motility, and the CEN-value paradox. *Cell* **1993**, *73*, 621–624.

77. Rattner, J.B.; Rao, A.; Fritzler, M.J.; Valencia, D.W.; Yen, T.J. CENP-F is a ca 400 kDa kinetochore protein that exhibits a cell-cycle dependent localization. *Cell Motil. Cytoskeleton* **1993**, *26*, 214–226.

78. Bernat, R.L.; Borisy, G.G.; Rothfield, N.F.; Earnshaw, W.C. Injection of anticentromere antibodies in interphase disrupts events required for chromosome movement at mitosis. *J. Cell Biol.* **1990**, *111*, 1519–1533.

79. Simerly, C.; Balczon, R.; Brinkley, B.R.; Schatten, G. Microinjected kinetochore antibodies interfere with chromosome movement in meiotic and mitotic mouse embryos. *J. Cell Biol.* **1990**, *111*, 1491–1504.

80. Wise, D.A.; Bhattacharjee, L. Antikinetochore antibodies interfere with prometaphase but not anaphase chromosome movement in living PtK₂ cells. *Cell Motil. Cytoskeleton* **1992**, *23*, 157–167.

81. Tomkiel, J.E.; Cooke, C.A.; Saitoh, H.; Bernat, R.L.; Earnshaw, W.C. CENP-C is required for maintaining proper kinetochore size and for a timely transition to anaphase. *J. Cell Biol.* **1994**, *125*, 531–545.

82. Wordeman, L.; Earnshaw, W.C.; Bernat, R.L. Disruption of CENP antigen function perturbs dynein anchoring to the mitotic kinetochore. *Chromosoma* **1996**, *104*, 551–560.

83. Cooke, C.A.; Heck, M.M.S.; Earnshaw, W.C. The INCENP antigens: movement from the inner centromere to the midbody during mitosis. *J. Cell Biol.* **1987**, *105*, 2053–2067.

84. Rattner, J.B.; Kingwell, B.G.; Fritzler, M.J. Detection of distinct structural domains within the primary constriction using autoantibodies. *Chromosoma* **1988**, *96*, 360–367.

85. Holland, K.A.; Kereső, J.; Zákány, J.; et al. A tightly bound chromosome antigen is detected by monoclonal antibodies in a ring-like structure on human centromeres. *Chromosoma* **1995**, *103*, 559–566.

86. Di Nardo, S.; Voelkel, K.; Sternglanz, R. DNA topoisomerase II mutant of *Saccharomyces cerevisiae:* topoisomerase II is required for segregation of daughter molecules at the termination of DNA replication. *Proc. Natl. Acad. Sci. USA* **1984**, *81*, 2616–2620.

87. Uemura, T.; Ohkura, H.; Adachi, Y.; Morino, K.; Shiozaki, K.; Yanagida, M. DNA topoisomerase II is required for condensation and separation of mitotic chromosomes in *S. pombe. Cell* **1987**, *50*, 917–25.

88. Buchenau, P.; Saumweber, H.; Arndt-Jovin, D.J. Consequences of topoisomerase II inhibition in early embryogenesis of *Drosophila* revealed by *in vivo* confocal laser scanning microscopy. *J. Cell Sci.* **1993**, *104*, 1175–1185.

89. Downes, C.S.; Mullinger, A.M.; Johnson, R.T. Inhibitors of DNA topoisomerase II prevent chromatid separation in mammalian cells but do not prevent exit from mitosis. *Proc. Natl. Acad. Sci. USA* **1991**, *88*, 8895–8899.

90. Clarke, D.J.; Johnson, R.T.; Downes, C.S. Topoisomerase II inhibition prevents anaphase chromatid segregation in mammalian cells independently of the generation of DNA strand breaks. *J. Cell Sci.* **1993**, *105*, 563–569.

91. Gorbsky, G.J. Cell cycle progression and chromosome segregation in mammalian cells cultured in the presence of the topoisomerase II inhibitors ICRF-187 [(+)-1,2-bis(3,5-dioxopiperazinyl-1-yl)propane; ADR-529] and ICRF-159 (Razoxane). *Cancer Res.* **1994**, *54*, 1042–1048.

92. Sumner, A.T. Inhibitors of topoisomerase II delay progress through mitosis and induce a doubling of the DNA content in CHO cells. *Exp. Cell Res.* **1995**, *217*, 440–447.

93. Williams, B.C.; Karr, T.L.; Montgomery, J.M.; Goldberg, M.L. The *Drosophila l(1)zw10* gene product, required for accurate mitotic chromosome segregation, is redistributed at anaphase onset. *J. Cell Biol.* **1992**, *118*, 759–773.

94. Holloway, S.L.; Glotzer, M.; King, R.W.; Murray, A.W. Anaphase is initiated by proteolysis rather than by the inactivation of maturation-promoting factor. *Cell* **1993**, *73*, 1393–1402.

95. Stratmann, R., Lehner, C.F. Separation of sister chromatids in mitosis requires the *Drosophila pimples* product, a protein degraded after the metaphase/anaphase transition. *Cell* **1996**, *84*, 25–35.

96. Yanagida, M. Frontier questions about sister chromatid separation in anaphase. *Bioessays* **1995**, *17*, 519–526.

97. Rattner, J.B. The structure of the mammalian centromere. *Bioessays* **1991**, *13*, 51–56.

98. Earnshaw, W.C.; Mackay, A.M. Role of nonhistone proteins in the chromosomal events of mitosis. *FASEB J.* **1994**, *8*, 947–956.

99. Sunkel, C.E., Coelho, P.A. The elusive centromere: sequence divergence and functional conservation. *Curr. Opin. Genet. Dev.* **1995**, *5*, 756–767.

100. Scheer, U.; Thiry, M.; Goessens, G. Structure, function and assembly of the nucleolus. *Trends Cell Biol.* **1993**, *3*, 236–241.

101. Wachtler, F.; Stahl, F. The nucleolus: a structural and functional interpretation. *Micron* **1993**, *24*, 473–505.

102. Derenzini, M.; Sirri, V.; Trerè, D. Nucleolar organizer regions in tumour cells. *Cancer J.* **1994**, *7*, 71–77.

103. Scheer, U.; Weisenberger, D. The nucleolus. *Curr. Opin. Cell Biol.* **1994**, *6*, 354–359.

104. Risueño, M.C.; Testillano, P.S. Cytochemistry and immunocytochemistry of nucleolar chromatin in plants. *Micron* **1994**, *25*, 331–360.

105. Mélèse, T.; Xue, Z. The nucleolus: an organelle formed by the act of building a ribosome. *Curr. Opin. Cell Biol.* **1995**, *7*, 319–324.

106. Sumner, A.T. *Chromosome Banding.* Unwin Hyman, London, 1990.

107. Roussel, P.; Hernandez-Verdun, D. Identification of Ag-NOR proteins, markers of proliferation related to ribosomal gene activity. *Exp. Cell Res.* **1994**, *214*, 465–472.

108. Schwarzacher, H.G.; Mikelsaar, A-V.; Schnedl, W. The nature of the Ag-staining of nucleolus organizer regions. *Cytogenet. Cell Genet.* **1978**, *20*, 24–39.

109. Miller, D.A.; Dev, V.G.; Tantravahi, R.; Miller, O.J. Suppression of human nucleolar organizer activity in mouse-human somatic hybrid cells. *Exp. Cell Res.* **1976**, *101*, 235–243.

110. Crocker, J. Nucleolar organiser regions. *Curr. Top. Pathol.* **1990**, *82*, 91–149.

111. Derenzini, M.; Ploton, D. Interphase nucleolar organizer regions in cancer cells. *Int. Rev. Cytol.* **1991**, *32*, 149–192.

112. Wilson, G.N. The structure and organization of human ribosomal genes. In: *The Cell Nucleus*, Vol. 10 (Busch, H.; Rothblum, L.; eds.). Academic Press, New York, 1982, pp 287–318.

113. Hecht, F.; Ramesh, K.H.; Lockwood, D.H. A guide to fragile sites on human chromosomes. *Cancer Genet. Cytogenet.* **1990**, *44*, 37–45.

114. Simonic, I.; Gericke, G.S. The enigma of common fragile sites. *Hum. Genet.* **1996**, *97*, 524–531.

115. Elder, F.F.B., Robinson, T.J. Rodent common fragile sites: are they conserved? Evidence from mouse and rat. *Chromosoma* **1989**, *97*, 459–464.

116. Smeets, D.F.C.M., van de Klundert, F.A.J.M. Common fragile sites in man and three closely related primate species. *Cytogenet. Cell Genet.* **1990**, *53*, 8–14.

117. Stone, D.M.; Jacky, P.B.; Hancock, D.D.; Prieur, D.J. Chromosomal fragile site expression in dogs. I. Breed specific differences. *Am. J. Med. Genet.* **1991**, *40*, 214–222.

118. Rønne, M. Putative fragile sites in the horse karyotype. *Hereditas* **1992**, *117*, 127–136.

119. Riggs, P.K.; Kuczek, T.; Chrisman, C.L.; Bidwell, C.A. Analysis of aphidicolin-induced chromosome fragility in the domestic pig (*Sus scrofa*). *Cytogenet. Cell Genet.* **1993**, *62*, 110–116.

120. Sutherland, G.R.; Hecht, F. *Fragile Sites on Human Chromosomes.* Oxford University Press, Oxford, 1985.

121. Oostra, B.A.; Verkerk, A.J.H.M. The fragile X syndrome: isolation of the FMR-1 gene and characterization of the fragile X mutation. *Chromosoma* **1992**, *101*, 381–387.

122. Hecht, F. Fragile sites, cancer chromosome breakpoints and oncogenes all cluster in light G-bands. *Cancer Genet. Cytogenet.* **1988**, *31*, 17–24.

123. Popescu, N.C. Chromosome fragility and instability in human cancer. *Crit. Rev. Oncog.* **1994**, *5*, 121–140.

124. Jones, C.; Penny, L.; Mattina, T.; et al. Association of a chromosome deletion syndrome with a fragile site within the proto-oncogene *CBL2*. *Nature* **1995**, *376*, 145–149.

125. Rocchi, A.; di Castro, M.; Prantera, G. Effects of DAPI on human leukocytes in vitro. *Cytogenet. Cell Genet.* **1979**, *23*, 250–254.

126. Zakharov, A.F.; Baranovskaya, L.I.; Ibraimov, A.I.; Benjusch, V.A.; Demintseva, V.S.; Oblapenko, N.G. Differential spiralization along mammalian metaphase chromosomes. II. 5-Bro-

modeoxyuridine and 5-bromodeoxycytidine-revealed differentiation in human chromosomes. *Chromosoma* **1974**, *44*, 343–359.

127. Laird, C.; Jaffe, E.; Karpen, G.; Lamb, M.; Nelson, R. Fragile sites in human chromosomes as regions of late-replicating DNA. *Trends Genet.* **1987**, *3*, 274–281.

128. Richards, R.I.; Sutherland, G.R. Heritable unstable DNA sequences. *Nature Genet.* **1992**, *1*, 7–9.

129. Oostra, B.A.; Willems, P.J. A fragile gene. *Bioessays* **1995**, *17*, 941–947.

130. De Vries, B.B.A.; Wiegers, A.M.; de Graaff, E.; et al. Mental status and fragile X expression in relation to FMR-1 gene mutation. *Eur. J. Hum. Genet.* **1993**, *1*, 72–79.

131. Heitz, E. Das Heterochromatin der Moose. I. *Jahrb. Wiss. Bot.* **1928**, *69*, 762–818.

132. Brown, S.W. Heterochromatin. *Science* **1966**, *151*, 417–425.

133. Beridze, T. *Satellite DNA.* Springer, Berlin, 1986.

134. Schmid, M.; Grunert, D.; Haaf, T.; Engel, W. A direct demonstration of somatically paired heterochromatin of human chromosomes. *Cytogenet. Cell Genet.* **1983**, *36*, 554–561.

135. Arrighi, F.E.; Hsu, T.C.; Pathak, S.; Sawada, H. The sex chromosomes of the Chinese hamster: constitutive heterochromatin deficient in repetitive DNA sequences. *Cytogenet. Cell Genet.* **1974**, *13*, 268–274.

136. Kalscheuer, V.; Singh, A.P.; Nanda, I.; Sperling, K.; Neitzel, H. Evolution of the gonosomal heterochromatin of *Microtus agrestis*: rapid amplification of a large, multimeric, repeat unit containing a 3.0-kb $(GATA)_{11}$- positive, middle repetitive element. *Cytogenet. Cell Genet.* **1996**, *73*, 171–178.

137. James, T.C.; Elgin, S.C.R. Identification of a nonhistone chromosomal protein associated with heterochromatin in *Drosophila melanogaster* and its gene. *Mol. Cell Biol.* **1986**, *6*, 3862–3872.

138. Eissenberg, J.C.; Ge, Y.-W.; Hartnett, T. Increased phosphorylation of HP1, a heterochromatin-associated protein of *Drosophila*, is correlated with heterochromatin assembly. *J. Biol. Chem.* **1994**, *269*, 21315–21321.

139. Wreggett, K.A.; Hill, F.; James, P.S.; Hutchings, A.; Butcher, G.W.; Singh, P.B. A mammalian homologue of *Drosophila* protein 1 (HP1) is a component of constitutive heterochromatin. *Cytogenet. Cell Genet.* **1994**, *66*, 99–103.

140. Nicol, L.; Jeppesen, P. Human autoimmune sera recognize a conserved 26 kD protein associated with mammalian heterochromatin that is homologous to heterochromatin protein 1 of *Drosophila*. *Chromosome Res.* **1994**, *2*, 245–253.

141. Gerasimova, T.I.; Gdula, D.A.; Gerasimov, D.V.; Simonova, O.; Corces, V.G. A Drosophila protein that imparts directionality on a chromatin insulator is an enhancer of position-effect variegation. *Cell* **1995**, *82*, 587–597.

142. Lewis, J.D.; Meehan, R.R.; Henzel, W.J.; et al. Purification, sequence, and cellular localization of a novel chromosomal protein that binds to methylated DNA. *Cell* **1993**, *69*, 905–914.

143. Disney, J.E.; Johnson, K.R.; Magnuson, N.S.; Sylvester, S.R.; Reeves, R. High-mobility group protein HMG-I localizes to G/Q- and C-bands of human and mouse chromosomes. *J. Cell Biol.* **1989**, *109*, 1975–1982.

144. Jeppesen, P.; Mitchell, A.; Turner, B.; Perry, P. Antibodies to defined histone epitopes reveal variations in chromatin conformation and underacetylation of centric heterochromatin in human metaphase chromosomes. *Chromosoma* **1992**, *101*, 322–332.

145. Belyaev, N.D.; Keohane, A.M.; Turner, B.M. Histone H4 acetylation and replication timing in Chinese hamster chromosomes. *Exp. Cell Res.* **1996**, *225*, 277–285.

146. Lavender, J.S.; Birley, A.J.; Palmer, M.J.; Kuroda, M.I.; Turner, B.M. Histone H4 acetylated at lysine 16 and proteins of the *Drosophila* dosage compensation pathway co-localize on the male X chromosome through mitosis. *Chromosome Res.* **1994**, *2*, 398–404.

147. Pimpinelli, S.; Bonaccorsi, S.; Gatti, M.; Sandler, L. The peculiar organization of *Drosophila* heterochromatin. *Trends Genet.* **1986**, *2*, 17–20.

148. Rudkin, G.T. Non-replicating DNA in *Drosophila*. *Genetics* **1969**, *61*, (Suppl.) 227–238.

149. Müller, F.; Bernard, V.; Tobler, H. Chromosome diminution in nematodes. *Bioessays* **1996**, *18*, 133–138.

150. Beerman, S. The diminution of heterochromatic chromosomal segments in *Cyclops* (Crustacea, Copepoda). *Chromosoma* **1977**, *60*, 297–344.

151. Nakai, Y.; Kubota, S.; Goto, Y.; Ishibashi, T.; Davison, W.; Kohno, S-I. Chromosome elimination in three Baltic, south Pacific and north-east Pacific hagfish species. *Chromosome Res.* **1995**, *3*, 321–330.

152. Bobrow, M. Heterochromatic chromosome variation and reproductive failure. *Exp. Clin. Immunogenet.* **1985**, *2*, 97–105.

153. Hsu, L.Y.F.; Benn, P.A.; Tannenbaum, H.L.; Perlis, T.E.; Carlson, A.D. Chromosomal polymorphisms of 1, 9, 16 and Y in 4 major ethnic groups: a large prenatal study. *Am. J. Med. Genet.* **1987**, *26*, 95–101.

154. Miklos, G.L.G.; John, B. Heterochromatin and satellite DNA in man: properties and prospects. *Am. J. Hum. Genet.* **1979**, *31*, 264–380.

155. Loidl J. Further evidence for a heterochromatin-chiasma correlation in some *Allium* species. *Genetica* **1982**, *60*, 31–35.

156. Berger, R.; Grielhuber, J. C-bands and chiasma distribution in *Scilla siberica* (Hyacinthaceae). *Genome* **1991**, *34*, 179–189.

157. Attia, T.; Lelley, T. Effects of constitutive heterochromatin and genotype on frequency and distribution of chiasmata in the seven individual rye bivalents. *Theor. Appl. Genet.* **1987**, *74*, 527–530.

158. Baker, W.K. Position-effect variegation. *Adv. Genet.* **1968**, *14*, 133–169.

159. Spofford, J.B. Position-effect variegation in *Drosophila*. In: *The Genetics and Biology of Drosophila*, Vol. 1c (Ashburner, M.; Novitski, E.; eds.). Academic Press, London, 1976, pp. 955–1018.

160. Karpen, G.H. Position-effect variegation and the new biology of heterochromatin. *Curr. Biol.* **1994**, *4*, 281–291.

161. Elgin, S.C.R. Heterochromatin and gene regulation in *Drosophila*. *Curr. Opin. Genet. Devel.* 1996, 6, 193–202.

162. Reuter, G.; Spierer, P. Position effect variegation and chromatin proteins. *Bioessays* **1992**, *14*, 605–612.

163. Belyaeva, E.S.; Demakova, O.V.; Umbetova, G.H.; Zhymulev. I.F. Cytogenetic and molecular aspects of position-effect variegation in *Drosophila melanogaster*. V. Heterochromatin-associated protein HP1 appears in euchromatic chromosomal regions that are inactivated as a result of position-effect variegation. *Chromosoma* **1993**, *102*, 583–590.

164. John, B. The biology of heterochromatin. In: *Heterochromatin: Molecular and Structural Aspects* (Verma, R.S.; ed.). Cambridge University Press, Cambridge, 1988, pp. 1–147.

165. Gatti, M.; Pimpinelli, S. Functional elements in *Drosophila melanogaster* heterochromatin. *Annu. Rev. Genet.* **1992**, *26*, 239–275.

166. Doshi, P.; Kaushal, S.; Benyajati, C.; Wu, C-I. Molecular analysis of the Responder satellite DNA in *Drosophila* melanogaster: DNA bending, nucleosome structure, and Rsp-binding proteins. *Mol. Biol. Evol.* **1991**, *8*, 721–741.

167. Hawley, R.S.; Irick, H.; Zitron, A.E.; et al. There are two mechanisms of achiasmate segregation in Drosophila, one of which requires heterochromatic homology. *Dev. Genet.* **1993**, *13*, 440–467.

168. Radic, M.Z.; Lungren, K.; Hamkalo, B. Curvature of mouse satellite DNA and condensation of heterochromatin. *Cell* **1987**, *50*, 1101–1108.

169. Martinez-Balbas, A.; Rodriguez-Campos, A.; Garcia-Ramirez, M.; et al. Satellite DNAs contain sequences that induce curvature. *Biochemistry* **1990**, *29*, 2342–2348.

170. Plohl, M.; Borstnik, B.; Ugarkovic, D.; Gamulin, V. Sequence-induced curvature of *Tenebrio molitor* satellite DNA. *Biochimie* **1990**, *72*, 665–670.

171. Garrido-Ramos, M.A.; Jamilena, M.; Lozano, R.; Ruiz Rejon, C.; Ruiz Rejon, M. The EcoRI centromeric satellite DNA of the Sparidae family (Pisces, Perciformes) contains a sequence motive common to other vertebrate centromeric satellite DNAs. *Cytogenet. Cell Genet.* **1995**, *71*, 345–351.

172. Bajer, A.; Molé-Bajer, J. Architecture and function of the mitotic spindle. *Adv. Cell Mol. Biol.* **1971**, *1*, 213–266.

173. Pathak, S.; Wurster-Hill, D.H. Distribution of constitutive heterochromatin in carnivores. *Cytogenet. Cell Genet.* **1977**, *18*, 245–254.

174. Vig, B.K. Sequence of centromere separation: a possible role for repetitive DNA. *Mutagenesis* **1987**, *2*, 155–159.

175. Macgregor, H.C.; Sessions, S.K. The biological significance of variation in satellite DNA and heterochromatin in newts of the genus *Triturus*; an evolutionary perspective. *Philos. Trans. R. Soc. Lond. Biol.* **1986**, *312*, 243–259.

176. Kurnit, D.M. Satellite DNA and heterochromatin variants: the case for unequal mitotic crossing over. *Hum. Genet.* **1979**, *47*, 169–186.

177. Baker, B.S.; Gorman, M.; Marin, I. Dosage compensation in *Drosophila melanogaster*. *Annu. Rev. Genet.* **1994**, *24*, 491–521.

178. Lyon, M.F. Gene action in the X chromosome of the mouse (*Mus musculus* L.). *Nature* **1961**, *190*, 372–373.

179. Takagi, N.; Sasaki, M. Preferential inactivation of the paternally derived X chromosome in XX mouse blastocysts. *Nature* **1975**, *256*, 640–642.

180. West, J.D.; Frels. W.I.; Chapman, V.M.; Papaioannou, Y. Preferential expression of the maternally derived X-chromosome in the mouse yolk sac. *Cell* **1977**, *12*, 873–882.

181. Grant, S.G.; Chapman, V.M. Mechanisms of X-chromosome regulation. *Annu. Rev. Genet.* **1988**, *22*, 199–233.

182. Sharman, G.B. Late DNA replication in the paternally derived X chromosome of female kangaroos. *Nature* **1971**, *230*, 231–232.

183. Barr, M.L.; Bertram, E.G. A morphological distinction between neurons of the male and female, and the behaviour of the nucleolar satellite during accelerated nucleoprotein synthesis. *Nature* **1949**, *163*, 676–677.

184. Penny, G.D.; Kay, G.F.; Sheardown, S.A.; Rastan, S.; Brockdorff, N. Requirement for *Xist* in X chromosome inactivation. *Nature* **1996**, *379*, 131–137.

185. Brown, C.J.; Hendrich, B.D.; Rupert, J.L.; et al. The human *XIST* gene: analysis of a 17 kb inactive X-specific RNA that contains conserved repeats and is highly localized within the nucleus. *Cell* **1992**, *71*, 527–542.

186. Clemson, C.M.; McNeil, J.A.; Willard, H.F.; Lawrence, J.B. XIST RNA paints the inactive X chromosome at interphase: evidence for a novel RNA involved in nuclear/chromosome structure. *J. Cell Biol.* **1996**, *132*, 259–275.

187. Migeon, B.R. X-chromosome inactivation: molecular mechanisms and genetic consequences. *Trends Genet.* **1994**, *10*, 230–235.

188. Disteche, C.M. Escape from inactivation in human and mouse. *Trends Genet.* **1995**, *11*, 17–22.

189. Jeppesen, P.; Turner, B.M. The inactive X chromosome in female mammals is distinguished by a lack of histone H4 acetylation, a cytogenetic marker for gene expression. *Cell* **1993**, *74*, 281–289.

190. Brown, S.W.; Nur, U. Heterochromatic chromosomes in the Coccids. *Science* **1964**, *145*, 130–136.

191. Nur, U. Nonreplication of heterochromatic chromosomes in a mealy bug, *Planococcus citri* (Coccoidea: Homoptera). *Chromosoma* **1966**, 19, 439–448.

192. Nur, U. Reversal of heterochromatinization and the activity of the paternal chromosome set in the male mealy bug. *Genetics* **1967**, *56*, 375–389.

193. Miller, G.; Berlowitz, L.; Regelson, W. Chromatin and histones in mealy bug cell explants: activation and decondensation of facultative heterochromatin by a synthetic polyanion. *Chromosoma* **1971**, *32*, 251–261.

194. Khosla, S.; Kantheti, P.; Brahmachari, V.; Chandra, H.S. A male-specific nuclease resistant fraction in the mealybug *Planococcus lilacinus*. *Chromosoma* **1996**, *104*, 386–392.

195. Lohe, A.R.; Hilliker, A.J. Return of the H-word (heterochromatin). *Curr. Opin. Genet. Devel.* **1995**, *5*, 746–755.

196. Spofford, J.B. Position-effect variegation in Drosophila. In: *The Genetics and Biology of Drosophila*, Vol. 1c (Ashburner, M.; Novitski, E.; eds.). Academic Press, London, 1976, pp. 955–1018.

197. Holmquist, G.P. Chromosome bands, their chromatin flavors, and their functional features. *Am. J. Hum. Genet.* **1992**, *51*, 17–37.

198. Drouin, R.; Holmquist, G.P.; Richer, C-L. High-resolution replication bands compared with morphologic G- and R-bands. *Adv. Hum. Genet.* **1994**, *22*, 47–115.

199. Okada, T.A.; Comings, D.E. Mechanisms of chromosome banding. III. Similarity between G-bands of mitotic chromosomes and chromomeres of meiotic chromosomes. *Chromosoma* **1974**, *48*, 65–71.

200. Jagiello, G.; Fang, J-S. A pachytene map of the mouse oocyte. *Chromosoma* **1980**, *77*, 113–121.

201. Hungerford, D.A.; Hungerford, A.M. Chromosome structure and function in man. VI. Pachytene chromomere maps of 16, 17 and 18; pachytene as a reference standard for metaphase banding. *Cytogenet. Cell Genet.* **1978**, *21*, 212–230.

202. Ambros, P.F.; Sumner, A.T. Correlation of pachytene chromomeres and metaphase bands of human chromosomes, and distinctive properties of telomeric regions. *Cytogenet. Cell Genet.* **1987**, *44*, 223–228.

203. Ganner, E.; Evans, H.J. The relationship between patterns of DNA replication and of quinacrine fluorescence in the human chromosome complement. *Chromosoma* **1971**, *35*, 326–341.

204. Latt, S.A. Microfluorimetric detection of deoxyribonucleic acid replication in human metaphase chromosomes. *Proc. Nat. Acad. Sci. USA* **1973**, *70*, 3395–3399.

205. Vogel, W.; Autenreith, M.; Speit, G. Detection of bromodeoxyuridine-incorporation in mammalian chromosomes by a bromodeoxyuridine-antibody. I. Demonstration of replication patterns. *Human. Genet.* **1986**, *72*, 129–132.

206. Kuroiwa, T. Asynchronous condensation of chromosomes from early prophase to late prophase as revealed by electron microscopic autoradiography. *Exp. Cell Res.* **1971**, *69*, 97–105.

207. Drouin R, Lemieux N, Richer C-L. Chromosome condensation from prophase to late metaphase: relationship to chromosome bands and their late replication time. *Cytogenet. Cell Genet.* **1991**, *57*, 91–99.

208. Kerem, B-S.; Goitein, R.; Diamond, G.; Cedar, H.; Marcus, M. Mapping of DNase I sensitive regions on mitotic chromosomes. *Cell* **1984**, *38*, 493–499.

209. de la Torre, J.; Sumner, A.T.; Gosalvez, J.; Stuppia, L. The distribution of genes on human chromosomes as studied by in situ nick translation. *Genome* **1992**, *35*, 890–894.

210. Manuelidis, L., Ward, D.C. Chromosomal and nuclear distribution of the Hind III 1.9-kb human DNA repeat segment. *Chromosoma* **1984**, *91*, 28–38.

211. Korenberg, J.R.; Rykowski, M.C. Human genome organisation: Alu, Lines, and the molecular structure of metaphase chromosome bands. *Cell* **1988**, *53*, 391–400.

212. Morad, M.; Jonasson, J.; Lindsten, L. Distribution of mitomycin C induced breaks on human chromosomes. *Hereditas* **1973**, *74*, 273–282.

213. San Roman, C.; Bobrow, M. The sites of radiation induced-breakage in human lymphocyte chromosomes, determined by quinacrine fluorescence. *Mutat. Res.* **1973**, 18, 325–331.

214. Buckton, K.E. Identification with G and R banding of the position of breakage points induced in human chromosomes by in vitro X-irradiation. *Int. J. Radiat. Biol.* **1976**, *29*, 475–488.

215. Nakagome, Y.; Chiyo, H. Non-random distribution of exchange points in patients with structural rearrangements. *Am. J. Hum. Genet.* **1976**, *28*, 31–41.

216. Chandley, A.C. A model for effective pairing and recombination at meiosis based on early replicating sites (R-bands) along chromosomes. *Hum. Genet.* **1986**, *72*, 50–57.

217. Müller, U.; Schempp, W. Homologous early replication patterns of the distal short arms of prometaphasic X and Y chromosomes. *Hum. Genet.* **1982**, *60*, 274–275.

218. Miller, O.J.; Erlanger, B.F. Immunochemical probes of human chromosome organization. *Pathobiol. Annu.* **1975**, 71–103.

219. Tobia, A.M.; Schildkraut, C.L.; Maio, J.J. Deoxyribonucleic acid replication in synchronized cultured mammalian cells. I. Time of synthesis of molecules of different average guanine + cytosine content. *J. Mol. Biol.* **1970**, *54*, 499–515.

220. Bostock, C.J.; Prescott, D.M. Shift in buoyant density of DNA during the synthetic period and its relation to euchromatin and heterochromatin in mammalian cells. *J. Mol. Biol.* **1971**, *60*, 151–162.

221. Yunis, J.J.; Kuo, M.T.; Saunders, G.F. Localization of sequences specifying messenger RNA to light-staining G-bands of human chromosomes. *Chromosoma* **1977**, *61*, 335–344.

222. Goldman, M.A.; Holmquist, G.P.; Gray, M.C. Caston, L.A.; Nag, A. Replication timing of genes and middle repetitive sequences. *Science* **1984**, *224*, 686–692.

223. Bernardi, G. The isochore organization of the human genome. *Annu. Rev. Genet.* **1989**, *23*, 637–661.

224. Bernardi, G. The isochore organization of the human genome and its evolutionary history—a review. *Gene* **1993**, *135*, 57–66.

225. Saccone, S.; de Sario, A.; della Valle, G.; Bernardi, G. The highest gene concentrations in the human genome are in telomeric bands of metaphase chromosomes. *Proc. Natl. Acad. Sci. USA* **1992**, *89*, 4913–4917.

226. Saccone, S.; de Sario, A.; Wiegant, J.; Raap, A.K.; della Valle, G.; Bernardi, G. Correlations between isochores and chromosomal bands in the human genome. *Proc. Natl. Acad. Sci. USA* **1993**, *90*, 11929–11933.

227. Craig, J.M., Bickmore, W.A. Chromosome bands - flavours to savour. *Bioessays* **1993**, *15*, 349–354.

228. Craig, J.M. The functional significance of mammalian chromosome banding. Ph.D thesis, University of Edinburgh, 1995.

229. Turner, B.M. Histone acetylation and control of gene expression. *J. Cell Sci.* **1991**, *99*, 13–20.

230. Breneman, J.W.; Yau, P.; Teplitz, R.L.; Bradbury, E.M. A light microscope study of linker histone distribution in rat metaphase chromosomes and interphase nuclei. *Exp. Cell. Res.* **1993**, *206*, 16–26.

231. Johnson, K.R.; Lehn, D.A.; Elton, T.S.; Barr, P.J.; Reeves, R. Complete murine cDNA sequence, genomic structure, and tissue expression of the high mobility group protein HMG-I(Y). *J. Biol. Chem.* **1988**, *263*, 18338–18342.

232. Goldman, M.A. The chromatin domain as a unit of gene regulation. *Bioessays* **1988**, *9*, 50–55.

233. Wang, H.C.; Kao, K.N. G-banding in plant chromosomes. *Genome* **1988**, *30*, 48–51.

234. Zhang, Z.; Yang, X. Effects of protein extraction on chromosome G-banding in plants. *Cytobios* **1990**, *62*, 87–92.

235. Kakeda K, Yamagata H, Fukui K, et al. High resolution bands in maize chromosomes by G-banding methods. *Theor. Appl. Genet.* **1990**, *80*, 265–272.

236. Peffley, E.B.; de Vries, J.N. Giemsa G-banding in *Allium*. *Biotech. Histochem.* **1993**, *68*, 83–86.

237. Cuny, G.; Soriano, P.; Macaya, G.; Bernardi, G. The major components of the mouse and human genomes. I. Preparation, basic properties and compositional heterogeneity. *Eur. J. Biochem.* **1981**, *115*, 227–233.

238. Stahl, A.; Luciani, J.M.; Devictor, M.; Capodano, A.M.; Hartung, M. Heterochromatin and nucleolar organizers during first meiotic prophase in quail oocytes. *Exp. Cell Res.* **1974**, *91*, 365–371.

239. Carlenius, C.; Ryttman, H.; Tegelstrom, H.; Jannson, H. R-, G- and C-banded chromosomes in the domestic fowl (*Gallus domesticus*). *Hereditas* **1981**, *94*, 61–66.

240. Schmid, M.; Guttenbach, M. Evolutionary diversity of reverse (R) fluorescent chromosome bands in vertebrates. *Chromosoma* **1988**, *97*, 101–114.

241. Bull, J. Sex chromosome differentiation: an intermediate stage in a lizard. *Can. J. Genet. Cytol.* **1978**, *20*, 205–209.

242. Yonenaga-Yassuda, Y.; Kasahara, S.; Chu, T.H.; Rodrigues, M.T. High-resolution RBG-banding pattern in the genus *Tropidurus* (Sauria: Iguanidae). *Cytogenet. Cell Genet.* **1988**, *48*, 68–71.

243. Thiery, J-P.; Macaya, G.; Bernardi, G. An analysis of eukaryotic genomes by density gradient centrifugation. *J. Mol. Biol.* **1976**, *108*, 219–235.

244. Sessions, S.K.; Kezer, K. Cytogenetic evolution in the plethodontid salamander genus *Aneides*. *Chromosoma* **1987**, *95*, 17–30.

245. Cuny, R.; Malcinski, G.M. Banding differences between tiger salamander and axolotl chromosomes. *Can. J. Genet. Cytol.* **1985**, *27*, 510–514.

246. Schempp, W.; Schmid, M. Chromosome banding in Amphibia. IV. BrdU-replication patterns in Anura and demonstration of XX/XY sex chromosomes in *Rana esculenta*. *Chromosoma* **1981**, *83*, 697–710.

247. Herrero, P.; de la Torre, J.; Arano, B.; Gosálvez.; Sumner, A.T. Patterns of DNase sensitivity in the chromosomes of *Rana perezi* (Amphibia: Anura). *Genome* **1995**, *38*, 339–343.

248. Schmid, M.; Steinlein, C. Chromosome banding in Amphibia. XVI. High-resolution replication banding patterns in *Xenopus laevis*. *Chromosoma* **1991**, *101*, 123–132.

249. Stock, A.D.; Mengden, G.A. Chromosome banding pattern conservation in birds and non-homology of chromosome banding patterns between birds, turtles, snakes and amphibians. *Chromosoma* **1975**, *50*, 69–77.

250. Schmid, M.; Löser, C.; Schmidtke, J.; Engel, W. Evolutionary conservation of a common pattern of activity of nucleolus organizers during spermatogenesis in vertebrates. *Chromosoma* **1982**, *86*, 149–179.

251. Delany, M.E.; Bloom, S.E. Replication banding patterns in the chromosomes of the rainbow trout. *J. Hered.* **1984**, *75*, 431–434.

252. Gold, J.R.; Li, Y.C.; Shipley, N.S.; Powers, P.K. Improved methods for working with fish chromosomes with a review of metaphase chromosome banding. *J. Fish Biol.* **1990**, *37*, 563–575.

253. Medrano, L.; Bernardi, G.; Couturier, J.; Dutrillaux, B.; Bernardi, G. Chromosome banding and genome compartmentalization in fishes. *Chromosoma* **1988**, *96*, 178–183.

254. Hudson, A.P.; Cuny, G.; Cortadas, J.; Haschemeyer, A.E.V.; Bernardi, G. An analysis of fish genomes by density gradient centrifugation. *Eur. J. Biochem.* **1980**, *112*, 203–210.

255. Wiberg, U.H. Sex determination in the European eel (*Anguilla anguilla*, L.). *Cytogenet. Cell Genet.* **1983**, *36*, 589–598.

256. Bernardi, G.; Bernardi, Jr., G. Compositional constraints and genome evolution. *J. Mol. Evol.* **1986**, *24*, 1–11.

257. Lima-da-Faria, A. The relationship between chromosomes, replicons, operons, transcription units, genes, viruses and palindromes. *Hereditas* **1975**, *81*, 249–284.

258. de la Torre, J.; Herrero, P.; Garcia de la Vega, C.; Sumner, A.T.; Gosálvez, J. Patterns of DNase I sensitivity in the chromosomes of the grasshopper *Chorthippus parallelus* (Orthoptera). *Chromosome Res.* **1996**, *4*, 56–60.

259. Cortes, F.; Escalza, P. Analysis of different banding patterns and late replicating regions in the chromosomes of *Allium cepa, A. sativum,* and *A. nigrum. Genetica* **1986**, *71*, 39–46.

260. Salinas, J.; Matassi, G.; Montero, L.M.; Bernardi, G. Compositional compartmentalization and compositional patterns in the nuclear genomes of plants. *Nucleic Acids. Res.* **1988**, *16*, 4269–4285.

261. Holm, P.B. The C and Q banding patterns of the chromosomes of *Lilium longiflorum* (Thunb.). *Carlsberg Res. Commun.* **1976**, *41*, 217–224.

262. Schweizer, D.; Strehl, S.; Hagemann, S. Plant repetitive DNA elements and chromosome structure. In: *Chromosomes Today*, Vol. 10. (Fredga, K.; Kihlman, B.A.; Bennett, M.D.; eds.). Unwin Hyman, London, 1990, pp. 33–43.

263. Moore, G.; Abbo, S.; Cheung, W.; et al. Key features of cereal genome organization as revealed by the use of cytosine methylation-sensitive restriction endonucleases. *Genomics* **1993**, *15*, 472–482.

264. Cavalier-Smith, T. Nuclear volume control by nucleoskeletal DNA, selection for cell volume and cell growth rate, and the solution of the DNA C-value paradox. *J. Cell Sci.* **1978**, *34*, 247–278.

265. van Dekken, H.; Pinkel, D.; Mullikin, J.; Trask, B.; van den Engh, G.; Gray J. Three-dimensional analysis of the organization of human chromosome domains in human and human-hamster hybrid interphase nuclei. *J. Cell Sci.* **1989**, *94*, 229–306.

266. Rawlins, D.J.; Highett, M.I.; Shaw, P.J. Localization of telomeres in plant interphase nuclei by in situ hybridization and 3D confocal microscopy. *Chromosoma* **1991**, *100*, 424–431.

267. Hutchison, N.; Weintraub, H. Localization of DNase I-sensitive sequences to specific regions of interphase nuclei. *Cell* **1985**, *43*, 471–482.

268. Puck, T.T.; Bartholdi, M.; Krystosek, A.; Johnson, R.; Haag, M. Confocal microscopy of genome exposure in normal, cancer and reverse-transformed cells. *Somatic Cell Mol. Genet.* **1991**, *17*, 489–503.

Advances in Genome Biology

Edited by **Ram S. Verma,** *Institute of Molecular Biology and Genetics Interscience, Brooklyn, NY*

Volume 1, Unfolding the Genome
1992, 392 pp. $109.50/£69.50
ISBN 1-55938-349-6

CONTENTS: Introduction to the Series, *Ram S. Verma.* Getting Techniques for Mapping and Sequencing the Genome: An Overview, *Ram S. Verma.* Cloning Defined Regions of the Human Genome by Microdissection of Banded Chromosomes and Enzymatic DNA Amplification, *Uwe Claussen, Gabriele Senger, Herman-Josef Lüdecke, and Bernhard Horsthemke.* In Situ Hybridization with Radiolabeled Probes, *Paul Szabo.* Analysis of Genome DNA by Southern Blotting, *Peter ten Dijke and Kees Stam.* Analysis of RNA by Northern Hybridization, *Ashok Kumar and M.A.Q. Siddiqui.* Slot-Blot Technique: Principles and Applications, *Geetha Vasanthakumar.* Spot-Blot Gene Mapping and Bivariate Flow Cytogenetics, *Roger V. Lebo.* Polymerase Chain Reaction: Principles and Applications, *Stephen M. Carleton.* Protocols for Pulsed Field Gel Electrophoresis: Separation and Detection of Large DNA Molecules, *Robert M. Gemmill, Jane Coyle-Morris, Patrick Scott, Sylvie Paulien, Michael Mendez, and Marileila Varella-Garcia.* Detection of Single Base Changes in Nucleic Acid, *Richard G.H. Cotton.* Identification of Chromosome-Specific Satellite DNA Using Non-Isotopic In Situ Hybridization, *Matteo Adinolif.* Construction and Use of Lambda and Cosmid Linkage Libraries, *Alan R. Kimmel.* Index.

Volume 2, Morbid Anatomy of the Genome
1993, 464 pp. $109.50/£69.50
ISBN 1-55938-583-9

CONTENTS: Preface, *Ram S. Verma.* Diagnosis of Human Genetic Diseases Using Recombinant DNA Techniques: An Overview, *Jörg Schmidtke and David N. Cooper.* Mitochondrial DNA and Disease, *S.R. Hammans and A.E. Harding.* The Cystic Fibrosis Gene, *Michael Wagner and Jochen Reiss.* Marfan Syndrome: Molecular Biology, *Brendan Lee, Maurice Godfrey, Petros Tsipouras, and Francesco Ramirez.* Marfan Syndrome: Molecular Pathogenesis, *Katariina Kainulainen and Leena Peltonen.* Molecular Genetics of Fragile X Syndrome, *Grant R. Sutherland and Robert I. Richards.* Molecular Genetics of Huntington Disease, *Catrin Pritchard, Ning Zhu, David R. Cox, and Richard M. Myers.* Gene for Von Recklinghausen Neurofibromatosis Type 1, *Dalal M. Jadayel.* Molecular Biology of Duchenne and Becker Muscular Dystrophies, *Jamel Chelly and Jean Claude Kaplan.* Molecular Genetics of Thalassemia, *J. Eric Russell and Stephen A. Liebhaber.* Applications of Molecular Genetics for Identity, *Zvi G. Loewy and Howard J. Baum.* Current Status and Future Directions in Human Gene Therapy, *Paul Tolstoshev.* Prelude to Reverse Genetics, *Ram S. Verma.* Index.

J A I

P R E S S

JAI PRESS INC.
55 Old Post Road No. 2 - P.O. Box 1678
Greenwich, Connecticut 06836-1678
Tel: (203) 661- 7602   Fax: (203) 661-0792

J
A
I

P
R
E
S
S

Advances in Structural Biology

Edited by **Sudarshan K. Malhotra**,
Department of Zoology, University of Alberta

Advances in Structural Biology as a new serial publication may require some explanation. First, structural biology now has a very broad range and is highly diversified; and second, research papers on structure, biochemistry, and function often appear in widely disparate journals. Consequently, investigators in specialized areas often find it difficult to maintain an integrated view of their own field of work, let alone an overview of advances in the field. The editor and publisher recognize their responsibility to specialists and generalists in providing a dedicated forum for a wide range of representative current subjects in structural biology.

— From the Preface

Volume 4, Biology and Physics of Cytoskeleton
In preparation, Summer 1998
ISBN 1-55938-967-2   Approx. $109.50/£69.50

Edited by **Sudarshan K. Malhotra** and
J. A. Tuszynski, *Department of*
Biological Sciences, University of Alberta

TENTATIVE CONTENTS: Preface, *S.K. Malhotra and J.A. Tuszynski.* Structure of Gamma Tubulin, *K. Wolf.* GTPase Activity of Tubulin Isotypes, *R. Luduena.* Statistical Mechanics of Biopolymers and Macromolecular Networks, *E. Frey and E. Sackmann.* Models of Motor Protein Motility, *G. Tsironis.* Dielectric Dipoles in Proteins of the Cytoskeleton, *A. Tsvilik.* Self-Organization of the Cytoskeleton, *J. Tabony.* Tensegrity in Mitosis, *A. Maniotis and Ingber.* Microtubule Oscillations *In Vitroi, M.F. Carlier.* Actin Filaments in Plant Cells and Light Absorption, *P. Malec.* Biological Ferroelectricity, *N. Mavromatos and K. Zioutas.* Quantum Cavity Effects in Proteins, *D. Nanopoulos and N. Mauromatos.* Biological Conduction at an Intracellular Level, *E. Insinna.* Data Processing in Studies of Microtubule Polymerization Processes, *J.A. Tuszynski and H. Bolterauer.* Index.

Also Available:
Volumes 1-3 (1991-1994) $109.50/£69.50 each

J
A
I

P
R
E
S
S

Advances in Organ Biology

Edited by **E. Edward Bittar,** *Department of Physiology, University of Wisconsin Medical School*

Volume 3, Retinoids: Their Physiological Function and Therapeutic Potential
1997, 312 pp. $128.50/£82.50
ISBN 0-7623-0285-2

Edited by **G.V. Sherbet,** *Cancer Research Unit, Medical School, University of Newcastle-upon-Tyne*

CONTENTS: Preface, *G. V. Sherbet.* Retinoid Structure, Chemistry and Biologically Active Derivatives, *Robert W. Curley and Michael J. Robarge.* Molecular Mechanisms of Retinoid Function, *Christopher P.F. Redfern.* Retinoids in Mammalian Embryonic Development, *Gillian M. Morriss-Kay.* The Role of Retinoids in Patterning Fish, Amphibian, and Chick Embryos, *Malcolm Maden and John Pizzey.* Retinoid and Growth Factor Signal Transduction, *G.V. Sherbet and M.S. Lakshmi.* Retinoids and Apoptosis, *Lin-Xin Zhang and Anton M. Jetten.* Retinoids in Tumor Cell Adhesion Invasion, and Metastasis, *Michael Edward.* Retinoid Receptors and Cancer, *Joseph A. Fontana and Arun K. Rishi.* Retinoids in the Management of Central Nervous System (CNS) Tumors, *M.E. Westarp.* Retinoids and Lung Cancer, *Andrew M. Arnold and Richard G. Tozer.* Index.

Also Available:
Volumes 1-2 (1996-1997) $128.50/£82.50 each

FACULTY/PROFESSIONAL discounts are available in the U.S. and Canada at a rate of 40% off the list price when prepaid by personal check or credit card and ordered directly from the publisher.

JAI PRESS INC.

55 Old Post Road No. 2 - P.O. Box 1678
Greenwich, Connecticut 06836-1678
Tel: (203) 661- 7602   Fax: (203) 661-0792

Printed and bound by CPI Group (UK) Ltd, Croydon, CR0 4YY

13/10/2024

01773499-0001